AF835428

Information and Communication Technologies in Public Administration

Innovations from Developed Countries

PUBLIC ADMINISTRATION AND PUBLIC POLICY
A Comprehensive Publication Program

EDITOR-IN-CHIEF

DAVID H. ROSENBLOOM
Distinguished Professor of Public Administration
American University, Washington, DC

Founding Editor

JACK RABIN

RECENTLY PUBLISHED BOOKS

Information and Communication Technologies in Public Administration: Innovations from Developed Countries, Christopher G. Reddick and Leonidas Anthopoulos

Creating Public Value in Practice: Advancing the Common Good in a Multi-Sector, Shared-Power, No-One-Wholly-in-Charge World, edited by John M. Bryson, Barbara C. Crosby, and Laura Bloomberg

Digital Divides: The New Challenges and Opportunities of e-Inclusion, Kim Andreasson

Living Legends and Full Agency: Implications of Repealing the Combat Exclusion Policy, G.L.A. Harris

Politics of Preference: India, United States, and South Africa, Krishna K. Tummala

Crisis and Emergency Management: Theory and Practice, Second Edition, Ali Farazmand

Labor Relations in the Public Sector, Fifth Edition, Richard C. Kearney and Patrice M. Mareschal

Democracy and Public Administration in Pakistan, Amna Imam and Eazaz A. Dar

The Economic Viability of Micropolitan America, Gerald L. Gordon

Personnel Management in Government: Politics and Process, Seventh Edition, Katherine C. Naff, Norma M. Riccucci, and Siegrun Fox Freyss

Public Administration in South Asia: India, Bangladesh, and Pakistan, edited by Meghna Sabharwal and Evan M. Berman

Making Multilevel Public Management Work: Stories of Success and Failure from Europe and North America, edited by Denita Cepiku, David K. Jesuit, and Ian Roberge

Public Administration in Africa: Performance and Challenges, edited by Shikha Vyas-Doorgapersad, Lukamba-Muhiya. Tshombe, and Ernest Peprah Ababio

Available Electronically
PublicADMINISTRATION*netBASE*
http://www.crcnetbase.com/page/public_administration_ebooks

Information and Communication Technologies in Public Administration

Innovations from Developed Countries

Edited by
Christopher G. Reddick
University of Texas at San Antonio, USA
Leonidas Anthopoulos
Technological Education Institute (TEI) of Thessaly, Greece

CRC Press
Taylor & Francis Group
Boca Raton London New York

CRC Press is an imprint of the
Taylor & Francis Group, an **informa** business

CRC Press
Taylor & Francis Group
6000 Broken Sound Parkway NW, Suite 300
Boca Raton, FL 33487-2742

© 2015 by Taylor & Francis Group, LLC
CRC Press is an imprint of Taylor & Francis Group, an Informa business

No claim to original U.S. Government works

Printed on acid-free paper
Version Date: 20150223

International Standard Book Number-13: 978-1-4822-3929-4 (Hardback)

This book contains information obtained from authentic and highly regarded sources. Reasonable efforts have been made to publish reliable data and information, but the author and publisher cannot assume responsibility for the validity of all materials or the consequences of their use. The authors and publishers have attempted to trace the copyright holders of all material reproduced in this publication and apologize to copyright holders if permission to publish in this form has not been obtained. If any copyright material has not been acknowledged please write and let us know so we may rectify in any future reprint.

Except as permitted under U.S. Copyright Law, no part of this book may be reprinted, reproduced, transmitted, or utilized in any form by any electronic, mechanical, or other means, now known or hereafter invented, including photocopying, microfilming, and recording, or in any information storage or retrieval system, without written permission from the publishers.

For permission to photocopy or use material electronically from this work, please access www.copyright.com (http://www.copyright.com/) or contact the Copyright Clearance Center, Inc. (CCC), 222 Rosewood Drive, Danvers, MA 01923, 978-750-8400. CCC is a not-for-profit organization that provides licenses and registration for a variety of users. For organizations that have been granted a photocopy license by the CCC, a separate system of payment has been arranged.

Trademark Notice: Product or corporate names may be trademarks or registered trademarks, and are used only for identification and explanation without intent to infringe.

Visit the Taylor & Francis Web site at
http://www.taylorandfrancis.com

and the CRC Press Web site at
http://www.crcpress.com

MIX
Paper from responsible sources
FSC www.fsc.org FSC® C013056

Printed and bound in Great Britain by
TJ International Ltd, Padstow, Cornwall

Contents

Foreword

Government and Information and Communications Technology Innovation: A Cliché Relation or a Universal Purpose?

Today, more and more government leaders are discussing innovation, and they recognize its contribution to economic growth, to enhancement of national competitiveness, to increment of local knowledge-based society, and so forth (Organisation for Economic Co-operation and Development, 2005; Porter & Schwab, 2008). Many could question that although governments talk about innovation, they do not transform themselves to more innovative forms. For instance, do governments update the ways they interact with their people? Do they utilize innovation to enhance their potential to deliver goods and services to their local societies?

To this end, information and communications technologies (ICTs) were seen as an innovative tool and were adopted quite early by governments as a means to become more efficient and effective in delivering their jobs. Internet-based government, for instance, was such an innovation in the early 1990s. Since then, many deliverables have been observed around the world: one-stop e-government portals that offer sophisticated public services, dashboards that illustrate government performance, online public procurement systems, maps and geo-data, publicly available information and documents (named open data), back-office information systems, online consultations, and others. All these products could be easily named "innovative" products for the governments, or maybe not?

Let us define the terms: *Innovation* concerns the practical refinement and development of an original invention into a usable technique or product, or a process, in which creativity is applied to every facet of an organization's value chain, from beginning to end, to develop new and better ways of creating value for customers (Fagerberg, 2004; Howells, 2005; Maital & Seshadri, 2007). According to this definition, all the above products concerned—in their initial appearance—ICT innovation; they are updated incrementally, while governments fund their development with extensive programs as open innovation (Anthopoulos & Fitsilis, 2014). Each of the above products concerns technological innovation, whereas all together, they are expected to perform extensive change to the ways that people interact with their governments and to become a social innovation. And we can say that "it is expected to deliver such a radical change" because despite its ambitious vision, e-government appears to become rather a single channel for citizens to access their governments, instead of performing an enormous change with regard to transactions between governments and citizens (Reddick & Anthopoulos, 2014).

The previous discussion validates that ICT innovation is something trivial for governments, but they have not achieved a radical change yet. On the other hand, little has happened and even less can be observed in improving the ways that governments perform their mission. Wars and poverty are still a reality, public spending is not absolutely transparent, citizens still prefer traditional channels to access their governments for custom services, and yet other myths still exist (Bekkers & Homburg, 2007). To this end, the value that the ICT innovative products deliver to citizens can be questioned, and it remains an unestablished objective.

All these justify the significant contribution of this book: This book views government ICT innovation from the lens of corresponding theoretical analysis and practical implementations in the developed world, which lead the innovation arena. To this end, chapter contributors examine various underlying theories and perspectives in the first section of this book. Moreover, exemplars and their added value have to be monitored and illustrated properly, and to this end, enough corresponding practices are examined in this first section. On the other hand, existing challenges and unaccomplished investments probably depict the future of ICT for governments. In this context, the second section of this book addresses various corresponding topical issues and demonstrates perhaps what governments can expect from ICTs or from changes that they trigger inside the public sector.

References

Anthopoulos, L., & Fitsilis, P. (2014). Trends in e-strategic management: How do governments transform their policies? *International Journal of Public Administration in the Digital Age, 1*(1), 15–38.

Bekkers, V., & Homburg, V. (2007). The myths of e-government: Looking beyond the assumptions of a new and better government. *The Information Society, 23*(5), 373–382.

Fagerberg, J. (2004). Innovation: A Guide to the literature. In Fagerberg, J., Mowery, D. C., & Nelson, R. R. (Eds.) *The Oxford Handbook of Innovation*. Oxford: Oxford University Press.

Howells, J. (2005). *The Management of Innovation and Technology: The Shaping of Technology and Institutions of the Market Economy*. London: Sage Publications.

Maital, S., & Seshadri, D. V. R. (2007). *Innovation Management: Strategies, Concepts and Tools for Growth and Profit*. New Delhi, India: Response Books, Sage Publications.

Organisation for Economic Co-operation and Development. (2005). *Oslo Manual: Guidelines for Collecting and Interpreting Innovation Data, 3rd Edition*. Paris: OECD Publishing.

Porter, M., & Schwab, K. (2008). *Global Competitiveness Report 2008–2009*. Geneva: World Economic Forum.

Reddick, C. G., & Anthopoulos, L. (2014). Interactions with e-government, new digital media, and traditional channel choices: Citizen-initiated factors. *Transforming Government: People, Process and Policy*, *8*(3), 398–419.

Vishanth Weerakkody
Brunel Business School
Brunel University
Uxbridge, United Kingdom

Preface

This book examines how information technology can be used to promote innovations in public administration. Information technology has a long history in the public sector, with claims that it will produce meaningful changes to public service delivery. It has been said to improve relations with citizens in their interaction with the government, improve the delivery of public services, and enhance the experience of businesses when working with government. This book examines public administration and information technology with a focus on developing countries. These countries are typically the most advanced in the development of information technology and public service delivery. This preface will provide a summary of the chapters in this book and demonstrate their main contributions to the field.

The first section of this book examines global perspectives on public administration and information technology innovations. The chapters in this section of the book cover important issues such as open government, social media, democracy, and management challenges. In Chapter 1, Sáez Martín, de Rosario, and Caba Pérez examine open government, which can be used by a democracy to facilitate more transparent and collaborative governments. The specific focus of their chapter is on open government and data disclosures by the central government of the European Union member countries. Their major findings are that countries of the European Union are making little use of the feedback possibilities offered by these open government portals.

In Chapter 2, Holzer and Zheng evaluate best practices in the websites of the largest municipalities in the 100 most wired nations. These nations have widely accepted, and more governments continuously adopt, e-governance to provide better services. These authors found that municipalities vary substantially in e-governance development. There is a substantial gap between municipalities in the Organisation for Economic Co-operation and Development (OECD) countries and non-OECD countries. The authors suggest four strategies to achieve success in e-governance, which are (1) upgrading information and communication technology infrastructure, (2) improving human resource management, (3) creating a corresponding political environment, and (4) promoting administrative performance.

In Chapter 3, Zheng and Manoharan used a cross-country analysis with structural equation modeling statistical techniques to test the impact of environmental factors (economy, democracy, education, Internet usage, and peer pressure) on e-government development and explored the relationships among these factors. The results of their chapter indicate that the overall external environment does affect e-government development at the national level.

In Chapter 4, Sgueo examines Italian public administration's use of web-based experiments in participatory democracy designed to address citizens' questions and needs. The findings of this chapter indicate that the use of large-scale Internet projects did not fix the democratic deficit of Italian contemporary politics and public administrations; the chapter also aims at speculating on possible future evolutions of web-based participatory initiatives in Italy.

In Chapter 5, Bolívar and Muñoz analyze social media implementation in public administration, examining the theories that support the use of social media, the main use of this technology for public administrations, and the trends and research innovations in this area in the future. Knowledge gaps and research opportunities are identified from these observations, which reveal changes in the research methods applied, reinforcing the development of a theoretical framework so that the application of social media may efficiently contribute to improving management in the public sector.

In Chapter 6, Anthopoulos, Reddick, Giannakidou, and Mavridis consider whether e-government is an "innovative product" for the public sector and suggest e-government deployment with the means of innovation management. In this chapter, respective theories are discussed and speculations are made for such an e-government alignment. Moreover, data from an analysis of Twitter feeds after the failed roll out of the Healthcare.gov website, or the Obamacare website in the United States, are used to examine this theory.

In Chapter 7, Caffrey and McDonagh provide insights into the strategic alignment of information technology and public administration as it relates to public organizations based on an assessment of empirical works. These authors note that although much has been discovered in relation to strategic alignment, a good deal continues to elude researchers, particularly in relation to public organizations, and some potential opportunities to advance further inquiry into this important management domain are presented.

The second section of this book examines topical issues in public administration and information technology innovation. In Chapter 8, Van Cauter, Snoeck, and Crompvoets test an updated Delone and McLean model on information systems failures. Their updated model did not detect a lack of trust, path dependency on information systems infrastructure, insufficient organizational competence, unfulfilled user expectations, and a lack of local management support, as other research has found.

In Chapter 9, de Vries examines innovations in interorganizational e-government projects in the public sector that tend to be shaped by discretionary decisions of information and communication technology professionals. Public sector managers who are insufficiently knowledgeable about the implications of such discretions can therefore not fully control the innovation. This chapter's findings justify the need for a better understanding of discretions to decrease the dependency on these in information and communication technology projects in the future, so that managing these projects effectively improves.

In Chapter 10, Manrique and Manrique examine a case study of both the promise and the difficulties associated with the increased use and acceptance of virtual currencies such as Bitcoin. The widespread use of virtual currencies will have implications for public administration, laws and regulations, business, and economic policy. Bitcoin has the potential to redefine not only business transactions but also transactions between citizens and government agencies as well.

In Chapter 11, Roy examines the relationship between mobility, Gov 2.0, and digital inclusion in Canada. This chapter examines the Province of Ontario, which has used several interrelated and relevant initiatives in this area. This chapter shows how digital inclusion is likely to be addressed within Ontario's evolving parameters of mobility, electronic service delivery, and wider Gov 2.0-stylized reform efforts.

In Chapter 12, Lodi, Maccioni, and Tortorelli show that governments have started sharing a large volume of public data sets on the Web. The issue is that to fully reuse the data and meet semantic interoperability requirements, it is crucial to provide them in a standard and machine-readable form. This paper aims at giving a broad overview of an e-government data project in Italy and of the initiatives that derived from it, including the development of three private sector applications.

In Chapter 13, Schade, Granell, and Perego show that scientific domains and the public sector increasingly open their data repositories. The chapter explores the concept of a cross-domain open data ecosystem by (a) explaining the current situation, (b) exploring innovative approaches and recent trends on information and communication technologies, and (c) outlining potential pathways to couple public sector information and public-funded research data.

In Chapter 14, Zacharis, Ipsilandis, and O'Kane show that for the last two decades, information technology professionals and public servants have witnessed valuable help in managing everyday operations of the organization but at the same time incubate a whole mindset of innovation and novelty. Especially in the educational area, the combination of large-scale automation projects with prolific didactic theory interventions shape contemporary high-level priorities in education. Their chapter examines innovation in the context of Greek educational institutions.

As a group, the chapters in this book offer both the innovations in important issues of information technology and public administration innovations and offer topical areas that need future exploration.

Christopher G. Reddick
Leonidas Anthopoulos

Contributors

Leonidas Anthopoulos
School of Business and Economics
TEI of Thessaly, Greece
Larissa, Greece

Manuel Pedro Rodríguez Bolívar
University of Granada
Department of Accounting and Finance
Faculty of Business Studies
University of Granada
Granada, Spain

Carmen Caba Pérez
University of Almería
Almeria, Spain

Eamonn Caffrey
School of Business
Trinity College Dublin
Dublin, Ireland

Joep Crompvoets
University of Leuven
Public Governance Institute
Leuven, Belgium

Arturo Haro de Rosario
University of Almería
Almeria, Spain

Walter Timo de Vries
Faculty of Geo-Information Science and Earth Observation
University of Twente
Enschede, the Netherlands

Irene Giannakidou
Department of Informatics
Aristotle University of Thessaloniki, Greece
Thessaloniki, Greece

Carlos Granell
Institute of New Imaging Technologies (INIT)
Universitat Jaume I
Castellón de la Plana, Spain

Marc Holzer
School of Public Affairs and Administration
Rutgers University–Newark
Newark, New Jersey

Pandelis Ipsilandis
Department of Project Management
School of Business and Economics
Technological Education Institute of Larissa
Thessalia, Greece

and

Board of Directors Greek Network of Project Managers
Nikaia Larissas, Greece

Giorgia Lodi
Agency for Digital Italy
Rome, Italy

Antonio Maccioni
Agency for Digital Italy
Rome, Italy

Aroon Manoharan
Department of Political Science
Kent State University
Kent, Ohio

Cecilia G. Manrique
Political Science/Public Administration
University of Wisconsin–La Crosse
La Crosse, Wisconsin

Gabriel Manrique
Economics
Winona State University
Winona, Minnesota

Nikolaos Mavridis
NCSR Demokritos
Institute of Informatics and Telecommunications
Attiki, Greece

Joe McDonagh
School of Business
Trinity College Dublin
Dublin, Ireland

Laura Alcaide Muñoz
University of Granada
Department of Accounting and Finance
Faculty of Business Studies
University of Granada
Granada, Spain

James O'Kane
Business School
Edge Hill University
Ormskirk, United Kingdom

Andrea Perego
European Commission–DG Joint Research Centre
Ispra, Italy

Christopher G. Reddick
Department of Public Administration
The University of Texas at San Antonio
San Antonio, Texas

Jeffrey Roy
School of Public Administration
Dalhousie University
Halifax, NS, Canada

Alejandro Sáez Martín
University of Almería
Almeria, Spain

Sven Schade
European Commission–DG Joint Research Centre
Ispra, Italy

Gianluca Sgueo
Center for Social Studies
University of Coimbra
Coimbra, Portugal

Monique Snoeck
University of Leuven
Research Centre for Management Informatics
Leuven, Belgium

Francesco Tortorelli
Agency for Digital Italy
Rome, Italy

Lies Van Cauter
University of Leuven
Public Governance Institute
Leuven, Belgium

Konstantinos Zacharis
Secondary Education Authority
Karditsa, Greece

and

Staffordshire University
Staffordshire, United Kingdom

Yueping Zheng
School of Public Affairs and Administration
Rutgers University–Newark
Newark, New Jersey

I

GLOBAL PERSPECTIVES ON PUBLIC ADMINISTRATION AND INFORMATION TECHNOLOGY INNOVATIONS

Chapter 1

Open Government Data: A European Perspective

Alejandro Sáez Martín, Arturo Haro de Rosario, and Carmen Caba Pérez

Contents

Abstract

The aim of the open government movement is for democratic governments to become more transparent, facilitating greater collaboration and participation among the administration, politicians, business representatives, and the general public. The concept of Open Government Data (OGD) denotes the provision of public sector information, to make public data available for stakeholders to store, use, and redistribute without restriction or charge. Many initiatives worldwide have been undertaken to develop web portals facilitating access to public sector data. The aim of this chapter is to examine the growing implementation of OGD, analyzing the data disclosed by the central governments of the European Union (EU) member countries through their official OGD websites. The methodology used is based on four analytical approaches: functions, semantics, quality, and content. The results obtained show, first, considerable heterogeneity among EU portals and, second, that the implementation of OGD portals in Europe is at an intermediate level of development. At present, the results obtained suggest that the countries of the EU are making little use of the feedback possibilities offered by their portals and, in many, cases do not allow the introduction of new data.

Keywords: public sector information (PSI), e-government, linked data, metadata, EU

1.1 Introduction

In recent years, public sector bodies have been urged by stakeholders to provide greater transparency about the use of resources, greater participation in the management of these resources, greater accountability by public managers, and greater access to the information processed and used by government (Sonáglio, 2013). In response to these demands, since the early 1990s, new forms of public management have been applied, and one of the most commonly adopted means of modernizing public administration is to make use of technological advances such as the Internet, thus enabling electronic public administration, or e-government (Khan, Young, & Park, 2012).

With the emergence of diverse forms of social media and with technological advances, in recent years, Web 2.0 tools have been developed, based on concepts such as Government 2.0 or open government (Osimo, Campbell, Kerr-Stevens, Bishop, & Bryant, 2009). This has led to the provision of online services based on interactivity, using new technologies such as social networking, mashups, open data platforms, interactive television, and interactive mobile platforms (Chun, Shulman, & Sandoval, 2010; Harrison et al., 2011; Khan et al., 2012; Mergel, 2010). The aim of the open government movement is for democratic governments to become more transparent, facilitating greater collaboration and participation among the administration, politicians, business representatives, and the general public (Lathrop & Ruma, 2010).

One strategy applied to achieve these goals is known as Open Government Data (OGD), which arises from the conjunction of open government and open data. The latter concept refers to the dissemination of information that can be used, reused, and redistributed by anyone, freely and without charge (Ayers, 2007; Bauer & Kaltenböck, 2011). Open Government Data is viewed as the availability of public sector information (PSI) through web portals, permitting its use by the general public by means of open, accessible, machine-readable formats (Alonso et al., 2009). The disclosure of data by government organizations has opened up new possibilities in terms of greater transparency, more extended reuse of government information, more efficient provision of public services, and greater participation by the population (Bennet & Harvey, 2009).

National and international initiatives have been developed to make governments more transparent and accountable to their citizens by making public data available and enabling reuse of these data. In this respect, national governments have pursued various strategies related to OGD; thus, the United States adopted the Open Government Directive in 2009 (Orszag, 2009) and the United Kingdom published the document "Power of Information Taskforce Report" (HM Government, 2009). In the international context, the Open Government Partnership (OGP) was created in September 2011 by eight countries (a group that has now expanded to 63), the G8 approved its Open Data Charter in June 2013 (OGP, 2014; Tinati, Halford, Carr, & Pope, 2012), and the Organization for Economic Co-operation and Development Council has recommended enhanced access to and more effective use of public information (Organization for Economic Co-operation and Development, 2008).

The European Union (EU) has, for some time, been aware of the benefits to society of greater openness and reuse of public data. In 2003, it issued PSI Directive 2003/98/EC (EC, 2003), aimed at establishing a basic level of harmonization of rules and procedures throughout the EU to facilitate the reuse of PSI. This idea was reinforced in 2010 with the establishment of the European Digital Agenda, in the context of Europe 2020. One year later, the EU published an official statement on OGD, entitled "Open Data: An Engine for Innovation, Growth and Transparent Governance" (EC, 2011). The 2003 Directive was recently revised and replaced by Directive 2013/37/UE (European Parliament, 2013).

The fact that EU countries are subject only to undemanding, minimum standards with respect to OGD may influence the decisions taken in each case regarding the disclosure of public data for reuse by society. A study conducted by the Spanish Ministry of Finance and Public Administration (2012) showed that stakeholders are calling for greater coordination with respect to the disclosure of PSI and that little such information is currently provided. Therefore, the challenge to be faced is that of ensuring the public the availability of the same type of information, in the same format, and within a similar time frame, to enable its reuse throughout the EU. The EU is aware of this lack of harmonization, and in response to the problem, it is funding Project HOMER (Harmonising open data in the Mediterranean through better access and reuse of PSI), the main objective of which is to establish an effective strategy for harmonizing open data policies and portals in Mediterranean countries, to be implemented by both regional and local governments (HOMER, 2014).

Taking into account the above considerations, this study examines the growing implementation of OGD, analyzing the data disclosed by the central governments of the EU member countries through their official OGD websites. We hope that this research gives rise to greater awareness of the potential benefits to society and European governments from the use of OGD and of the need to create a common framework for access to public data and for their reuse within the EU.

This chapter is organized in seven sections. After this introduction, Section 1.2 provides an overview of the OGD initiative. Section 1.3 then analyzes the different ways in which PSI is provided, produced, and classified. In Section 1.4, we describe the research methodology applied and show how the sample selection was obtained. In Sections 1.5 and 1.6, respectively, we present and discuss the results obtained, and in Section 1.7, the main conclusions are presented.

1.2 The OGD Initiative

The OGD movement emerged in 2004–2005 as the expression of transparency and open government, with particular focus on improvement of democratic processes (Davies, 2010; OKFN and Access-Info, 2010), greater availability of knowledge and information about the public sector (OKFN and Access-Info, 2010), the reuse of this information by sectors of society and business (Davies, 2010), and technological advances in the field of linked data, through the Semantic Web (Tinati et al., 2012).

Coleman (2011) and Davies (2010) have argued that OGD initiatives are largely the result of demands made by a growing minority of civil society and business interests for the government to disclose public information and to allow its reuse under "Open Definition" terms (http://opendefinition.org/okd). This would mean the data being published online proactively, in technically open formats, allowing access free of charge for citizens and at a lower marginal cost for governments, making use of the Conformant License. However, OGD initiatives originated in the academic world, as a development-oriented movement created by scientific communities to ensure free access to the data published in online academic libraries (Murray-Rust, 2008).

One of the key factors in the development of the OGD initiative, according to Alexopoulos, Spiliotopoulou, and Charalabidis (2013), has been the use of technological resources, particularly the transition from Web 1.0 to Web 2.0. The first initiatives in the greater availability of public data, based on Web 1.0, were characterized by the use of infrastructure by means of which data were published in formats that were not processable by machines (e.g., PDF), lacking context information or means of linking with other data. As a consequence, the functionality of first initiatives was limited to the simple uploading and downloading of data. However, the emergence of Web 2.0 dramatically boosted possibilities, offering users the opportunity to exploit the full potential of PSI, focusing on the development of applications to enhance the data provided and on its use and reuse, to benefit scientific research, economic growth, confidence among stakeholders, and broad-based participation (Maio, 2009; Wattal, Schuff, Mandviwalla, & Williams, 2010). Furthermore, the reuse of PSI by means of Web 2.0 enables society to benefit from the collective knowledge of the population, sourcing data via social media and mobile applications (Khan et al., 2012) and interactive data platforms (Mergel, 2010).

The EU has developed various programs and initiatives with respect to open data, notably the creation of the European Public Sector Information (EPSI) Platform, Legal Aspects of Public Sector Information 2.0, and the European data portal. The first (the EPSI Platform) focuses on the importance of the reuse of PSI and on promotion of dynamic reuse of PSI market throughout the EU (EPSI Platform, 2014). The objective of Legal Aspects of Public Sector Information 2.0 is to identify the remaining legal barriers and obstacles to access and reuse of PSI on the European content market and to propose measures and tools to stimulate the progress of the European market toward open data. The European data portal was created in 2010 as part of the LOD2 project "Creating Knowledge out of Interlinked Data." This is a research project on information and communication technology funded by the 7th Framework Programme of the European Commission, that provides access to freely reusable databases provided by local, regional, and national public authorities throughout Europe (PublicData.eu, 2014).

National initiatives by EU countries regarding OGD are very diverse. The following are some of the initiatives and projects carried out by EU countries in the field of open data. In 2012, the Netherlands established an open data knowledge network to develop best practices and share insights; in addition, legal and technical barriers to the reuse of government information were eliminated. Other initiatives have been implemented in Sweden, Slovakia, and Italy. Sweden promotes an enhanced use of the open format (application programming interface), which enables consumers to reuse data; Slovakia has developed its Openness Barometer, which measures the data provided by the open data portals of government ministries and other official bodies (Government Office of the Slovak Republic, 2012); and Italy has developed the program "Simplify Italy" to measure administrative costs and to streamline the consultation of public data. In this respect, the document "Promoting an Annual National Contest for Creative Re-use of Public Data" was published to encourage participation in the area of open data (Italian OGP Action Plan, 2012). Finally, in 2013, the U.K. Government announced the creation of a publicly accessible central register of U.K. company beneficial ownership. This register will contain information on individuals who ultimately own or control more than 25% of a company's shares or voting rights or who otherwise exercise control over the company and its management. The register will cover private and public companies and limited liability partnerships, but not companies that already comply with disclosure rules under the U.K. Disclosure and Transparency Rules (OGP, 2014).

1.3 Making PSI Available

According to the research conducted as part of the ENGAGE (2011) project, PSI can be provided in the following ways:

- OGD catalogue aggregators. These websites provide a single access point from which to locate all other PSI portals and initiatives that publish and maintain lists of PSI catalogues. A good example of this is the European project "Public Sector Information," which compiles the data sets released by government bodies (Peset, Ferrer-Sapena, & Subirats-Coll, 2011).
- Web-based platforms for sharing data source URLs. These websites function as a record of PSI sources in a hierarchically organized machine-processable format. For example, the CKAN generalized data hub portal and the OGD Initiative apply a Microsoft online tool to publish and use a broad range of public data supplied by government agencies.
- Public Sector Sources and Open Data initiatives. This is the main category of OGD portals and contains public websites and other initiatives that publish original information. In most cases, these initiatives assume responsibility for the entire lifecycle of the data and are also the legal owners.

One of the keys to realizing the full potential of the PSI made available is the user's ability to locate the necessary information. This is often achieved by means of a data catalogue tool containing structured OGD records and enabling the use of metadata (Cyganiak & Maali, 2012; Kučera, Chlapek, & Nečaský, 2013). On the other hand, Cunningham (2012) argued that if the metadata are inadequate, they will be difficult to manage, the PSI will not be best used, and public money will be wasted.

For a data catalogue to perform satisfactorily, and to be of sufficient quality, it must meet certain requirements, such as accuracy, completeness, consistency, and timeliness (Batini, Cappiello, Francalanci, & Maurino, 2009). In accordance with these demands,

- All the information recorded in the catalogue should match the data described.
- The entire set of OGD published must be recorded, with no omissions.
- All catalogue records must be consistent, with no contradictions or discrepancies; the same terms or concepts should be used to classify data of the same type or category, and unknown information should be treated in the same way.
- All the information recorded in the catalogue must be current, not only in time but also in its location in the catalogue.

With respect to how PSI records should be opened and shared, OGD initiatives are fairly recent, and thus far, little has been done to establish a classification scheme. However, studies conducted by the World Wide Web Consortium (W3C, 2009) and by Berners-Lee (2009, 2010) are taken as benchmarks for determining levels of openness and data availability.

The W3C, an e-government interest group, has established three stages in which public administrations should provide and share their data (W3C, 2009): first, the publication of data as files in well-known, open formats (i.e., not requiring a user's license), such as CSV or XML; second, the creation of online catalogues; and finally, making the published data machine readable.

Berners-Lee (2009) suggests that governments publish information in accordance with the principles of linked data. The main difference between the hypertext web and the semantic web is that the former links pages or documents by HTML, whereas the latter goes beyond the "document" concept to link structured data (Berners-Lee, 2009). This author also proposed a five-star maturity model with which to measure the ways in which public data are published, according to the reuse and linkage characteristics observed (see Table 1.1).

Shadbolt and O'Hara (2013) claim that data linking enhances the benefits of open data, because with less effort, users can access more information content. Moreover, data linking makes it possible to create a data network with decentralized, heterogeneous sources, and RDF data can be interconnected via URI links (Hausenblas, 2009).

Both and Schieferdecker (2012) point out the important role played by users as a valuable source of OGD feedback, by notifying the editors of an OGD catalogue of records with missing information and by detecting an incorrect description of a data set. In this respect, open data can help establish a favorable

Table 1.1 Five Stars of Linked Data Classification for Linked OGD

*	Data are published online under an open license (e.g., PDF).
**	Data are published in machine-readable format (e.g., as an Excel spreadsheet).
***	Data are published in machine-readable, nonproprietary format, using open standards such as CSV.
****	Data are published in accordance with linked-data principles, such as RDF.
*****	Available data are linked to other RDF data.

Source: Berners-Lee, T. (2010). *Design issues: Linked data.* Retrieved April 2, 2014, from http://www.w3.org/DesignIssues/LinkedData.html.

environment for proactive participation, providing users with the opportunity to create applications using the data sets made available on the web portal. In turn, these applications could promote collaboration and cooperation by stakeholders in supplying additional information necessary to keep the applications up to date (Kassen, 2013).

1.4 Methodology

In designing the methodology to be applied, to analyze the use of OGD in the EU and to determine to what extent these initiatives are being used and which countries have the most complete OGD portals, we first reviewed the literature on OGD. In this respect, Kassen (2013) examined the Chicago Open Data project using a study framework based on an analysis of the corresponding legal, political, and economic environments and on an evaluation of the open data project itself. In another context, Peset et al. (2011) discussed, first, the global outlook for Open Data and Linked Open Data and, second, the particular situation in Spain. Shadbolt and O'Hara (2013) reviewed the state of linked OGD in the UK in the context of the potential for the organizations publishing the data and the Linked Data web itself, as well as administrative and political issues.

Another important study in this field was presented by Bertot, Jaeger, and Grimes (2010), who described the practical realities of measuring U.S. Federal agency transparency and open government initiatives. Alexopoulos et al. (2013) considered OGD in Greece from three standpoints: functional, semantic, and technical. Kučera et al. (2013) described current OGD activities in the Czech Republic and discussed the quality of the catalogue records and proposed techniques for their improvement. Finally, Kalampokis, Tambouris, and Tarabanis (2011) proposed a classification scheme for OGD initiatives and applied it to a sample of 24 OGD initiatives produced by disparate public agencies in terms of governance level (federal, national, regional, and local). In summary, many diverse methodologies and types of analysis have been applied, but few studies have specifically analyzed OGD portals in the EU.

Therefore, taking into account the above-cited texts, the methodology applied in the present study is based on four types of analysis: functional, semantic, quality, and content.

1.4.1 Functional Analysis

In line with Alexopoulos et al. (2013) and Kalampokis et al. (2011), we examined OGD from a functional standpoint. For this purpose, an analysis of the tools and functions available for users to acquire information from OGD portals was conducted. This analysis was composed of the following dimensions (see Figure 1.1):

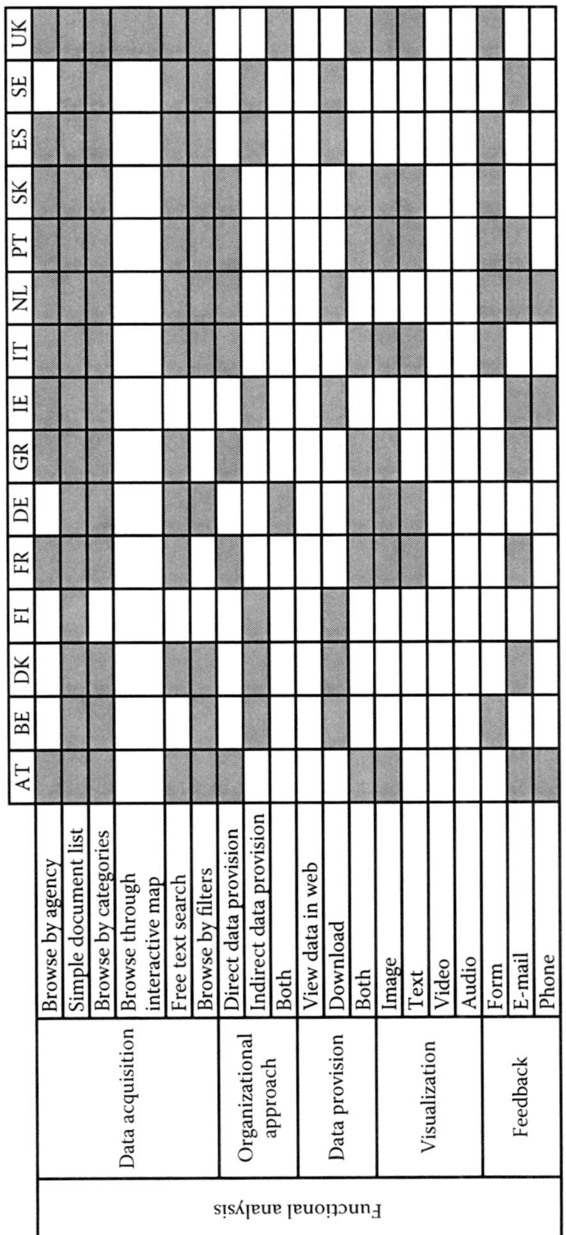

Figure 1.1 European functional analysis results.

- Data acquisition: The tools and utilities available for data search, such that the user can obtain information by agency, by category, or by key words, among other criteria (Alexopoulos et al., 2013).
- Organizational approach for providing data: We distinguish between two approaches: direct data provision, when data from various government agencies are published in a single OGD portal, and indirect data provision, when the data are published in a decentralized way (with each organization or agency using its own website), while the OGD portal only offers a link to these sites (Kalampokis et al., 2011).
- Data provision: OGD portals, apart from supplying data directly or indirectly, may offer the option to view information online, to download it, or both (Kalampokis et al., 2011).
- Visualization: If the OGD portal allows online data visualization, this can be done using various display modes, including text, image, video, and audio (Alexopoulos et al., 2013).
- Feedback: The possibilities offered by OGD for stakeholders to communicate with government (by forms, e-mail or telephone).

1.4.2 Semantic Analysis

A semantic analysis was performed by examining the representation and structure of the information available on the OGD portal. The data thus made available can be in the form of simple, flat structures or that of comprehensive, linked ones. In this analysis, we followed the methodology proposed by Alexopoulos et al. (2013) and analyzed the following dimensions (see Figure 1.2):

- Level of metadata: Metadata is structured information that describes other information and enables us to find, manage, and preserve it. In line with Interoperability Solutions for European Public Administrations (2011), we distinguish between metadata ignorance, closed, open, and reusable metadata.
- Level of open data: Using the Five Stars of Linked Data classification proposed by Berners-Lee (2010), but reduced to four stars because from the point of view of a user, it is not known whether available data are linked to other RDF data.
- Multilingualism: The multilingual capabilities of an OGD portal are measured, taking into account, therefore, the language in which the data are presented. For this purpose, we distinguish between official language and another language (Alexopoulos et al., 2013).
- Data format: In accordance with the formats used by Alexopoulos et al. (2013) and those found in the OGD portals analyzed, we examined the most common formats and analyzed a total of 19 types of formatting.

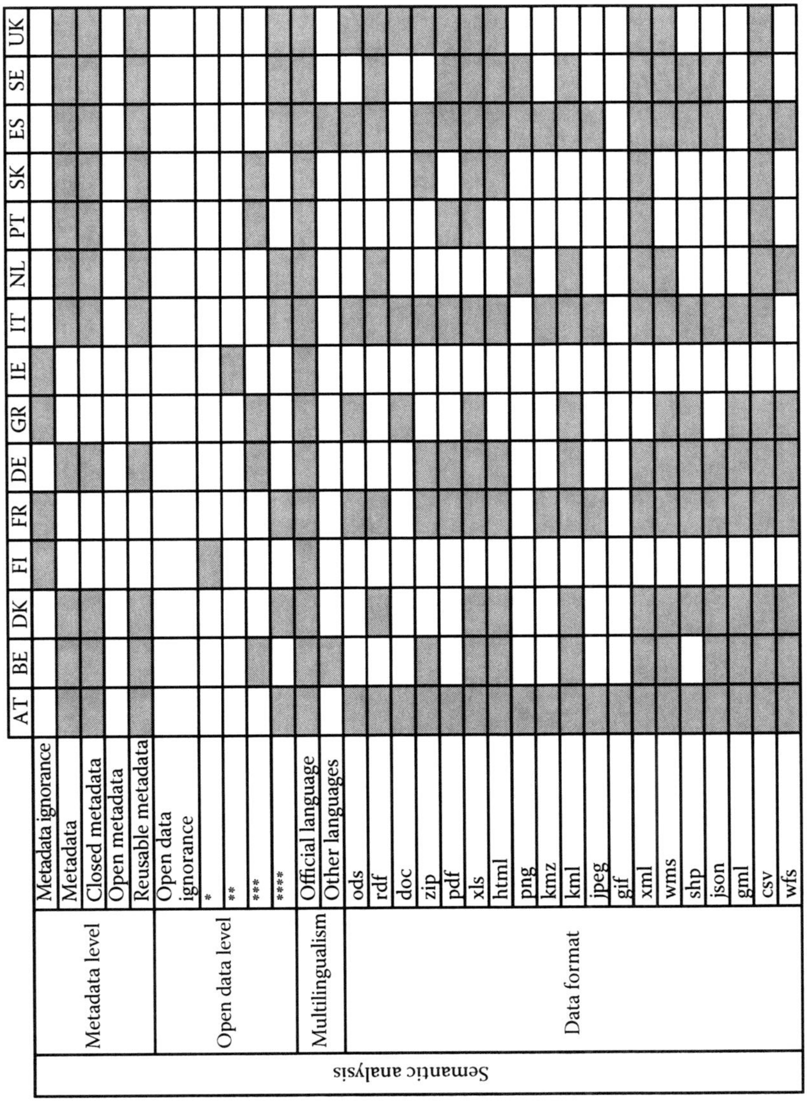

Figure 1.2 **European semantic analysis results.**

1.4.3 Quality Analysis

Following Kučera et al. (2013), we analyzed the quality of OGD catalogues. Taking as a model the concept of quality defined by Batini et al. (2009), OGD catalogues were examined to determine whether they make data searching easier and more precise, that is, whether the information thus provided is accurate, complete, consistent, and timely. Under these premises, the following dimensions were analyzed (see Figure 1.3):

- Free information: One of the premises of any open data is that the information should be made available to users free of charge. In the context of OGD, this premise is even more relevant because the data are public. In addition, the fact that society has unrestricted access to all public information can be interpreted as a favorable indication of the quality of the OGD catalogue.
- Data accuracy: The content of the data must match the information described in the catalogue (Kučera et al., 2013).
- Timeliness: The data must be up to date (Kučera et al., 2013).
- Glossary or list of terms: A list should be provided of the words/abbreviations most commonly used and their definitions (Kučera et al., 2013).
- Possibility of incorporating new data: Kassen (2013) noted that OGD can be used to promote public participation. Thus, allowing users to incorporate additional information into OGD catalogues can enhance their quality, provided that this new information is of value to society. Therefore, apart from considering feedback channels in the functional analysis, it is also necessary to determine whether there exists an application allowing users to add new data to the OGD.

1.4.4 Content Analysis

The studies by Alexopoulos et al. (2013) and Kučera et al. (2013) examined how data are provided in OGD portals and the quality of the catalogues presented. However, they did not take into consideration the type of information disclosed, the categories into which the information is grouped, the filters used to locate the information, the total volume of data, or the number of agencies involved. Accordingly, and to determine how the information provided by each OGD is grouped and filtered, we analyzed the content of OGD catalogues, examining the following dimensions (see Figure 1.4):

- Categories: In accordance with the provisions of PSI Directive 2003/98/EC (EC, 2003) and following Alexopoulos et al. (2013), the PSI made available in OGD portals for use, processing, and reuse is classified into the following main fields (see Table 1.2): business, geographic, legal, meteorological, social, transport, and other.
- Number of catalogues: This indicates the total number of catalogues available in each OGD portal.

		AT	BE	DK	FI	FR	DE	GR	IE	IT	NL	PT	SK	ES	SE	UK
Quality analysis	Free information															
	Accuracy of data															
	Timeliness (days)	518	n/a	1661	n/a	n/a	384	1230	n/a	656	n/a	1529	n/a	888	461	1189
	Glossary/list of terms															0
	Ability to add data															0

Figure 1.3 European quality analysis results.

			AT	BE	DK	FI	FR	DE	GR	IE	IT	NL	PT	SK	ES	SE	UK
Content analysis	Categories	Business data	2	1	1	0	1	1	0	7	7	0	1	0	5	3	2
		Geographic data	2	0	1	0	1	1	2	0	6	18	1	0	1	1	1
		Legal data	0	1	1	0	0	2	0	1	1	2	1	2	1	0	1
		Meteorological data	0	0	0	0	0	1	0	1	0	0	0	0	0	0	0
		Social data	8	6	8	0	5	7	6	10	23	7	8	4	10	3	5
		Transport data	1	1	1	0	1	1	0	1	0	4	1	0	1	1	0
		Other	1	4	1	0	2	1	3	5	2	7	3	0	4	3	1
	Number of catalogues		14	13	13	0	10	14	11	25	39	38	15	6	22	11	10
	Number of data		1182	n/a	n/a	n/a	12,896	6755	n/a	1220	6659	6009	484	205	1495	42	17,849
	Number of agencies		1	0	0	0	265	0	28	59	62	50	19	10	15	0	10
	Number of filters		4	3	3	0	0	4	0	0	4	5	4	3	4	5	6
	Filters	Category	■	■	■			■			■	■	■	■	■	■	■
		Subcategory			■											■	
		Agency	■								■	■	■	■	■		■
		Tag														■	
		Format	■					■			■			■	■	■	■
		License						■			■	■				■	■
		Linked			■							■					■
		Geography		■								■					
		Date											■		■		
		Keyword						■					■				
		Other	■	■													■

Figure 1.4 European content analysis results.

Table 1.2 Classification Categories for Catalogues

Business information: includes Chamber of Commerce information, official business registers, patent and trademark information, and public tender databases
Geographic information: includes address information, aerial photos, buildings, cadastral information, geodetic networks, geology, hydrographic data, and topographic information
Legal information: includes decisions of national, foreign, and international courts; national legislation; and treaties
Meteorological information: includes climate data and models and weather forecasts
Social data: include various types of statistics (economic, public spending, employment, health, population, public administration, social)
Transport information: includes information on traffic congestion, road works, public transport, and vehicle registration

Source: EC (2003). *Directive 2003/98/EC of the European Parliament and of the Council of 17 November 2003 on the re-use of public sector information.* Retrieved July 2, 2014, from http://ec.europa.eu/information_society/policy/psi/docs/pdfs/directive/psi_directive_en.pdf.

- Number of data: This indicates the total volume of data available in each OGD portal.
- Number of agencies: This indicates the total number of agencies providing data to the OGD portal.
- Number of filters: This indicates the total number of filters that can be used simultaneously to refine an information search in an OGD portal.
- Types of filters: Alexopoulos et al. (2013) observed that an OGD portal should enable searching with filters but did not analyze the different types of filters available. Therefore, in this book, we examined the 11 filters most commonly applied: category, subcategory, agency, tag, format, license, linked, geography, date, keyword, and other.

1.5 Sample

The European Digital Agenda, which forms part of the Europe 2020 strategy, is a landmark event in the area of policies for open data and the reuse of PSI in the EU (EC, 2011). In this vein, in 2011, the European Commission published the communication "Open Data: An Engine for Innovation, Growth and Transparent Governance" (EC, 2011). Accordingly, and taking into account the commitment made by the EU in the field of open data, we decided to analyze the present state of OGD in Europe.

To select the EU countries for the study sample, we first examined the list of official OGD portals offered by the Data.gov platform in the "International Countries" category. Data.gov is the main site for the U.S. government's Open Data and also the world's largest open government website (Hendler, Holm, Musialek, & Thomas, 2012). Of the 16 EU countries included in Data.gov, one was rejected because its OGD website did not meet the basic standards for an open data portal. Therefore, 15 countries, which represent 80.06% of the total EU population, were selected for the final study sample: Austria (AT), Belgium (BE), Denmark (DK), Finland (FI), France (FR), Germany (DE), Greece (GR), Ireland (IE), Italy (IT), Netherlands (NL), Portugal (PT), Slovak Republic (SK), Spain (ES), Sweden (SE), and United Kingdom (UK).

1.6 Results

Following the methodology described at the outset, we now present the results of the functional, semantic, quality, and content analysis, showing both individual-level data for each OGD portal analyzed and data for the whole EU. These results evidence diversity and the lack of homogeneity in the production and availability of data within the EU.

In the functional analysis, and starting with the question of data acquisition, all the countries analyzed present a simple document list, that is, a list of the data made available to users, usually ordered alphabetically, but under no special criterion or search field (see Figure 1.5). Most of them also provide the option of searching for data by category and by free text, in which key words are entered to locate data. Less common is the option to browse by filters and/or by agency, which is provided by around half of the countries studied, see, for example, Austria and Italy (see Figure 1.1). On the contrary, searching by interactive maps is available only in the U.K. portal, which is one of the most complete OGD systems. An interactive map displays the information in question in map form, incorporating a digital mapping tool to optimize geoinformation search by users.

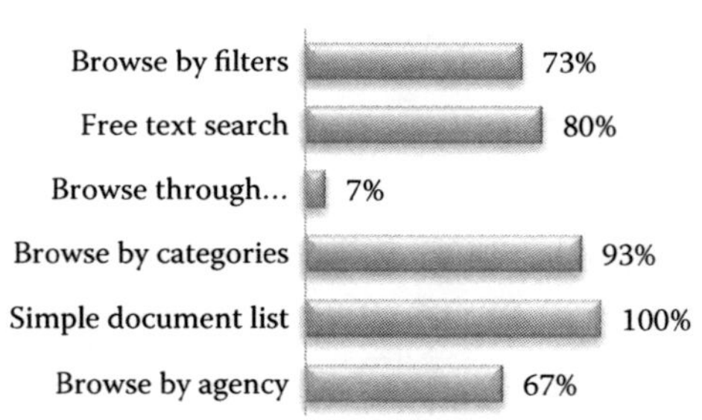

Figure 1.5 Data acquisition.

Examination of how these countries provide information on their OGD portals reveals that seven do so directly, 47% opt for indirect data provision, and 13% combine these two approaches (see Figure 1.6). Conversely, 40% of the countries, including Belgium and Spain, provide data only indirectly (see Figure 1.1).

Regarding data provision, 53% of the OGD portals analyzed allow data to be viewed online and also downloaded; this is the case in Denmark and France. In contrast, 47% allow data to be downloaded only, without previewing it on the web; this is the case for the Netherlands and Finland, for example (see Figure 1.7). In this respect, of the eight portals where information can be viewed online, it is offered as an image in six cases and as text in all of them (see Figure 1.8).

The final dimension of functional analysis concerns the feedback options available. The main ones are forms (39%) and e-mail (44%), with telephone contact being less common (17%) (see Figure 1.9). The leader in this respect is the Netherlands portal, where users have telephone, e-mail, and forms available as feedback channels. In contrast, countries like Finland and Denmark do not provide any feedback channel to users (see Figure 1.1).

The results of the semantic analysis are summarized in Figure 1.2. The level of metadata is high, with 73% of OGD portals applying this system, which, moreover, provides

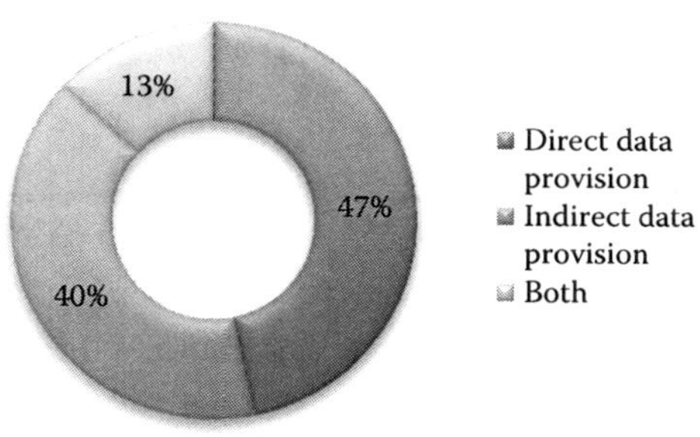

Figure 1.6 Organizational approach.

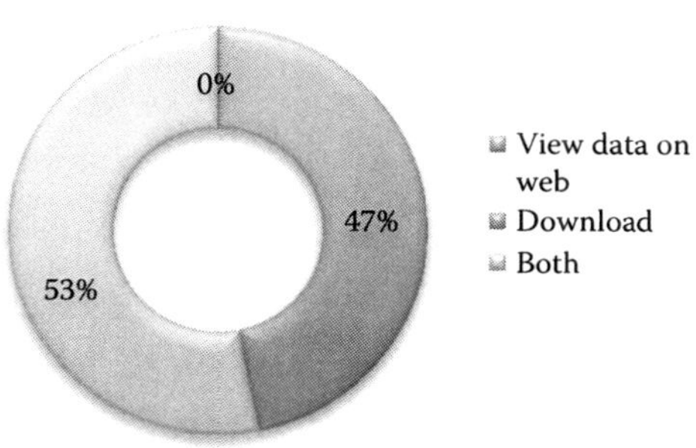

Figure 1.7 Data provision.

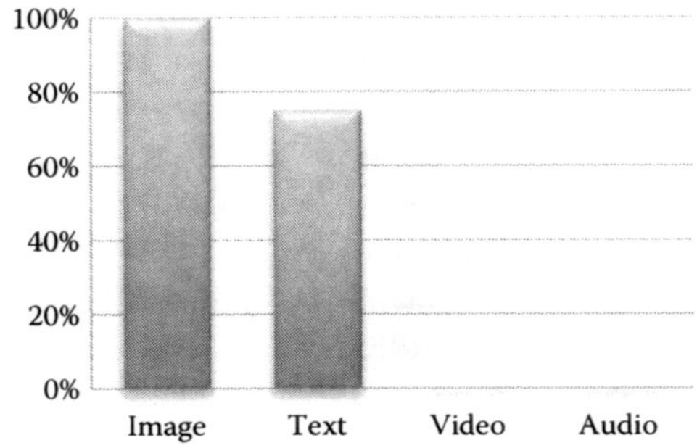

Figure 1.8 Visualization.

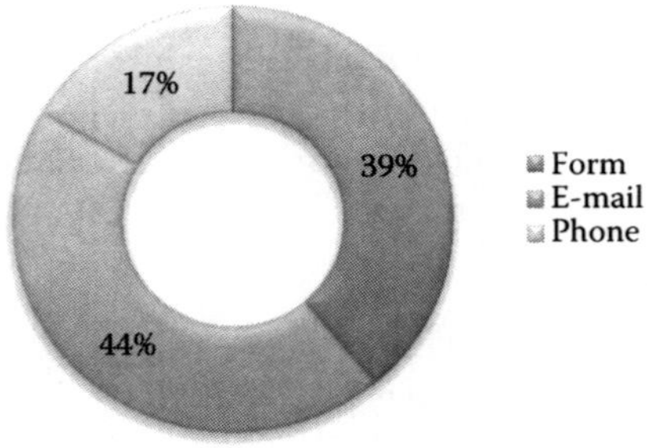

Figure 1.9 Feedback.

reusable metadata (see Figure 1.10). Among the countries that do not have metadata, the cases of Finland and Ireland are striking (see Figure 1.2), as their results are poor not only in the semantic analysis but also, in general, in all the aspects considered. On the other hand, the overall level of open data can be considered acceptable, with 53% of the OGD portals meeting the highest levels of the classification used (the Berners-Lee Five Stars of Linked Data, reduced to four stars) (see Figure 1.11), whereas Ireland and Finland present the lowest levels (see Figure 1.2).

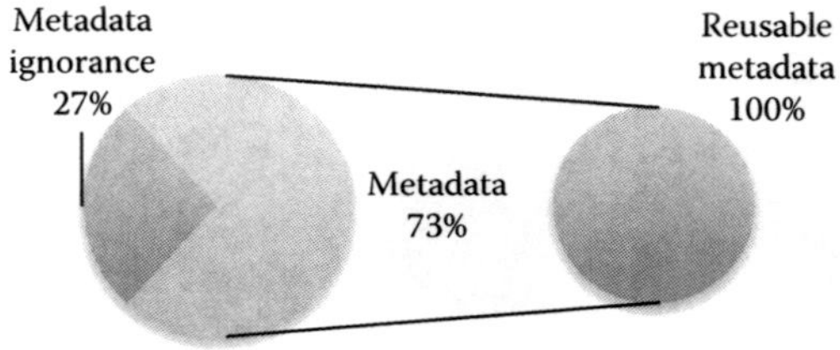

Figure 1.10 Metadata level.

On the other hand, only two countries offer the possibility of consulting their OGD in a language other than the official one; this is the case for Spain and Finland (see Figure 1.2). Although some sites do enable other languages for browsing, information can be downloaded only in the official language, although in Belgium, there is more than one official language.

Of the formats used to present the data, the most widely used is CSV, which is found in 13 portals, followed by XML and XLS (see Figure 1.12). At the other extreme, the least used formats are those that are least editable, such as image types (GIF and JPEG). Austria provides the largest number of formats for users, followed by Italy, Spain, and France (see Figure 1.2).

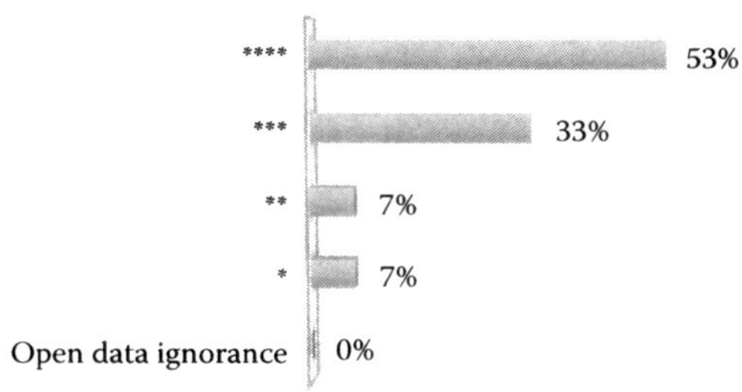

Figure 1.11 Open data level.

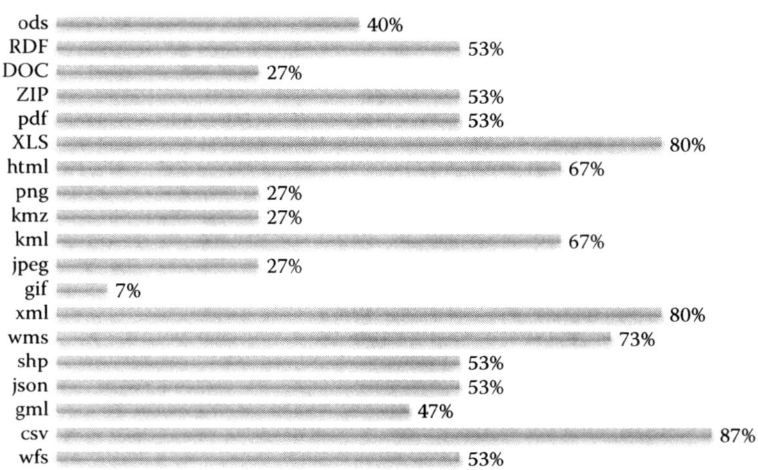

Figure 1.12 Data format.

The analysis of data quality (see Figure 1.3) shows that OGD is available free of charge in all cases. The data accuracy requirement, that is, that the data content matches the information described in the catalogue, is also met in all the portals analyzed. As for the timeliness of the data, that is, the requirement that the data provided in OGD portals be up to date and not merely historical, this question is measured by the number of days the least up-to-date information has remained on the portal; in this respect, the results show that Germany is the country with the most up-to-date OGD. However, it should be clarified that we were not able to measure this parameter in all cases (see Figure 1.3).

Only Italy, Portugal, and Slovakia provide a glossary of relevant terms in their OGD portals, and only six portals enable external users to add new data. Therefore, it seems that these OGD portals fail to make best use of the possibilities offered by the platform as a tool for public participation and communication.

The fourth analytic approach concerns the content of the OGD portals. With regard to the categories into which the catalogues are grouped (see Figure 1.4), most catalogues contain information of a social nature; this is true of 46.4% of the OGD portals, whereas meteorological content is the least commonly found, in only 0.8% (see Figure 1.13). However, the number of catalogues provided in an OGD portal is not necessarily conclusive because an OGD that has a catalogue for each category (provided that each catalogue contains an adequate quantity of information) can be considered more complete than one with many catalogues in all possible categories but which provides little significant information in each catalogue.

The total volume of data was the dimension that presented the greatest dispersion. In this respect, the UK and France were at the forefront, with 17,849 and 12,896 data files, respectively, whereas Sweden had only 42, and in another four portals, it was impossible to determine the total number of files available (see Figure 1.4). As for the number of agencies providing the data, this information could not be determined in six cases. France was outstanding in this respect, deriving its OGD from 265 agencies (see Figure 1.4).

Finally, we analyzed the types of filters that can be applied in performing a data search. The most commonly found were those enabling the user to browse by category

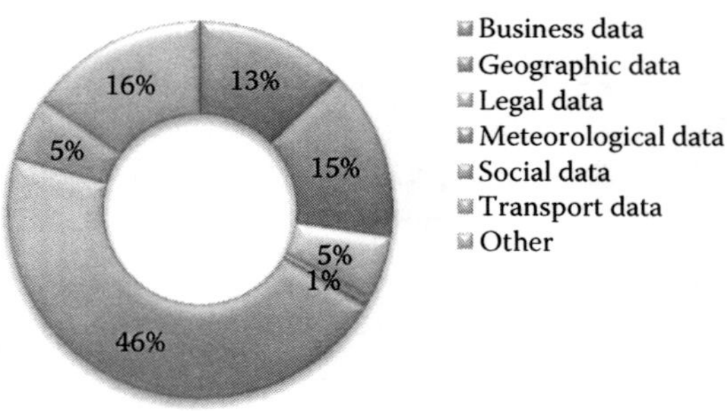

Figure 1.13 Categories.

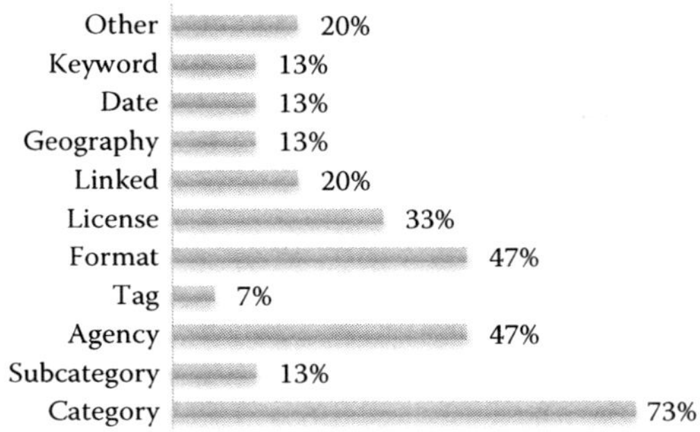

Figure 1.14 Types of filters.

(73%), followed by agency and format (47%) (see Figure 1.14). Conversely, tag filters were least often provided (7%). In this dimension, too, the United Kingdom excelled, enabling the use of up to six filters simultaneously, followed by the Netherlands and Sweden, with five types of filters each. In contrast, France, Finland, Greece, and Ireland do not enable the use of search filters to locate data (see Figure 1.4).

1.7 Conclusions

Open data has revolutionized access to online data. In the field of public administration, the movement initiated by the United States has rapidly spread to most developed countries. However, despite the increasing prevalence of OGD in Europe, as yet, little research attention has been paid to the situation in the EU.

As observed throughout this study, if OGD is to be developed to its maximum potential, public data should be supplied in a standard form so that it can be reused. However, our analysis reveals considerable heterogeneity among EU portals, which present widely varying structures, with different data search processes and different formats for presenting information. Therefore, further initiatives are required to promote the development and improvement of international models or standards, to harmonize and coordinate access to PSI through these portals.

In addition, it would be advisable for governments to make use of OGD to create a favorable context for proactive public participation, inviting stakeholders to cooperate in the provision of further data. In addition to providing information, users could also create applications (apps), making use of the content presented in OGD portals. In short, it would be desirable for governments to establish a new type of environment for cooperation with society. At present, however, the results obtained suggest that the countries of the EU are making little use of the feedback possibilities offered by their portals and, in many cases, do not allow the introduction of new data.

Open Government Data portals are intended to make useful information available to society. Although the present study was not designed to analyze the use made of the public information thus made available, we find that most portals have not implemented any system to measure the popularity of their data sets, although a system of this type could be used to identify and clarify the information needs of the community and thus enable more such data to be published.

In summary, the results obtained in the present study contribute to our understanding of how public data are presented, how accessible these data are, and how these data are processed. We also determined the extent to which the OGD initiative is being developed and implemented in different EU countries. In future research in this field, it would be useful to examine a larger study sample and to analyze the factors that may influence the adoption and implementation of OGD initiatives.

Author Biographies

Alejandro Sáez Martín is a researcher in government administration in the Department of Economics and Business, University of Almería (Spain). His research interests are focused on the financial information disclosures on the web (e-government) and on the management system and financial information in the federal and local governments. He is an author of numerous articles in national and international journals, including *Retos*, *Transforming Government: People, Process and Policy*, and *Local Government Studies*.

Arturo Haro de Rosario, who has a PhD in governmental accounting, is a senior lecturer at the Department of Economics and Business, University of Almería (Spain). His research interests are focused on the financial information disclosures on the web (e-government) and on the management system and financial information in the federal and local governments. He is an author of numerous articles in national and international journals, including *Small Business Economics*, *Transforming Government: People, Process and Policy*, *Revista Finanzas y Politicas Públicas*, *Revista de Ciencias Sociales*, and *Local Government Studies*.

Carmen Caba Pérez, who has a PhD in governmental accounting, is an associate professor at the Department of Accounting and Finance, University of Almería (Spain). She has been accredited as a full professor by ANECA, Spain's National Agency for Quality Assessment and Accreditation. Her research interests are focused on the financial information disclosures on the web (e-government) and on the management system and financial information in the federal and local governments. She is an author of numerous articles in national and international journals, including *Government Information Quarterly*, *Online Information Review*, *International Review of Administrative Sciences*, *American Review of Public Administration*, *Public Administration and Development*, *Local Government Studies*, and *The International Journal of Public Sector Management*.

Acknowledgment

This research was carried out with financial support from the Regional Government of Andalusia (Spain), Department of Innovation, Science and Enterprise (Research Projects P10-SEJ-06628 and P11-SEJ-7700).

References

Alexopoulos, C., Spiliotopoulou, L., & Charalabidis, Y. (2013). Open data movement in Greece: A case study on open government data sources. *17th Panhellenic Conference on Informatics*, 279–286.

Alonso, J., Azañón, O., Ambur, O., Bennett, D., Novak, K., & Sheridan, J. (2009). *Improving access to government through better use of the web*. W3C Working Draft 2009. Retrieved April 4, 2014, from http://www.w3.org/TR/2009/NOTE-egov-improving-20090512.

Ayers, D. (2007). Evolving the link. *IEEE Internet Computing, 11*(3), 94–96.

Batini, C., Cappiello, C., Francalanci, C., & Maurino, A. (2009). Methodologies for data quality assessment and improvement. *ACM Computing Surveys, 41*(3), 1–52.

Bauer, F., & Kaltenböck, M. (2011). *Linked open data: The essentials*. Vienna: DGS-Druck.

Bennet, D., & Harvey, A. (2009). *Publishing open government data*. W3C Working Draft 2009. Retrieved April 4, 2014, from http://www. w3.org/TR/2009/WD-gov-data-20090908.

Berners-Lee, T. (2009). *Putting government data online*. Retrieved April 2, 2014, from http://www.w3.org/DesignIssues/GovData.html.

Berners-Lee, T. (2010). *Design issues: Linked data*. Retrieved April 2, 2014, from http://www.w3.org/DesignIssues/LinkedData.html.

Bertot, J. C., Jaeger, P. T., & Grimes, J. M. (2010). Using ICTs to create a culture of transparency: E-government and social media as openness and anti-corruption tools for societies. *Government Information Quarterly, 27*(3), 264–271.

Both, W., & Schieferdecker, I. (2012). *Berliner Open Data-Strategie. Organisatorische, rechtliche und technische Aspekte offener Daten in Berlin*. Berlin: Fraunhofer Verlag.

Chun, S. A., Shulman, S., & Sandoval, A. R. (2010). Government 2.0: Marking connections between citizens. *Data and Government. Information Polity, 15*(12), 1–9.

Coleman, E. (2011). *From new public management to open governance—A new future or the more things change the more they stay the same?* (MPA Thesis). Coventry, UK: University of Warwick.

Cunningham, A. (2012). *El uso de Metadatos en la Administración Pública*. Digital Preservation Europe. Retrieved April 2, 2014, from http://www.digitalpreservationeurope.eu/publications/briefs/es_metadata_for_public_administrations.pdf.

Cyganiak, R., & Maali, F. (2012). *Use cases and requirements for the data catalog vocabulary*. Retrieved April 2, 2014, from http://dvcs.w3.org/hg/gld/raw-file/default/dcat-ucr/index.html.

Davies, T. (2010). *Open data, democracy and public sector reform: A look at open government data use from data.gov.uk* (Master's thesis). Oxford, UK: University of Oxford.

EC. (2003). *Directive 2003/98/EC of the European Parliament and of the Council of 17 November 2003 on the re-use of public sector information*. Retrieved July 2, 2014, from http://ec.europa.eu/information_society/policy/psi/docs/pdfs/directive/psi_directive_en.pdf.

EC. (2011). *Open data: An engine for innovation, growth and transparent governance*. COM (2011) 882 final, Brussels. Retrieved April 2, 2014, from http://eur-lex.europa.eu/LexUriServ/LexUriServ.do?uri=COM:2011:0882:FIN:EN:PDF.

ENGAGE. (2011). *Analysis report of public sector data and knowledge sources*. Retrieved April 2, 2014, from http://www.engage-project.eu/engage/wp/wpcontent/plugins/download-monitor/dow nload.php?id=4.

EPSI Platform. (2014). Retrieved April 9, 2014, from http://www.epsiplatform.eu/.

European Parliament. (2013). *Directive 2013/37/UE, June 26, 2013*. Retrieved April 2, 2014, from http://www.boe.es/doue/2013/175/L00001-00008.pdf.

Government Office of the Slovak Republic. (2012). *Open government initiative*. Retrieved July 2, 2014, from http://www.otvorenavlada.gov.sk/slovakia-presented-its-open-government-partnership-acti on-plan-at-the-annual-ogp-conference-in-brasilia/.

Harrison, T. M., Santiago, G. G., Burke, B., Cook, M., Cresswell, A., Helbig, N., Hrdinová, J., & Pardo, T. (2011). Open government and e-government: Democratic challenges from a public value perspective. *Proceedings of the 12th Annual International Conference on Digital Government College Park*, College Park, MD.

Hausenblas, M. (2009). Exploiting linked data to build web applications. *IEEE Internet Computing, 13*(4), 68–73.

Hendler, J., Holm, J., Musialek, C., & Thomas, G. (2012). US government linked open data: Semantic.data.gov. *IEEE Intelligent Systems, 27*(3), 25–31.

HM Government. (2009). *Putting the frontline first: Smarter government*. Retrieved April 2, 2014, from http://www.hmg.gov.uk/media/52788/smarter-government-final.pdf.

HOMER. (2014). *Harmonising open data in the Mediterranean through better access and reuse of public sector information*. Retrieved April 8, 2014, from http://homerproject.eu/.

Interoperability Solutions for European Public Administrations. (2011). *Towards open government metadata*. European Commission. Retrieved April 2, 2014, from https://joinup.ec.europa.eu/sites/default/files/towards_open_government_metadata_0.pdf.

Italian OGP Action Plan. (2012). *Open government partnership Italian action plan*. Retrieved July 7, 2014, from http://www.funzionepubblica.gov.it/media/968941/piano%2011%20aprile%20%20opengovpartnership%20for%20consultation_en.pdf.

Kalampokis, E., Tambouris, E., & Tarabanis, K. (2011). A classification scheme for open government data: Towards linking decentralized data. *International Journal of Web Engineering and Technology, 6*(3), 266–285.

Kassen, M. (2013). A promising phenomenon of open data: A case study of the Chicago open data project. *Government Information Quarterly, 30*, 508–513.

Khan, G. F., Young, H. Y., & Park, H. W. (2012). Social media use in public sector: A comparative study of the Korean & US government agencies. *Proceedings of the 8th International Conference on Webometrics, Informatics and Scientometrics*, Seoul, South Korea, October 23–26.

Kučera, J., Chlapek, D., & Nečaský, M. (2013). Open government data catalogs: Current approaches and quality perspective. In A. Kö, C. Leitner, H. Leitold, & A. Prosser (Eds.), *Technology-enabled innovation for democracy, government and governance* (pp. 152–166). Praga (Czech Republic): Springer.

Lathrop, D., & Ruma, L. (2010). *Open government: Collaboration, transparency, and participation in practice*. Sebastopol, CA: O'Reilly Media.

Maio, A. D. (2009). *Web 2.0 in UK Government needs a director of digital engagement*. Homepage of Gartner. Retrieved July 5, 2014, from http://blogs.gartner.com/andrea_dimaio/2009/05/13/web-20-in-uk-government-needs-a-director-of-digital-engagement/.

Mergel, I. (2010). Government 2.0 revisited: Social media strategies in the public sector. *American Society for Public Administration, 33*(3), 7–10.

Murray-Rust, P. (2008). Open data in science. *Serials Review, 34*(1), 52–64.

OGP. (2014). *Open Government Partnership*. Retrieved April 5, 2014, from http://www.opengovpartnership.org/.

OKFN & Access-Info. (2010). *Beyond access: Open government data and the "right to reuse."* Retrieved April 2, 2014, from http://www.accessinfo.org/documents/Access_Docs/Advancing/Beyond_Access_7_January_2011_web.pdf.

Organization for Economic Co-operation and Development. (2008). *OECD recommendation of the Council for enhanced access and more effective use of public sector information*. Retrieved April 5, 2014, from http://www.oecd.org/internet/ieconomy/40826024.pdf.

Orszag, P. (2009). *Open government directive. Memorandum for the heads of executive departments and agencies*. Retrieved April 2, 2014, from http://www.whitehouse.gov/open/document ts/opengovernment-directive.

Osimo, D., Campbell, D., Kerr-Stevens, J., Bishop, C., & Bryant, L. (2009). *Public services 2.0: Web 2.0 from the periphery to the centre of public service delivery*. Brussels: European Commission Information Society and Media.

Peset, F., Ferrer-Sapena, A., & Subirats-Coll, I. (2011). Open data y linked open data: Su impacto en el área de bibliotecas o documentación. *El profesional de la información, 20*(2), 165–173.

PublicData.eu. (2014). Retrieved April 9, 2014, from http://publicdata.eu/.

Shadbolt, N., & O'Hara, K. (2013). Linked data in government. *IEEE Computer Society, 17*(4), 72–77.

Sonáglio, C. (2013). Open government data: A value chain model proposal. *Proceedings of the 14th Annual International Conference on Digital Government Research*, 285–286.

Tinati, R., Halford, S., Carr, L., & Pope, C. (2012). *Using mixed methods to track the growth of the web: Tracing open government data initiatives*. Retrieved April 2, 2014, from http://eprints.soton.ac.uk/335270/1/WWW12_Web_Science_Track__Tracing_Open_Government_Data_Initiatives.pdf.

W3C. (2009). *Publishing open government data*. W3C Working Draft. Retrieved April 2, 2014, from http://www.w3.org/TR/gov-data/.

Wattal, S., Schuff, D., Mandviwalla, M., & Williams, C. (2010). Web 2.0 and politics: The 2008 U.S. presidential election and an e-politics research agenda. *MIS Quarterly, 34*(4), 669–688.

Chapter 2

Best Practices in E-Governance: A Comparative Study Based on the Rutgers University Worldwide Digital Governance Survey

Marc Holzer and Yueping Zheng

Contents

Abstract

E-government has been widely accepted, and more governments continuously adopt it to provide better services. Great changes have occurred in the past decade, leading to the transformation from e-government to e-governance and emphasizing more citizen participation. Although increasing numbers of e-governance initiatives have been evident, e-governance development varies substantially both at the municipal and national levels. Following a model we have been developing and applying at Rutgers University–Newark since 2003 for evaluating the websites of the largest municipalities in the 100 most wired nations, our 2011–2012 Worldwide Digital Governance Survey indicates that municipalities vary substantially in e-governance. The average score for the top five municipalities (Seoul, Toronto, Madrid, Prague, and Hong Kong) is 66.54, whereas Lagos, Kuwait City, Baghdad, Asuncion, and Tashkent rank lowest at an average of only 12.05. A substantial gap also exists between municipalities in Organization for Economic Co-operation and Development (OECD) countries and those in non-OECD countries by a ratio of almost 2:1. Similar variances at the national level have been confirmed by e-government surveys conducted by the United Nations. To better illustrate the latest innovations and usage of information and communication technologies to improve e-governance, best practices need to be explored because they can contribute both to the academy and to practice. In this study, excellent examples from Seoul, Toronto, Berlin, and Hong Kong are highlighted to reflect the latest innovations and enlighten the development of e-governance in municipalities worldwide. We suggest four strategies to achieve success in e-governance, including (1) upgrading information and communication technology infrastructure, (2) improving human resource management, (3) creating a corresponding political environment, and (4) promoting administrative performance.

Keywords: citizen participation, e-government, government website

2.1 Introduction

In the past two decades, "information technology (IT) has been adopted and implemented around the world as a means of improving government performance" (Lee, Chang, & Berry, 2011, p. 444), particularly in e-government. The importance of e-government has been widely accepted, and more governments continuously adopt computer-based approaches to provide better services through websites. The sharp increase in Internet users has promoted change from "single and disorder" to "comprehensive and ordered"; citizens have pushed e-government to change from "providing information and services" to "enabling citizen participation," from "government oriented" to "citizen oriented." All of these changes transform e-government to e-governance, which emphasizes citizen participation and meeting public needs.

Although more and more e-governance initiatives have been evident, e-governance development varies substantially both at municipal and national levels. Best practices need to be explored to illustrate the latest innovations and usage of information and communication technologies (ICTs) to improve e-governance. This could contribute both to the academy and to practice. Following a model we have been developing and applying since 2003, in the 2011–2012 Worldwide Digital Governance Survey from Rutgers University–Newark, we have evaluated the websites of the largest municipalities in the 100 most wired nations in terms of five dimensions of digital governance: privacy/security, usability, content, services, and citizen participation. Our study indicates that municipalities vary substantially. Seoul, Toronto, Madrid, Prague, and Hong Kong rank as the top five municipalities, with an average score 66.54, whereas Lagos, Kuwait City, Baghdad, Asuncion, and Tashkent rank lowest, with an average score of 12.05. A substantial gap also exists between municipalities in Organization for Economic Co-operation and Development (OECD) countries and those in non-OECD countries by a ratio of almost 2:1.

This comparative study in e-governance across municipalities worldwide explores best practices, illustrating how ICTs are being used to better serve the public and promote citizen participation. On the basis of the 2011–2012 Worldwide Digital Governance Survey, several municipalities such as Seoul, Toronto, Berlin, and Hong Kong were selected for their excellent performance across multiple dimensions of e-governance. These examples reflect the latest innovations that are potentially applicable to the development of e-governance in other municipalities. More than 10 years' continual research in worldwide e-governance enables us to present a longitudinal study to better understand and reflect the progress in e-governance.

2.2 Transformation from E-Government to E-Governance

Information and communication technologies provide new ways for government to provide information and services, which is commonly called e-government. E-government emphasizes "the use of multiple information and communication technologies (ICTs) such as broadband, e-mail, fiber optics, text messaging, 4G, search engines, teleconferencing and WiFi to more rapidly deliver information and services to citizens, businesses and other governments" (Bhatnagar, 2004; Carter & Bélanger, 2005; Chadwick, 2006; Heeks, 1999; Hernon, Cullen, & Relyea, 2006; Ho, 2002; Holden, Norris, & Fletcher, 2003; Homberg, 2008; Milakovich, 2012, p. 10; Roucheleau, 2006, 2007; West, 2005). Milakovich (2012) viewed the development of e-government as several stages: presence, interaction, transaction, and transformation.

In the past decade, ICTs were gradually introduced by governments to better involve citizens. E-mails and online chatting are used by the public to contact governments; listservs are being used to distribute the latest news or updates; bulletin boards and discussion forums are designed to collect public opinions; online surveys and polls are being conducted to obtain public feedback toward government performance; social media such as YouTube, Facebook, and Twitter enable citizens to actively connect with the government; and e-petitioning allows the public to raise issues to gain the government's attention. The combination of ICTs and citizen participation creates e-participation, enabling citizens to play a better role in government running.

The rise of e-participation motivates the transformation from e-government to e-governance, which is more than just providing information and services. E-governance, meaning "electronic governance," is usage of ICTs at various levels of the public sectors and beyond for the purposes of enhancing governance (Bedi, Singh, & Srivastava, 2001; Holmes, 2001; Okot-Uma & London, 2000; Palvia & Sharma, 2007). It emphasizes a citizen-centric approach, "providing citizens with the ability to choose the manner in which they wish to interact with their government" (Milakovich, 2012, p. 17; Oliver & Sanders, 2004). The benefits and opportunities brought about by digital governance have been discussed by Milakovich (2012), such as enhancing direct access to a wider range of government services, attracting better participation, integrating operations inside the government, reducing costs and layers of organizational processes, and improving effectiveness, trust, and quality of services.

At the far end of the e-participation continuum, political participation includes activities "such as contacting political officials, attending a rally, or signing a petition" (Tolbert & McNeal, 2003, p. 177). Weber and Berman (2001) (see also Tolbert & McNeal, 2003) found that those individuals who engaged in Internet activities such as using e-mail and chat rooms were more likely to be engaged in a variety of political activities.

2.3 E-Governance Varies at the Municipal Level

Although more and more governments have realized the importance of e-governance and improve it, jurisdictions still vary substantially at municipal and national levels. Several studies have focused on e-government, e-participation, and e-governance, and among them, the Rutgers E-Governance Survey has an enduring history. Having done research in e-governance for more than 10 years, the E-Government Institute at the School of Public Affairs and Administration of Rutgers University–Newark developed one of the most comprehensive indexes, with 104 scaled items used to evaluate e-governance performance at the municipality level. By evaluating the websites of the largest cities of the most wired nations in the world, the 2011–2012 Rutgers E-Governance Survey reflects the development and variance in e-governance worldwide at the municipal level.

2.4 Methodology

2.4.1 Choosing 100 Largest Municipalities

Building on previous surveys since 2003, the 2011–2012 survey focused on the largest municipalities of the 100 most wired nations in the world (Table 2.1). These nations were chosen based on data from the International Telecommunication Union. Moreover, the rationale for selecting the largest municipalities stems from the e-governance literature, which suggests a positive relationship between population and e-governance capacity at the local level (Holzer & Manoharan, 2012; Moon, 2002; Moon & deLeon, 2001; Musso, Weare, & Hale, 2000; Weare, Musso, & Hale, 1999). These municipalities serve as a surrogate for the e-governance levels of their representative nations.

Table 2.1 100 Municipalities Selected by Continent (2011–2012)

Africa (12)	
Accra (Ghana)	Dar es Salaam (Tanzania)[a]
Algiers (Algeria)[a]	Kampala (Uganda)[a]
Cairo (Egypt)	Lagos (Nigeria)
Cape Town (South Africa)	Nairobi (Kenya)
Casablanca (Morocco)	Omdurman (Sudan)[a]
Dakar (Senegal)	Tunis (Tunisia)

(Continued)

Table 2.1 (Continued) 100 Municipalities Selected by Continent (2011–2012)

Asia (31)	
Aleppo (Syria)[a]	Kuala Lumpur (Malaysia)
Almaty (Kazakhstan)	Kuwait City (Kuwait)
Amman (Jordan)	Mumbai (India)
Baghdad (Iraq)	Muscat (Oman)
Baku (Azerbaijan)	Quezon City (Philippines)
Bangkok (Thailand)	Riyadh (Saudi Arabia)
Bishkek (Kyrgyzstan)[a]	Sana'a (Yemen)[a]
Colombo (Sri Lanka)	Seoul (Republic of Korea)
Dhaka (Bangladesh)	Shanghai (China)
Dubai (United Arab Emirates)	Singapore (Singapore)
Ho Chi Minh City (Vietnam)	Tashkent (Uzbekistan)
Hong Kong (Hong Kong)	Tbilisi (Georgia)
Jakarta (Indonesia)	Tehran (Iran)
Jerusalem (Israel)	Tokyo (Japan)
Karachi (Pakistan)	Yerevan (Armenia)
Kathmandu (Nepal)	
Europe (35)	
Amsterdam (Netherlands)	Chisinau (Moldova)
Athens (Greece)	Copenhagen (Denmark)
Belgrade (Serbia and Montenegro)	Dublin (Ireland)
Berlin (Germany)	Helsinki (Finland)
Bratislava (Slovak Republic)	Istanbul (Turkey)
Brussels (Belgium)	Kiev (Ukraine)
Bucharest (Romania)	Lisbon (Portugal)
Budapest (Hungary)	Ljubljana (Slovenia)

Table 2.1 (Continued) 100 Municipalities Selected by Continent (2011–2012)

London (United Kingdom)	Sofia (Bulgaria)
Madrid (Spain)	Stockholm (Sweden)
Minsk (Belarus)	Tallinn (Estonia)
Moscow (Russian Federation)	Tirane (Albania)
Oslo (Norway)	Vienna (Austria)
Paris (France)	Vilnius (Lithuania)
Prague (Czech Republic)	Warsaw (Poland)
Riga (Latvia)	Zagreb (Croatia)
Rome (Italy)	Zurich (Switzerland)
Sarajevo (Bosnia and Herzegovina)	
North America (10)	
Guatemala City (Guatemala)	Saint Joseph (Costa Rica)
Havana (Cuba)[a]	San Juan (Puerto Rico)
Mexico City (Mexico)	San Salvador (El Salvador)
New York (United States)	Santo Domingo (Dominican Republic)
Panama City (Panama)	Toronto (Canada)
South America (10)	
Asuncion (Paraguay)	Lima (Peru)
Buenos Aires (Argentina)	Montevideo (Uruguay)
Caracas (Venezuela)	Santa Fe De Bogota (Colombia)
Guayaquil (Ecuador)	Santiago (Chile)
La Paz (Bolivia)	Sao Paulo (Brazil)
Oceania (2)	
Auckland (New Zealand)	Sydney (Australia)

Source: Holzer, M., & Manoharan, A. (2012). *Digital governance in municipalities worldwide (2011–2012)*. Newark, NJ: E-Governance Institute, Rutgers University.

[a] Official city websites unavailable.

2.4.2 Rutgers E-Governance Index

For our survey, we used the Rutgers E-Governance Index to evaluate the official websites of these municipalities in terms of information and services provided. This index utilizes 104 measures belonging to five distinct categorical areas, including privacy/security, usability, content, services, and citizen and social engagement. Table 2.2 contains the measures in the index.

With the increase in e-government usage, "concerns regarding the extent to which 'information security' and 'user privacy' can be ensured are raised" (Lambrinoudakis, Gritzalis, Dridi, & Pernul, 2003, p. 1873). Protection to users'

Table 2.2 E-Governance Performance Measures

E-Governance Category	*Key Concepts*	*Raw Score*	*Weighted Score*	*Key Words*
Privacy/Security	19	27	20	Privacy policies, authentication, encryption, data management, cookies
Usability	20	32	20	User-friendly design, branding, length of homepage, targeted audience links or channels, and site search capabilities
Content	26	63	20	Access to current accurate information, public documents, reports, publications, and multimedia materials
Services	21	61	20	Transactional services—purchase or register, interaction between citizens, businesses, and government
Citizen and social engagement	18	48	20	Online civic engagement/ policy deliberation, social media applications, citizen-based performance measurement
Total	104	231	100	

Source: Holzer, M., & Manoharan, A. (2012). *Digital governance in municipalities worldwide (2011–2012).* Newark, NJ: E-Governance Institute, Rutgers University.

information security and privacy is the precondition for developing e-governance. That consideration is emphasized in the Rutgers E-Governance Survey. The Security/Privacy category pays attention to whether a statement or policy is available to explain the collection of personal data, how governments use these data, whether these data would be disclosed to third parties, how governments use "cookies" or "Web Beacons" to track users, and so on. This grouping examines the extent that municipalities protect website users' privacy and security.

Asiimwe and Lim (2010) believed that the benefits that government websites offer to citizens and governments could not be realized if websites are unusable. Along those lines, the Rutgers survey examines whether municipal websites are convenient to use. The second component, usability, focuses on the performance of a government website in enabling users to locate information or services conveniently, that is, whether municipalities' websites are user-friendly: whether the homepage is concise and consistent, targeted audience links and sitemaps are available, online forms are well designed, search tools are advanced and easy to use, and so on. Thus, this component of the survey tests the usability of municipal websites.

The third part is content, examining what kinds of information municipalities are providing to citizens. Website content is examined "in five key areas: access to contact information, public documents, disability access, multimedia materials, and time-sensitive information" (Holzer & Manoharan, 2012). The survey diagnoses whether municipal websites provide contact information for departments or public officials; newsletters or community updates are available to citizens; codes, charts, or policy priorities are supplied; public meeting minutes, budgets, and related documents are offered; and performance measurements and results are posted for citizens. This section of the survey aims at examining the website's content and focuses on what kinds of information municipalities are providing to citizens.

An important aspect of e-governance is providing services online. The Rutgers survey "examines two different types of services: (1) those that enable citizens to interact with the municipality and (2) those that allow users to register for events or services online" (Holzer & Manoharan, 2012). Specifically, it indicates whether governments enable citizens to pay utilities, taxes, fines, or tickets online; allow users to apply for permits or licenses online; enable citizens to file complaints or request services online; and allow users to report crimes, violations of administrative laws and regulations, corruption, and so on.

The importance of citizen participation has been emphasized by many scholars; it holds a sacrosanct role in U.S. political culture (Day, 1997; Irvin & Stansbury, 2004), serves as a vital component of democracy (Crow & Stevens, 2012), benefits government performance improvement (Holzer & Mullins, 2012), and enables citizens to learn more about the government, to gain skills for activist citizenship, and to influence the policy process (Irvin & Stansbury, 2004). E-governance is expected to use ICTs to improve citizen participation. Therefore, the last component of the Rutgers E-Governance Survey, citizen and social engagement, focuses on the online methods provided by the government for citizens to engage government. These methods include filling out online forms to provide comments or feedback, subscription options for newsletters or listservs, bulletin boards to gather citizen input, online discussion forms or chatting capabilities for citizens to discuss public issues, online surveys or polls, real-time chat or instant messaging, and others.

2.5 Results

Among all 100 municipalities, 92 were found to have official websites, with an increase from 87 in the 2009 survey, 86 in the 2007 survey, and 81 in the 2005 survey. This reflects an increase in e-governance adoption among municipalities worldwide. As Figure 2.1 shows, the average score for e-governance over all 92 municipalities is 33.76, somewhat of a decrease from 35.93 in 2009 versus increases from 33.37 in 2007, 33.11 in 2005, and 28.49 in 2003.

As Table 2.3 indicates, however, municipalities vary a lot in their e-governance performance. Only 7 municipalities score higher than 60 of 100, and only 19 score higher than 50. Twenty-one municipalities even score lower than 20, reflecting their poor performance in e-governance. The results indicate differences over five categories (as Figure 2.2 reflects). The average score for usability is 12.09, the highest among these five components, whereas that for content is 7.38 and that for service is only 5.78. The average score for privacy/security is 4.99, whereas that for citizen and social engagement is 3.53, which is the lowest. This indicates that governments pay more attention to technical aspects, such as whether their websites are user-friendly with advanced tools. However, they pay limited attention to the services and information that users need through their websites. Moreover, few of them are doing well in enabling citizens to participate online.

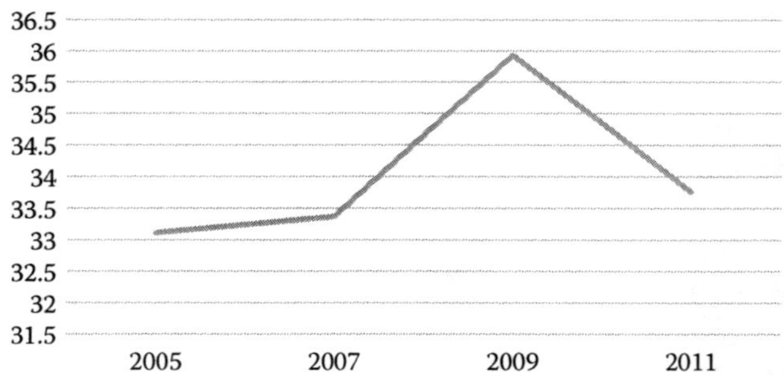

Figure 2.1 Average e-governance score 2005–2011. (From Holzer, M., & Manoharan, A. (2012). *Digital governance in municipalities worldwide (2011–2012)*. Newark, NJ: E-Governance Institute, Rutgers University.)

Table 2.3 2011–2012 E-Governance Survey Results

Final Rank	*Nations*	*Largest City (by Population)*	*Total Score*	*S/P Score*	*Usability Score*	*Content Score*	*Service Score*	*C&S Score*
1	Korea (Rep.)	Seoul	82.23	13.33	18.44	16.67	17.55	16.25
2	Canada	Toronto	64.31	10.74	16.88	16.83	12.79	7.09
3	Spain	Madrid	63.63	12.22	16.88	15.08	12.79	6.67
4	Czech Republic	Prague	61.72	12.59	16.25	13.02	8.20	11.67
5	Hong Kong, China	Hong Kong	60.81	11.11	17.82	13.65	13.44	4.80
6	United States	New York	60.49	11.11	15.94	13.81	12.13	7.50
7	Sweden	Stockholm	60.26	17.41	13.13	12.54	11.15	6.04
8	Slovak Republic	Bratislava	56.74	13.33	16.26	10.64	9.02	7.50
9	United Kingdom	London	56.19	12.22	15.63	11.75	11.81	4.79
10	China	Shanghai	55.49	7.78	13.44	12.07	12.62	9.58
11	Lithuania	Vilnius	55.35	10.74	16.57	11.59	10.00	6.46
12	Austria	Vienna	54.79	11.11	13.44	12.38	8.69	9.17
13	Finland	Helsinki	54.22	13.33	13.75	11.11	8.52	7.50
14	New Zealand	Auckland	53.19	12.22	13.13	13.18	12.79	1.88
15	United Arab Emirates	Dubai	53.18	12.60	15.32	7.94	12.13	5.21

(Continued)

Table 2.3 (Continued) 2011–2012 E-Governance Survey Results

Final Rank	*Nations*	*Largest City (by Population)*	*Total Score*	*S/P Score*	*Usability Score*	*Content Score*	*Service Score*	*C&S Score*
16	Singapore	Singapore	52.21	4.82	15.00	12.70	13.45	6.25
17	Russia	Moscow	51.77	3.34	16.57	11.27	11.64	8.96
18	Denmark	Copenhagen	50.06	11.11	14.69	10.80	7.21	6.25
19	Armenia	Yerevan	49.97	4.45	16.26	12.38	13.77	3.13
20	France	Paris	48.65	9.26	11.88	12.54	6.23	8.75
21	Germany	Berlin	47.16	14.08	12.82	9.21	6.07	5.00
22	Slovenia	Ljubljana	46.25	6.30	14.38	14.61	6.39	4.59
23	Japan	Tokyo	45.35	7.04	14.38	11.59	9.02	3.33
24	Croatia	Zagreb	44.43	5.93	16.25	8.25	7.54	6.46
25	Brazil	Sao Paulo	44.22	4.82	15.32	9.05	9.84	5.21
26	Ireland	Dublin	43.76	9.63	14.69	6.51	9.18	3.75
27	Norway	Oslo	42.60	5.19	15.01	10.64	8.04	3.75
28	Estonia	Tallinn	41.69	2.22	12.19	14.13	10.66	2.50
29	The Netherlands	Amsterdam	40.73	4.45	15.63	8.89	6.56	5.21
30	Switzerland	Zurich	39.90	9.63	13.44	9.53	3.78	3.54

31	Colombia	Bogota	39.88	5.93	15.63	7.15	6.40	4.80
32	Kazakhstan	Almaty	37.76	5.56	13.76	5.56	6.23	6.67
33	Bolivia	La Paz	37.16	2.59	14.07	7.78	9.18	3.54
34	Malaysia	Kuala Lumpur	37.09	6.30	14.07	7.46	6.56	2.70
35	Mexico	Mexico City	36.98	5.74	12.51	6.19	7.54	5.00
36	Belgium	Brussels	36.78	4.45	13.44	12.22	4.59	2.09
37	Portugal	Lisbon	36.49	2.96	11.88	7.46	6.07	8.13
38	Italy	Rome	35.06	8.89	11.26	7.15	5.90	1.88
39	South Africa	Johannesburg	34.03	3.33	11.25	9.21	6.07	4.17
40	Iran (I.R.)	Tehran	33.09	7.78	10.32	6.03	6.89	2.09
41	Vietnam	Ho Chi Minh	32.95	5.93	11.88	4.76	7.05	3.34
42	Israel	Jerusalem	32.83	0.00	12.82	8.89	7.38	3.75
43	Belarus	Minsk	32.11	1.11	14.38	6.35	3.61	6.67
44	Argentina	Buenos Aires	31.15	1.85	13.75	7.46	4.76	3.33
45	Saudi Arabia	Riyadh	30.66	7.78	14.07	5.40	1.97	1.46
46	Australia	Sydney	30.52	7.04	13.44	6.99	1.81	1.25
47	Chile	Santiago	29.26	1.11	11.25	7.62	8.04	1.25

(*Continued*)

Table 2.3 (Continued) 2011–2012 E-Governance Survey Results

Final Rank	*Nations*	*Largest City (by Population)*	*Total Score*	*S/P Score*	*Usability Score*	*Content Score*	*Service Score*	*C&S Score*
48	Greece	Athens	29.20	8.52	12.19	4.45	1.97	2.09
49	India	Mumbai	28.99	8.15	11.25	3.81	5.57	0.21
50	Latvia	Riga	28.85	2.60	10.94	7.15	4.43	3.75
51	Oman	Muscat	28.72	4.45	11.25	4.45	2.96	5.63
52	Romania	Bucharest	28.12	4.82	12.19	5.56	1.81	3.75
53	Peru	Lima	27.80	1.11	11.57	7.46	5.58	2.09
54	Indonesia	Jakarta	27.07	3.71	11.88	5.87	3.12	2.50
55	Uruguay	Montevideo	26.98	2.59	12.19	7.46	3.28	1.46
56	Tunisia	Tunis	26.65	1.11	13.13	5.72	2.95	3.75
57	Bulgaria	Sofia	26.35	7.41	8.13	6.35	1.97	2.50
58	Turkey	Istanbul	25.81	1.85	5.63	6.83	9.84	2.03
59	Guatemala	Guatemala City	25.43	1.11	12.51	5.08	5.90	0.84
60	Ukraine	Kiev	25.01	0.00	12.19	7.14	4.43	1.25
61	Poland	Warsaw	24.94	7.04	10.94	4.44	1.48	1.04
62	Egypt	Cairo	24.64	7.04	12.50	2.06	2.63	0.42

63	Moldova	Chisinau	24.55	1.11	8.76	7.30	4.27	3.13
64	Jordan	Amman	23.70	1.11	11.26	4.45	3.77	3.13
65	Dominican Republic	Santo Domingo	23.27	4.07	9.69	4.76	2.46	2.29
66	Sri Lanka	Colombo	22.93	0.00	13.75	2.70	3.77	2.71
67	Hungary	Budapest	22.67	1.11	8.75	6.98	3.12	2.71
68	Philippines	Quezon City	22.48	0.00	14.38	4.76	2.30	1.04
69	Albania	Tirane	22.18	0.00	9.07	6.35	4.26	2.50
70	Serbia	Belgrade	22.04	0.00	11.26	4.92	2.95	2.92
71	Puerto Rico	San Juan	21.42	9.26	3.13	4.92	3.28	0.83
72	Ecuador	Guayaquil	19.69	0.37	9.07	5.40	3.61	1.25
73	Ghana	Accra	19.41	1.85	10.63	3.18	2.30	1.46
74	Thailand	Bangkok	18.53	1.48	6.25	3.65	3.61	3.54
75	Bosnia and Herzegovina	Sarajevo	18.31	0.00	10.00	5.24	0.99	2.09
76	Senegal	Dakar	18.20	0.37	11.57	2.38	1.81	2.09
77	Venezuela	Caracas	17.50	0.37	9.69	4.92	1.48	1.04
78	Nepal	Kathmandu	16.81	0.00	8.13	4.29	3.77	0.63
79	Bangladesh	Dhaka	16.79	2.22	9.69	1.59	2.46	0.84

(*Continued*)

Table 2.3 (Continued) 2011–2012 E-Governance Survey Results

Final Rank	*Nations*	*Largest City (by Population)*	*Total Score*	*S/P Score*	*Usability Score*	*Content Score*	*Service Score*	*C&S Score*
80	Morocco	Casablanca	16.77	0.00	11.88	3.49	0.98	0.42
81	Panama	Panama City	16.33	0.00	9.38	3.33	2.79	0.83
82	Pakistan	Karachi	16.25	0.00	11.26	2.06	1.48	1.46
83	Georgia	Tbilisi	15.78	0.00	7.82	4.76	2.79	0.42
84	Costa Rica	Saint Joseph	15.69	1.11	7.51	4.29	2.79	0.00
85	Azerbaijan	Baku	15.05	3.70	6.25	4.44	0.66	0.00
86	El Salvador	San Salvador	15.04	1.11	6.57	4.29	2.46	0.63
87	Kenya	Nairobi	14.48	0.37	9.69	1.43	2.79	0.21
88	Nigeria	Lagos	14.29	2.22	7.51	3.50	0.66	0.42
89	Kuwait	Kuwait City	14.22	1.11	9.07	2.07	1.15	0.84
90	Iraq	Baghdad	14.11	1.11	9.07	1.91	0.99	1.04
91	Paraguay	Asuncion	10.76	0.74	6.26	2.70	0.66	0.42
92	Uzbekistan	Tashkent	6.76	0.00	5.32	0.96	0.49	0.00

Source: Holzer, M., & Manoharan, A. (2012). *Digital governance in municipalities worldwide (2011–2012)*. Newark, NJ: E-Governance Institute, Rutgers University.

Note: C&S = citizen and social; S/P = security/privacy.

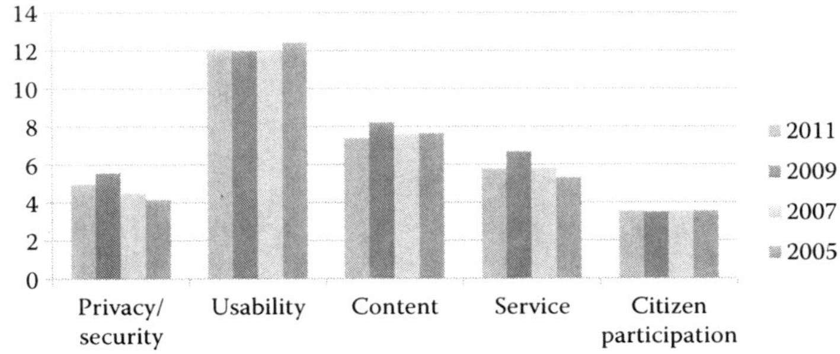

Figure 2.2 Average score by categories 2005–2011. (From Holzer, M., & Manoharan, A. (2012). *Digital governance in municipalities worldwide (2011–2012)*. Newark, NJ: E-Governance Institute, Rutgers University.)

This study has some limitations. To begin with, the survey includes only 100 municipalities in selected countries, a limited sample in terms of reflecting e-governance developments at the municipal level worldwide. In addition, the largest municipality in each country may not represent all municipalities in a country. The study focuses only on a municipality's official website, exploring government performance in providing e-governance services and engaging citizens. However, although the data for each were collected from a user's perspective, the study may not fully address users' experiences and provide limited information about public usage (width and depth) toward e-governance. In addition, examining only digital tools used via government websites is limited in terms of reflecting the entire e-governance level. For example, in the category citizen and social engagement, whether governments provide convenient ways for citizens to interact with governments is examined; however, citizen participation is more potentially more extensive. Whether and how governments make use of public feedback and engage citizens in policy making processes in a meaningful way also need to be tested.

2.6 Best Practices in E-Governance

Best practices are common in many disciplines. Exploring best practices is important because these municipalities provide excellent examples for others to learn. Similarly, e-governance needs to identify best practices as well, for better understanding e-governance and its achievement. In this study, on the basis of the 2011–2012 digital governance survey data, we select municipalities ranking high in various categories and use detailed examples to describe their performance in privacy and security, usability, content, services, and social engagement, thus contributing to a better understanding of good e-governance.

2.6.1 Privacy and Security

Among all the municipalities evaluated, Berlin is doing an excellent job in privacy and security. A privacy policy (Figure 2.3) is available at the bottom of the homepage, providing detailed information as to what kind of data are collected, used, and protected through the website. The policy covers 12 elements, including collection and processing of personal data; cookies; security; disclosure of personal data to third parties; social plugins; Google analytics; a privacy statement for using the scalable central measuring procedure; advertising; other data processing; links to other websites; changes to privacy statement; and questions, comments, and information. These parts compose a comprehensive statement, enabling users to clearly understand the level of protection of their privacy and security when using this website.

In the section on collection and processing personal data, the policy explicitly addresses name, data of birth, address, payment method, and payment data; other kinds of data such as personal interest, telephone number, gender, and others, are collected when registration is required for certain services or fee-based services (website links). As to cookies, the policy explains what cookies are and how Berlin uses cookies. It also teaches users how to disable or block cookies if they want to.

The policy also discusses how Berlin stores users' data and these data's safety. The disclosure of personal data to third parties, as Figure 2.4 shows, is clearly stated. With more and more people using social media, such as Facebook, Twitter, and Google+, Berlin offers users additional services. Within this policy, a section termed "Social Plugins" explains how social media work through Berlin's website and how users' data are collected. Also, Google Analytics and use of the Scalable

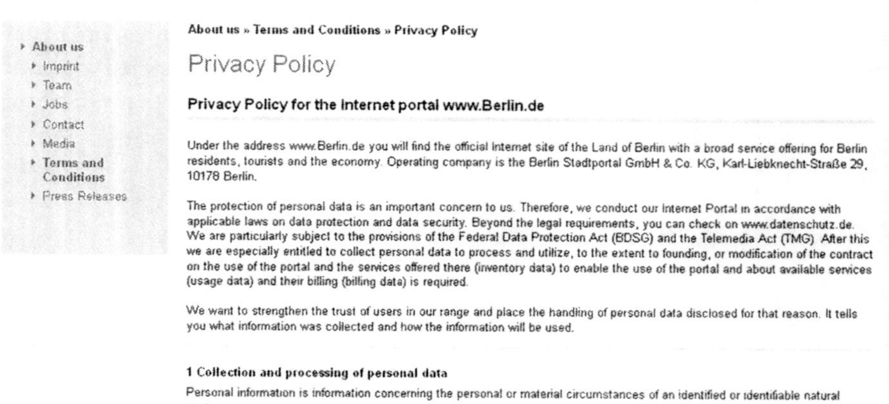
- About us
  - Imprint
  - Team
  - Jobs
  - Contact
  - Media
  - Terms and Conditions
  - Press Releases

About us » Terms and Conditions » Privacy Policy

Privacy Policy

Privacy Policy for the Internet portal www.Berlin.de

Under the address www.Berlin.de you will find the official Internet site of the Land of Berlin with a broad service offering for Berlin residents, tourists and the economy. Operating company is the Berlin Stadtportal GmbH & Co. KG, Karl-Liebknecht-Straße 29, 10178 Berlin.

The protection of personal data is an important concern to us. Therefore, we conduct our Internet Portal in accordance with applicable laws on data protection and data security. Beyond the legal requirements, you can check on www.datenschutz.de. We are particularly subject to the provisions of the Federal Data Protection Act (BDSG) and the Telemedia Act (TMG). After this we are especially entitled to collect personal data to process and utilize, to the extent to founding, or modification of the contract on the use of the portal and the services offered there (inventory data) to enable the use of the portal and about available services (usage data) and their billing (billing data) is required.

We want to strengthen the trust of users in our range and place the handling of personal data disclosed for that reason. It tells you what information was collected and how the information will be used.

1 Collection and processing of personal data

Personal information is information concerning the personal or material circumstances of an identified or identifiable natural person.

Figure 2.3 Privacy policy in Berlin official website.

3 Security

We store your data on specially protected servers in Germany. Access is restricted to authorized persons who are involved in the technical, commercial or editorial maintenance of the servers. We take precautions to ensure the safety of their personal data. Your data will be conscientiously protected against loss, destruction, falsification, manipulation and unauthorized access or unauthorized disclosure.

4 Disclosure of personal data to third parties

Some of the services offered on our site are provided by third parties ("Partners"). These are so-called ad-admin. In that regard, subject to the processing and use of these data to the data protection regulations of the respective partner. Create Our partners using cookies and dynamic IP addresses pseudonymous user profiles in order to inform the individual user through specially tailored to him, interesting offers and advertising can. For most partners you can collect and use of user profiles thus obtained contradict any time by clicking on the opt-out pages. We try on the page www.berlin.de/wir-ueber-uns-be/agb/datenschutz /optout/ to maintain such opt-out links. Once the servers and browsers support the so-called track-out process, we will also be documented there. We will transfer personal information without your consent to third parties who are not partners, unless we are forced to surrender due to mandatory statutory provisions. There is no transfer of personal data to third parties for advertising purposes.

5 Social Plugins

Our website offers you as an additional service, so-called social plugins that enable it to interact with the social services *Facebook*, *Twitter* and *Google+*. To an unintended transfer of your data usage to avoid (eg the currently visited page) to these services, the use of social plug-ins can only after activation by a targeted click (two-click method to take place). Previously, these buttons are not available and not functional. After activation of these buttons by clicking you can use this; but it is working in the same train that called social networks via the enabled plugins collect usage and possibly user data. Be collected and analyzed to what extent these data can not be influenced by us.

Figure 2.4 Berlin official website privacy policy.

Central Measuring procedure are explained in detail. In addition, Berlin notifies users of privacy statement changes by e-mail and provides contact methods for questions and comments. Overall, the policy is clear and comprehensive. Berlin is doing a good job in enabling users to know the collection, usage, and protection of their personal data.

2.6.2 Usability

Hong Kong (Figure 2.5) provides an example of building a user-friendly website. Target audience links are divided into four categories—residents, business and trade, nonresidents, and social groups—enabling users to efficiently locate information with their status. A link to the homepage is available and the navigation bar is consistent throughout the site. Information and service links are listed in the homepage alphabetically, such as "Communication & Technology, Culture," "Leisure & Sports," "Education & Training," and so on. In Hong Kong's website, these forms and services links are directly posted in the homepage. Municipality news is also available in the homepage. Different types of multimedia, mobile websites, and apps are listed on the homepage as well.

In addition, the Site Map (Figure 2.6) is available at the bottom of the homepage, which contains four main categories: Residents, Business & Trade, Non-Residents, and Miscellaneous. All the page links are under labeled items that are listed alphabetically within categories. For example, "Education & Training,"

Figure 2.5 Hong Kong official website.

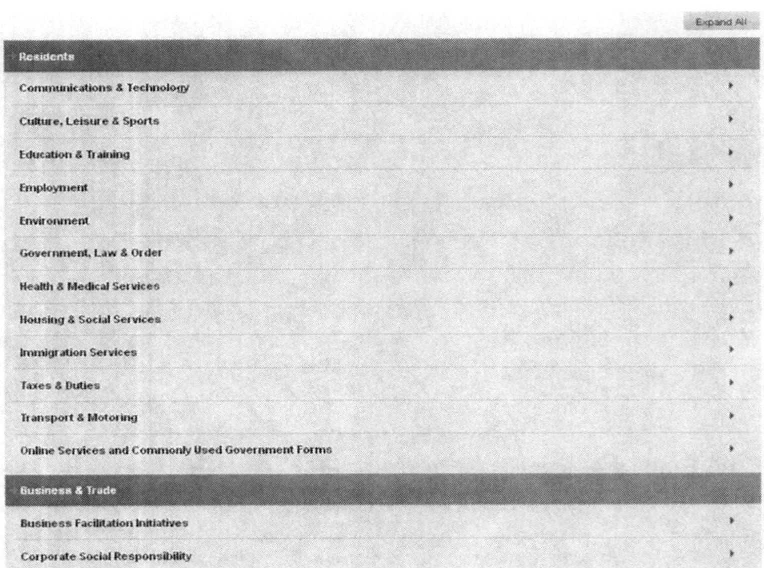

Figure 2.6 Hong Kong official website map.

"Employment," "Environment," and "Government, Law & Order" are listed under the category of "Residents," and a government structure link can easily be found under "Government." All the pages in Hong Kong's website are arranged clearly and enable the public to find information and services conveniently.

Online forms (Figure 2.7) provided on the website are well designed. It is convenient to fill in forms or to complete applications online. For example, if a resident wants to apply for a government job, he or she just needs to fill out the online form following the instructions. The column on the left indicates the steps to take as well as the current status. Clear instructions are available to help applicants complete forms, and required fields are clearly labeled. The system also provides additional information when users make errors, enabling them to identify and fix. Users do not need to reenter correct information.

Another example of the website's usability is its well-designed search tools. The search function (Figure 2.8) is located at the top of the homepage and is convenient to use. For example, to search "budget," users can sort the search results by date or by relevance. An advanced search function is also available that users can manage search scope by clearly inputting words or exact phrases, choosing the file format, selecting sections of the Hong Kong website and languages, modifying the time period, and so on. In addition, FAQs for search services are provided to help users familiar with the search tools. Overall, the well-designed search tools enable users to find information and services by themselves.

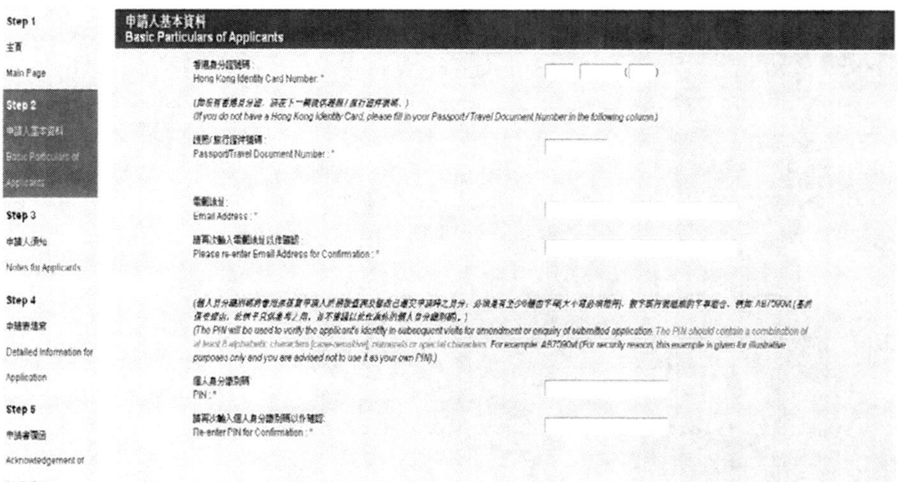

Figure 2.7 Hong Kong official website government job online application.

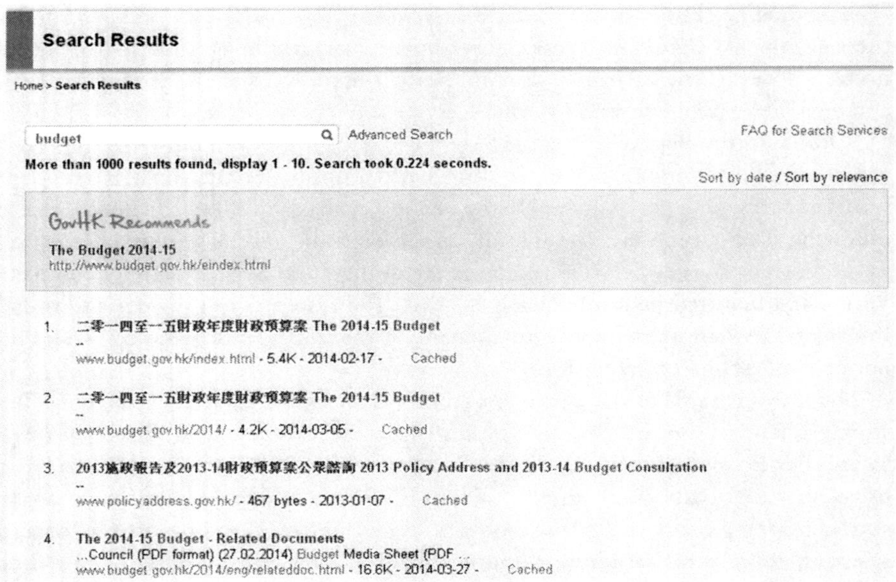

Figure 2.8 Hong Kong official website budget search results.

2.6.3 Content

Toronto ranks first in content among all the 100 municipalities. The homepage (Figure 2.9) is clear and the contents follow in four parts: living in Toronto, doing business, visiting Toronto, and accessing the city hall. For the residents of Toronto, information such as health, environment and energy, grants, awards, immigration, arts and culture, recreation, housing, and transportation are available. For people who want to do business, information is provided on topics such as how to start your business, operate and grow business in Toronto, do business or be a partner with Toronto, and others. And for people interested in visiting Toronto or accessing the city hall, information is supplied in a clear manner.

Offices' and public officials' contact information is available to all citizens, including their office locations, telephone numbers, and e-mail addresses. This information is highlighted or underlined for easy access. The staff directory is listed alphabetically by surname and can be downloaded as well. The municipality enables citizens to connect with the government and follow municipal updates in many ways. An e-newsletter is available for the public, stakeholders, media, City Council, and staff, focusing on the latest services, innovations, and accomplishments. As Figure 2.10 indicates, citizens can subscribe to Toronto E-Updates to receive updates or news. Social media tools, such as Facebook, Twitter, and YouTube, are provided to facilitate following city updates. Toronto utilizes its city blogs and Flickr to share its news and photos. Residents may also subscribe to podcasts to follow recent city initiatives and

Figure 2.9 Toronto official website.

Figure 2.10 Toronto official website "Connect with Your City."

services. In addition to public meeting schedules and minutes, government budgets (Figure 2.11) are also available to the public. The budget-at-a-glance report is brief and clearly shows the public how the municipal government obtains and uses money.

To help citizens better understand their running, governments are expected to provide performance information. Toronto enables residents to access government performance data in the "Performance Management and Benchmarking Report" (Figure 2.12) that is "produced every year and that provides detailed information

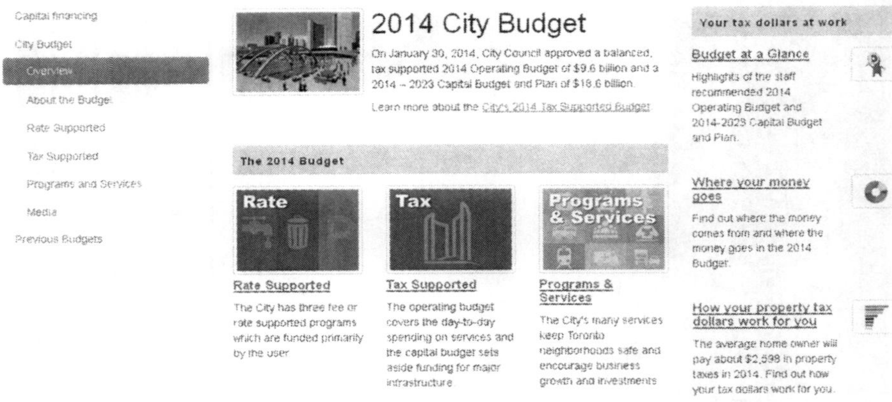

Figure 2.11 Toronto official website "2014 City Budget" overview.

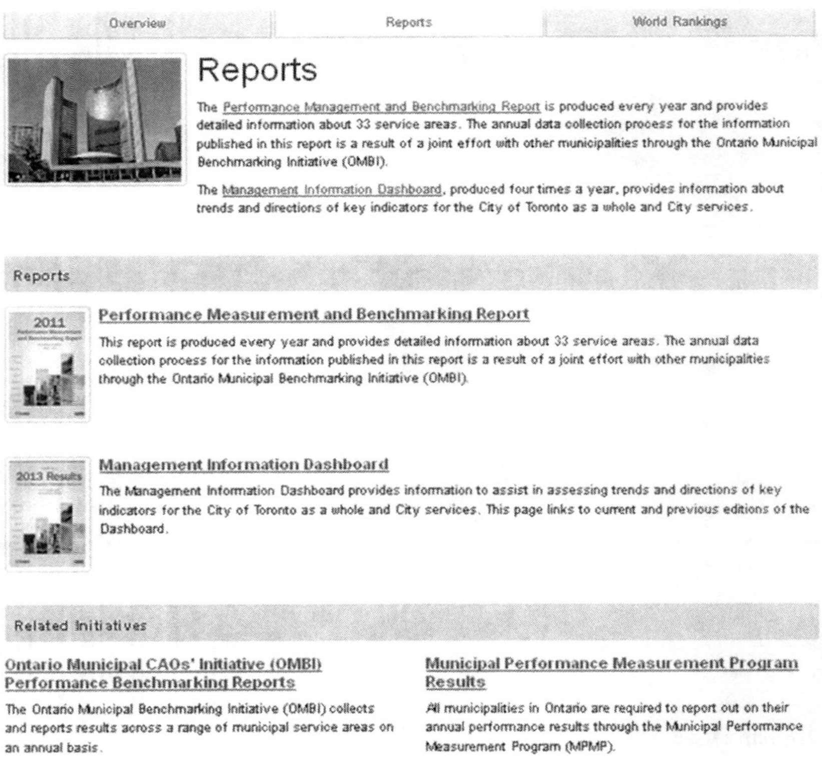

Figure 2.12 Toronto official website "Reports."

about 33 service areas" (Toronto Progress Portal, 2014), including building services, children's services, cultural services, and others. In the report on each service area, the indicator or measurement is listed and comparisons are made both internally and externally to better understand government performance. For example, the "number of paid parking spaces (all types) managed per 100,000 population" is used to measure the service level, and "parking services operating cost per paid parking space (all types) managed" is used to measure the efficiency level. Then, on the basis of the comparisons, citizens could know clearly whether the department of parking services has made improvements in its performance.

2.6.4 Services

Within online services, Seoul (Figure 2.13) ranks highest. Convenient methods are provided for the public to contact public officials and for government agencies to obtain information and services. Bulletin boards are available for the public to ask questions and get answers from public officials. To search for answers to common questions, FAQs with subcategories of topics are provided. Users can directly request information with online forms. The site allows users to report corruption, violations of administrative laws and regulations, and others. Complaints to the government via the website can be tracked as well. The website also allows users to pay their utilities, such as water and gas. Taxes, fines, or tickets may also be paid online. Permits and licenses could be applied and tracked online. The site also

Figure 2.13 Seoul official website.

enables users to look up property assessments used for levying taxes, and searchable databases are also provided on the website. Users can also establish their own account to access private information online (e.g., educational records, medical records, and point total of driving violations).

2.6.5 Citizen and Social Engagement

Seoul also ranks number 1 in citizen and social engagement, providing convenient means for citizens to engage, as indicated in Figure 2.14. A subscription option is available for citizens to follow updates. E-mails and online forms are provided by both departments and public officials to enable citizens to contact them. A link termed "Citizen Participation" is at the top of homepage, and a particular subwebsite can be located with this link, which provides a variety of active pathways for citizen participation. An online bulletin board is provided for citizens to post ideas, comments, and opinions. A discussion forum is available to citizens for dialogue on public issues.

Online forms (Figure 2.15) are provided to users to provide suggestions. Online surveys and polls are used to collect public opinions. Satisfaction surveys are conducted to provide feedback, and the results are available to citizens. In addition, an e-petition tool is available for the public to raise issues to attract the government's attention or concerns. The website also encourages citizens to post information, photos, and videos. All of these means enable citizens to connect with the government well. They could not only understand government operating but also actively participate by providing comments or feedback, completing online surveys or polls, discussing public issues with others, and raising public issues themselves.

Figure 2.14 Seoul official website "Citizen Participation."

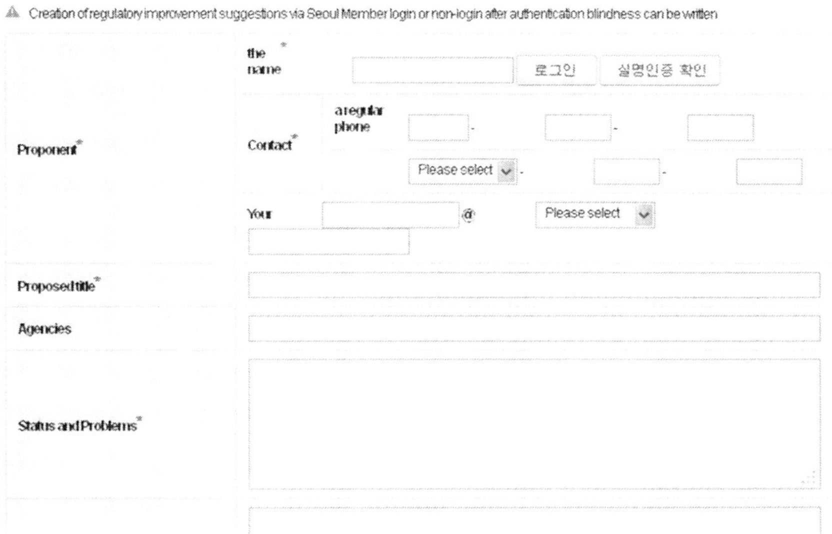

Figure 2.15 Seoul official website online form for providing comments.

2.7 Strategies for Successful E-Governance

The study of e-governance has emerged as a major new area of research in public administration (Holzer, Manoharan, & Van Ryzin, 2010). E-governance provides benefits such as improvement in government accountability, efficiency and effectiveness (Carter & Bélanger, 2005; Holzer et al., 2010), transparency, higher levels of public involvement, and others. The importance of e-governance and the benefits it brings increase the need for governments to transform from e-government to e-governance. Many factors both inside and outside governments impact e-governance adoption and development, such as economy, education, and culture. Success in e-governance requires government precise efforts and strategies to achieve the transformation. Four strategies include (1) upgrading ICT infrastructure, (2) improving human resource management, (3) creating a corresponding political environment that emphasizes citizen participation, and (4) promoting administrative performance.

Information and communication technology infrastructure plays a fundamental role in developing e-governance. Bwalya (2009) emphasized the importance of ICT infrastructure in e-government adoption and development. Similar to e-government, the achievement of e-governance goals strongly relates to the ICT infrastructure level. High Internet usage will result in high public pressure for developing e-governance, with requests for better online services as well as more e-participation opportunities. Better ICT infrastructure inside a government serves as a good foundation for the government to implement e-governance strategies and increase the possibilities for e-governance success. In Seoul, for example, based on International Telecommunication Union data, the percentage of individuals using the Internet is 84.1%, which is very high in comparison with other countries. Korea is at the leading edge in using ICTs. High ICT usage brings correspondingly high e-governance services usage, and better ICT infrastructure serves as an encouraging environment within which the government can develop its e-governance.

Successes in e-governance cannot be achieved without better human resource management. Professionals need to be recruited to manage information mobility, data collection and processing, and e-governance affairs. In New York City, a top municipality in e-governance, for example, a chief technology officer in the Department of Transportation is appointed to provide technical support and use technological resources to help the department and the entire municipality achieve their strategic objectives. That officer, by making better use of information technologies and resources, improves government performance and services provided. Besides, rapid development of ICT requires government officials and employees to continually update their knowledge. Knowledge management, with the ICT help, enables them to learn in a more collaborative and efficient manner. E-governance success indicates protecting citizens' privacy and security well, providing information and services conveniently, and actively engaging citizens. All of these require better human resource management to improve government's capacity in achieving e-governance goals.

A contrast between e-governance and e-government is the emphasis on citizen participation: e-governance moves beyond e-government in terms of seeking opportunities for more active citizen engagement in government decision-making processes. Seoul serves as a leading example in promoting citizen participation. Multiple channels offered by ICTs promote citizen-centric governance, motivating citizens to propose public agenda, express themselves during policy making, give feedbacks and comments on public policy implementation, and supervise the government more comprehensively. To better engage citizens and develop e-governance, technological changes need to be paired with a customer-focused strategy such as Citizen Relationship Management to facilitate citizen-focused service (Milakovich, 2012, p. 65). Citizen Relationship Management requires the government to be citizen-centric, providing information and services that citizens prefer, strengthening cooperative links between government and citizens, developing strategies from a 360-degree view of the citizen, and orienting to citizens' needs (Milakovich, 2012, p. 65). A political environment that emphasizes citizen participation and social engagement needs to be built, serving as the political foundation for transforming e-government to e-governance.

Developing e-governance asks for higher levels of government accountability and performance. This demand brings changes to expectations and challenges for governments in balancing its administrative values, such as efficiency and professionalism, with its political values, such as participation and democracy. To achieve that balance, "public agencies are expected to meet balanced performance standards with citizens, interests and elected officials to accommodate preferences of diverse individuals and groups" (Milakovich, 2012, p. 115). Toronto, another top municipality in e-governance, offers an example. With its "Performance Management and Benchmarking Report," Toronto provides the public with detailed information covering a wide range of services. The indicators and measurements listed enable the public to better understand and evaluate government performance. Feedback from citizens and other stakeholders helps government to create balanced performance standards to improve government performance. External feedback also helps to increase government transparency and accountability. All of these contribute to the further development of e-governance.

To sum up, transforming e-government to e-governance requires multifaceted government efforts, among which ICT infrastructure, human resource management, a political environment emphasizing citizen participation, and administrative performance cannot be ignored. Higher levels of ICT infrastructure both inside and outside the government serve as an incubating environment for e-governance development. Better human resource management supplies qualified employees for the adoption and implementation of e-governance initiatives and increases the possibility of e-governance success. A political environment that emphasizes citizen participation and social engagement reduces the barriers to implementing e-governance and gaining more political support for it. Moreover, higher levels of government performance benefit the transformation from e-government to e-governance.

Author Biographies

Marc Holzer is dean of the School of Public Affairs and Administration at Rutgers University–Newark. His work on e-government and e-governance spans the last decade and includes five worldwide surveys as well as state and municipal evaluations in the United States. His publications in this area include "Global Cities on the Web: An Empirical Typology of Municipal Websites," "Explaining E-participation Diffusion at the National Level: An Examination of External Environmental Influences," and others.

Yueping Zheng is a senior research associate at the E-Governance Institute and a PhD Candidate in the School of Public Affairs and Administration, Rutgers University–Newark. His research interests are e-government, performance management, and citizen participation.

References

Asiimwe, E. N., & Lim, N. (2010). Usability of government websites in Uganda. *Electronic Journal of e-Government, 8*(1), 1–12.

Bedi, K., Singh, P. J., & Srivastava, S. (2001). *Government@net: New governance opportunities for India*. New Delhi: Sage Publications.

Bhatnagar, S. (2004). *E-government: From vision to implementation, a practical guide with case studies*. Thousand Oaks, CA: Sage Publications.

Bwalya, K. J. (2009). Factors affecting adoption of e-government in Zambia. *The Electronic Journal of Information Systems in Developing Countries, 38*(4), 1–13.

Carter, L., & Bélanger, F. (2005). The utilization of e-government services: Citizen trust, innovation and acceptance factors. *Information Systems Journal, 15*(1), 5–25.

Chadwick, A. (2006). *Internet politics: States, citizens, and new communication technologies*. New York: Oxford University Press.

Crow, D. A., & Stevens, J. R. (2012). Citizen engagement in local environmental policy: Information, mobilization and media. In H. L. Schachter & K. Yang (Eds.), *The state of citizen participation in America* (pp. 131–162). Raleigh, NC: Information Age Publishing.

Day, D. (1997). Citizen participation in the planning process: An essentially contested concept? *Journal of Planning Literature, 11*(3), 421–434.

Heeks, R. (Ed.). (1999). *Reinventing government in the information age: International practice in IT-enabled public sector reform*. London: Routledge.

Hernon, P., Cullen, R., & Relyea, H. C. (Eds.). (2006). *Comparative perspectives on e-government: Servicing today and building for tomorrow*. Lanham, MD: The Scarecrow Press.

Ho, T.-K., A. (2002). Reinventing local governments and the e-government initiative. *Public Administrative Review, 62*(July/August), 434–444.

Holden, S. H., Norris, D. F., & Fletcher, P. D. (2003). Electronic government at the local level: Progress to date and future issues. *Public Productivity and Management Review, 26*(3), 1–20.

Holmes, D. (2001). *Egov: Ebusiness strategies for government*. London: Nicholas Brealey Publishing.

Holzer, M., & Manoharan, A. (2012). *Digital governance in municipalities worldwide (2011–2012)*. Newark, NJ: E-Governance Institute, Rutgers University.

Holzer, M., Manoharan, A., & Van Ryzin, G. (2010). Global cities on the web: An empirical typology of municipal websites. *International Public Management Review, 11*(3), 104–121.

Holzer, M., & Mullins, L. B. (2012). Citizen participation and performance: A model for citizen-based government and performance management. In H. L. Schachter & K. Yang (Eds.), *The state of citizen participation in America* (pp. 325–347). Raleigh, NC: Information Age Publishing.

Homberg, V. (2008). *Understanding e-government: Information systems in public administration*. London: Routledge.

Irvin, R. A., & Stansbury, J. (2004). Citizen participation in decision making: Is it worth the effort? *Public Administration Review, 64*(1), 55–65.

Lambrinoudakis, C., Gritzalis, S., Dridi, F., & Pernul, G. (2003). Security requirements for e-government services: A methodological approach for developing a common PKI-based security policy. *Computer Communications, 26*(16), 1873–1883.

Lee, C. P., Chang, K., & Berry, F. S. (2011). Testing the development and diffusion of e-government and e-democracy: A global perspective. *Public Administration Review, 71*(3), 444–454.

Milakovich, M. E. (2012). *Digital governance: New technologies for improving public service and participation*. New York: Routledge.

Moon, M. J. (2002). The evolution of e-government among municipalities: Rhetoric or reality? *Public Administration Review, 62*(4), 424–433.

Moon, M. J., & deLeon, P. (2001). Municipal reinvention: Municipal values and diffusion among municipalities. *Journal of Public Administration Research and Theory, 11*(3), 327–352.

Musso, J., Weare, C., & Hale, M. (2000). Designing web technologies for local governance reform: Good management or good democracy? *Political Communication, 17*(1), 1–19.

Okot-Uma, R. W. O., & London, C. S. (2000). *Electronic governance: Re-inventing good governance.* London: Commonwealth Secretariat.

Oliver, E. L., & Sanders, L. (2004). *E-government reconsidered: Renewal of governance for the knowledge age.* Regina, Canada: Canadian Plains Research Center and the Saskatchewan Institute of Public Policy.

Palvia, S. C. J., & Sharma, S. S. (2007). E-government and e-governance: Definitions/domain framework and status around the world. *International Conference on E-governance.*

Roucheleau, B. A. (2006). *Public management information systems.* Hershey, PA: Idea Publishing Group.

Roucheleau, B. A. (Ed.). (2007). *Case studies on digital government.* Hershey, PA: Idea Publishing Group.

Tolbert, C. J., & McNeal, R. S. (2003). Unraveling the effects of the Internet on political participation? *Political Research Quarterly, 56*(2), 175–185.

Toronto Progress Portal—City initiatives | City of Toronto. (n.d.). Retrieved April 10, 2014, from http://www1.toronto.ca/wps/portal/contentonly?vgnextoid=2258dce9014df310VgnVCM10000071d60f89RCRD.

Weare, C., Musso, J., & Hale, M. (1999). Electronic democracy and the diffusion of municipal web pages in California. *Administration and Society, 31*(1), 3–27.

Weber, L. M., & Bergman, J. (2001). Who participates and how? A comparison of citizens "online" and the mass public. Annual Meeting of the Western Political Science Association, Las Vegas, NV, March, pp. 15–17.

West, D. (2005). *Digital democracy: Technology and public sector performance.* Princeton, NJ: Princeton University Press.

Chapter 3

Does External Environment Affect E-Government? A Cross-Country Analysis

Yueping Zheng and Aroon Manoharan

Contents

Abstract

E-government has made tremendous progress and became a global phenomenon in the past two decades, attracting attention from both researchers and practitioners. A growing body of research continues to focus on explaining the e-government divide and to determine the factors impacting e-government development. To date, however, few studies have examined the impact of external environment on e-government development worldwide; this study intends to fill some of the gaps. With combined data at country level, we used structural equation modeling statistical techniques to test the impact of environmental factors (economy, democracy, education, Internet usage, and peer pressure) on e-government development and explored the relationships among these factors. The results indicate that the overall external environment does affect e-government development at the national level, with Internet usage, democracy, and education exhibiting significant influence. Economy can also indirectly impact e-government by affecting other factors. However, this study did not find any significant impact of peer pressure on e-government at the national level.

Keywords: external environment, e-government, cross-country analysis

3.1 Introduction

3.1.1 Definition of E-Government

There are many definitions of e-government—some are narrow and some are broad. Narrowly defined, e-government refers to the delivery of information and services online through the Internet or other digital means (West, 2004). Sakowicz (2004) also defined it as the use of information and communication technologies (ICTs) to transform government by making it more accessible, effective, and accountable. Srivastava and Teo (2007) defined e-government as the use of ICT and the Internet to enhance the access to and delivery of all facets of government services and operations for the benefit of citizens, businesses, employees, and other stakeholders. Simply stated, e-government is the use of technology to enhance the access to and delivery of government services to benefit citizens, business partners, and employees (Silcock, 2001). These definitions are very similar in that they all mention both the tools and functions of e-government while highlighting the technologies used to deliver services to public.

Viewing e-government broadly, Sakowicz (2004) argued that e-government can include all ICT used to support government operations, engage citizens, and provide government services. Therefore, from Sakowicz's viewpoint, e-government embraces the whole range of governance and administrative projects, including e-services, e-democracy, e-voting, e-justice, and in some ways, even e-education or e-healthcare. Jansen (2005) supports this viewpoint, whose definitions explain how e-government employs technologies as tools to achieve better government (Organization for Economic Co-operation and Development, 2003) and bring about improvements in government operations that may include increased effectiveness, efficiency, service quality, or transformation (U.S. Government, 2002). Silcock (2001) argued that e-government is not just technology; it is not just business processes; it is not just resources. Instead, she argued that it is all these areas combined.

3.1.2 Social Benefits of E-Government

E-government is becoming a global phenomenon that is consuming the attention of politicians and policy makers as well as ordinary citizens (Jansen, 2005). In the face of rising demands from demographic, economic, social, and global trends, e-government no longer appears to be a matter of choice, but a necessity for any country wishing to enter the 21st century as a competitive nation in the world arena (Kumar, Mukerji, Butt, & Persaud, 2007). Ho (2002) mentioned that, in the early 1990s, city governments began to use electronic mail, listserv, and the World Wide Web to deliver information and services. By the end of 1990s, web-based services were already an integral and significant part of the new e-government (Ho, 2002). Data from the United Nations (UN) show that over 173 countries had already developed government websites by 2003 (Jansen, 2005).

E-government presents a tremendous opportunity to move forward in the 21st century with higher quality and more cost-effective government services, as well as with a better relationship between citizens and government (Fang, 2002). Past research on e-government impact highlights some of the benefits that it offers citizens, businesses, and the government (Srivastava & Teo, 2007). With e-government, the quality of services provided to citizens and businesses can be improved significantly while simultaneously attaining greater efficiency for all participants (Kumar et al., 2007). Services can be provided 24 hours a day and with more flexible methods that enable citizens to better communicate with the government, benefit from government services, and participate. Kumar et al. (2007) mentioned that e-government provides citizens with more control over how and when they interact with the government than they had previously possessed. Bwalya (2009) argued that e-government has the potential to improve public service delivery by public institutions in the areas of transparency, accountability, and responsiveness, while also promoting collaborative and joint-up administrations in which other stakeholders in the government can access services through portals or "one-stop shops."

In fact, the benefits brought about by e-government not only include improvement in government services and performance but also exist in social, economic, and political impacts on the whole of society. Through websites, citizens have more opportunities to give feedback to government officials and employees. E-government enables citizens to enjoy greater participation, more ably express themselves, and better play roles in the policy-making process, all of which improve the functions of democracy. Several studies have noted that e-government has helped not only in improving service delivery (Kibsi, Boer, Mourshed, & Rea, 2001; Srivastava & Teo, 2007; Von Haldenwang, 2004; West, 2004) and increasing democratization (Von Haldenwang, 2004; West, 2004) but also in reducing corruption and increasing government transparency (Banerjee & Chau, 2004; Cho & Choi, 2004; Von Haldenwang, 2004; Wong & Welch, 2004). In addition, e-government has been proposed as a solution for increasing citizens' communication with government agencies and, ultimately, furthering political trust (Chadwick & May, 2003; Clift, 2000; Norris, 2001; Seifert & Peterson, 2002; Tapscott, 1997; Thomas & Streib, 2003; Tolbert & Mossberger, 2006; West, 2004). Therefore, the benefits of e-government can be seen as both diverse and broad, confirming the importance of e-government to both societies in general and the government itself.

3.1.3 E-Government Gap and External Environment at Country Level

Although the importance of e-government has been widely accepted, its adoption among nations worldwide occurs at varying levels (see Figure 3.1). Based on the data of the UN e-government Survey of 2010, the average score of the top 10 countries is 0.81 out of 1, whereas the lowest 10 countries scored 0.02. In addition, of the 191 countries evaluated, only 61 received a score higher than 0.5, with 56 of those countries scoring below 0.3. Differences also exist among various regions. For example, the average score for countries in Europe is 0.6277, whereas the region of

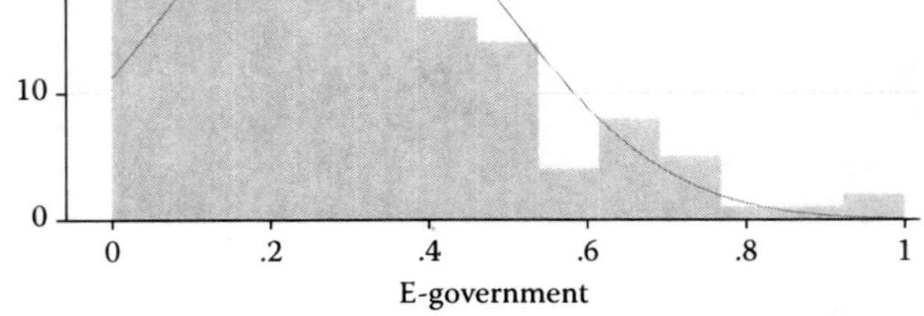

Figure 3.1 Adoption of e-governance among nations.

Africa scored an average of only 0.2733. With the average for the whole world set at 0.44, the data reflect a great divide existing at country level worldwide.

Researchers have attempted to outline the factors determining the development of e-government at country level. Some scholars focus on the factors inside the government, such as budgeting, technology, and human resources. In contrast, Basu (2004) summarized the external factors that impede the progress of an e-government in developing countries. These core factors include problems with institutional weakness, funding arrangements, local environment, technology and information changes, and legal inadequacy. Ho and Ni (2004) contend that perceived economic benefits, such as the potential for cost savings and cost recovery, might provide incentives for policy makers and managers to support the adoption of e-government. They also believed that resource availability plays a role in e-government adoption.

These arguments are based on the assumption that governments, if in possession of these resources, are eager to develop e-government. However, given sufficient funds, human resources, and technology, some might ask if South Korea or North Korea would try their best to develop an e-government—providing information, online services, and opportunities for citizens to participate. Many people would say no. Therefore, it seems that e-government development depends on a variety of factors, not only those of funding, human resources, and technology.

In fact, developing an e-government is a kind of government behavior impacted by both internal and external environments. Whether governments want to develop an e-government not only depends on the resources they have but also relates to whether they have the political will to develop it. Ke and Wei (2004) analyzed the successful experience of Singapore in e-government. They found that strong leadership and political will are very important. Even if the government has the proper technologies, human resources, and funding to improve its e-government, as soon as it lacks the will to continue, the process will become slow and inactive. Political issues not only relate to the internal environment but are also impacted by the external environment. With higher levels of democracy, Internet usage, education, and economy, public needs for e-government would likely rise, which would increase the political will of the government to improve its e-government development.

3.2 Literature Review and Conceptual Model

E-government development at the country level exists in the environment of economy, Internet usage, education, and politics, among other elements. These factors indirectly impact a country's e-government by serving as a base for the government to develop an e-government and influence the public's need for it. In this case, public needs can be viewed as pressures to push government to develop e-government to meet those needs. In addition, the e-government development of neighboring countries can be viewed as a form of peer pressure to a country.

3.2.1 Internet Usage and E-Government

A reasonable level of Internet usage is a necessary precondition for e-government development. Possessing sufficient Internet access makes it possible to provide services to citizens online while demonstrating public need for e-government development. Singh, Das, and Joseph (2007) argued that the maturity of e-government in a country could be reasonably expected to depend on the state of the ICT infrastructure because such infrastructure limits the proportion of the citizenry that can be served by e-government services. Bavec and Vintar (2007) also found that the level of Internet access impacts the development of e-government. Higher levels of Internet usage motivate more citizens to use information technologies to seek information, services, and participation in government affairs. Therefore, the public need of e-government should be higher in countries with higher levels of infrastructure. Thus, the following hypothesis can be drawn:

H1: Higher levels of Internet usage lead to higher levels of e-government.

3.2.2 Democracy and E-Government

Lee, Chang, and Berry (2011) emphasized the fundamental role of political and civic norms in the e-government and e-democracy success. A democratic society usually reflects greater levels of openness, freedom of expression, civil rights, and political stability (Das, DiRienzo, & Burbridge, 2009; Zheng & Holzer, 2013) and is generally more successful in e-government development. E-government not only uses information technologies to deliver services but also promotes the value of transparency, freedom, and democracy. Countries with higher levels of democracy are likely to have higher demand for e-government because they put more pressure on the government to provide government documents to citizens, give public opportunities to submit comments and feedback, and provide for participation in the policy-making process. Das et al. (2009) stated that with greater acceptance of civil liberties and freedom of expression comes a greater provision for any digital e-facility.

The development of mass communication media, such as the postal system, the newspaper, the telephone, television, and the Internet, has not only occurred in economic or regulatory contexts but has also come about in a political context (Guillén & Suárez, 2005). Thus, the influence of political environment on Internet usage cannot be ignored. Guillén and Suárez (2005) argued that the political regime prevalent in any given society acts as an exogenous influence on Internet development because it is a historical and causal variable existing prior to the emergence of the Internet as a popular means of communication during the 1990s (see also Norris, 2001; van Dijk, 1999). Internet development brings the decentralization in power of jurisdictions, which becomes a big concern to governments and policy makers. Countries with higher levels of democracy would be more likely to motivate the public to use the Internet to communicate, express, and discuss, which would generally increase the Internet usage level within these countries. Based on the arguments above, the following hypotheses can be developed:

H2: Higher levels of democracy result in higher levels of e-government development.

H3: Higher levels of democracy result in higher levels of Internet usage.

3.2.3 Education and E-Government

Higher levels of education serve as a great foundation for governments to improve its e-government level. Educated societies provide more qualified employees and increase the public need for e-government because more and more citizens realize the importance of e-government and can enjoy the services. In addition, higher levels of education can indirectly affect e-government by increasing levels of democracy and Internet usage. Singh et al. (2007) also emphasized that, besides the negative influence of government instability, pervasive corruption, arbitrary rule, and an ineffective government mandate, a voiceless citizenry and human capital make it unlikely for e-government to progress beyond basic information publishing, which consists mainly of propaganda.

The hypothesis that higher levels of education lead to more democratic politics (Glaeser, Ponzetto, & Shleifer, 2006; Lipset, 1959; Zheng & Holzer, 2013) has received a good deal of empirical support (Barro, 1999; Glaeser et al., 2006; Papaioannou & Siourounis, 2005). Glaeser et al. (2006) argued that schooling teaches people to interact with others and increases the benefits of civic participation, including voting and organizing. Education enables citizens to comprehend the mechanisms of government, understand their right to receive services from the government and express themselves, know their responsibilities to supervise and push the government in order to improve its performance, and increase their desire to participate in the government decision-making process. Consequentially, countries with higher levels of education should exhibit higher levels of democracy.

In addition, research shows that people with higher education are more likely to use a computer and the Internet. The average educational level in a country is a strong predictor of Internet usage (Guillén & Suárez, 2005; Zheng & Holzer, 2013). Specifically, educated individuals have more desire and ability to use the Internet for work and entertainment. Therefore, countries with a higher level of education should have higher Internet usage rates, which should result in higher levels of e-government.

H4: Countries with higher education levels have higher levels of e-government.

H5: Countries with higher education levels have higher levels of democracy.

H6: Countries with higher education levels have higher levels of Internet usage.

3.2.4 Economy and E-Government

The economy serves as a foundation for e-government development at the country level. Countries with higher per capita gross domestic product are better able to afford and support e-government (Das et al., 2009). A solid economy enables the government to provide consistent and adequate funds to use technology and human resources to develop e-government, as well as provides a sound base for the government to improve e-government, as a good economy is the precondition for the type of infrastructural development that provides citizens with more computers and Internet access. Countries with higher per-capita gross domestic product are in a better position to afford pervasive, high-quality physical ICT infrastructure (Singh et al., 2007; Zheng & Holzer, 2013), enabling citizens to have more access to e-government services. In summary, a solid economy should serve as a foundation to develop the type of infrastructure that promotes e-government development.

Lipset's observation that democracy is related to economic development, first advanced in 1959, has generated the largest body of research on any topic in comparative politics (Przeworski & Limongi, 1997). Although the relationship between economy and democracy is complex, many scholars still believe that rich countries are more likely to be democratic. In addition, it is widely accepted that educational strength and economic strength are linked. Higher levels of education should provide more qualified employees, wiser leaders, more innovative technologies, and better management, all of which would help economic development. At the same time, higher levels of economy would provide a strong base and adequate funding for education in better facilities and teaching environments while enabling more citizens to extend their education's duration. Therefore, based on the arguments above, the following hypotheses can be built:

H7: Countries with higher levels of economic development have higher levels of Internet usage.

H8: Countries with higher levels of economic development have higher levels of democracy.

H9: Economic development is positively correlated with education at the country level.

3.2.5 Peer Pressure and E-Government

The development of a country is impacted by that of the countries around it due to the effect of peer pressure. Ho and Ni (2004) used the case study of the Iowa County Treasurer's offices to explain the adoption of e-government features. They argued that peer influence is another external factor that may facilitate innovation adoption, especially when the net benefit of an innovation is difficult to assess.

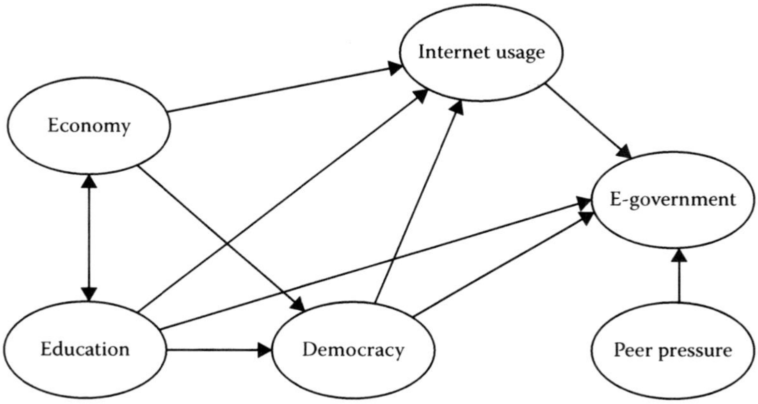

Figure 3.2 Conceptual model.

They believe that neighboring jurisdictions can easily influence each other in policy experiments, as they tend to share similar political and socioeconomic backgrounds and policy challenges (Figure 3.2). Consequentially, we may hypothesize the following:

H10: Countries whose neighbors have higher levels of e-government have higher levels of e-government.

3.3 Methodology

3.3.1 Data and Measurements

All data included are secondary data pulled from various sources. The observations were made at the country level and the e-government score came from the 2010 UN e-government survey. Online services were used to measure the e-government at the country level, which attempts to capture a country's performance in a single internationally comparable value using a four-stage model of online service maturity. This four-stage model consists of emerging information services, enhanced information services, transactional services, and connected services (Department of Economic and Social Affairs, 2010, p. 95). The sample size was 193.

Concerning the variable of education, expected education life was used to measure the education level in a country. The expected education life reflects the expected number of years of a student's schooling. The data were from the Human Development Index of 2010, a study conducted by the UN Development Programme, and the sample size was 173. In defining levels of economy, the gross

Table 3.1 Descriptive Statistics

Variable	*M*	*SD*	*Min*	*Max*
Dependent Variable				
E-government	0.31	0.21	0	1
Independent Variable				
Internet usage	32.4	27.6	0.72	93.27
Democracy	5.57	2.14	1.52	9.8
Education	0.63	0.21	0.18	0.99
Economy	12.25	13.52	0.32	61.24
Peer pressure	0.31	0.16	0.02	0.94

Note: The final sample size is 129.

national income (GNI) per capita (current $1,000) in a country was used as a unit of measure. The democracy score came from the Democracy Index of 2010 from the Economist Intelligence Unit. Based on the report, the index provides a snapshot of the state of democracy worldwide for 165 independent states and two territories (Economist Intelligence Unit, 2011). The Internet usage percentage from the International Telecommunications Union's 2010 data was used to measure Internet usage. The average e-government score of a country's land borders was used to measure peer pressure. For example, the level of peer pressure exerted upon Germany depended on the average of the e-government scores of Austria, Belgium, Czech Republic, Denmark, France, Luxembourg, the Netherlands, Poland, Sweden, Switzerland, and the United Kingdom. The descriptive statistics are in Table 3.1.

3.4 Results

To estimate the accuracy of the structural model, the maximum likelihood estimates for the model coefficients were first obtained with the STATA software. The model fit the statistics, showing that the model was accurate. The root mean square error of approximation was 0.053, which met the criteria set forth by Arbuckle (2006), who argued that a root mean square error of approximation between 0.05 and 0.08 is typically interpreted to represent between a "close" and a "reasonable" model fit. In addition, the comparative fit index was 0.998, and the Tucker–Lewis index was 0.991. Lastly, the coefficient of determination was 0.863, which is also very high. Overall, the model (as shown in Figure 3.3) proved to be strong, and most of the hypotheses were confirmed.

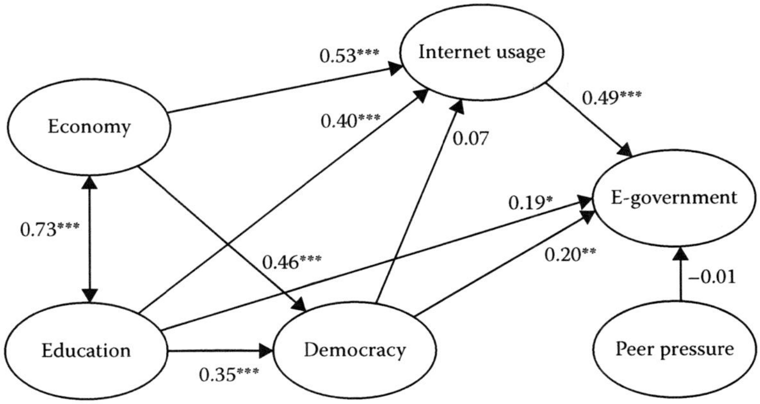

Figure 3.3 Final structural model. *p < .1; **p < .05; ***p < .01.

3.5 Findings and Discussion

As Table 3.2 indicates, the study supports the conclusion that e-government development at the country level is impacted by the external environment in which it exists. Countries with a better external environment—higher Internet usage, better education, more democracy, and a good economy—have higher levels of e-government.

Countries with higher levels of Internet usage are more likely to have a higher level of e-government, in agreement with Hypothesis 1. Higher levels of Internet usage provide a great precondition for e-government development, enabling citizens to enjoy the services brought by e-government and increasing public need for e-government. Thus, they can motivate a government to improve its e-government level.

The analyzed results also support the hypothesis that higher levels of democracy result in higher levels of e-government. This confirms the statement made by Das et al. (2009) that a democratic society is generally more successful with regard to the development of e-government. Countries with higher levels of democracy have more demand for the government to provide government documents to citizens, give public opportunities to submit comments and feedback, and allow for participation in the policy-making process. These demands would be transferred into the demand for e-government in the 21st century as more and more services are provided through e-government.

However, the study did not find a significant impact of democracy on Internet usage. It is believed that in countries with higher levels of democracy, citizens have more desire to express themselves and discuss government affairs. It is also accepted that the Internet becomes a more important vehicle for citizens to communicate

Table 3.2 Analytical Results

	E-Government	*Internet Usage*	*Democracy*	*Education*	*Economy*	*Peer Pressure*
Internet usage	0.49***					
Democracy	0.20**	0.07				
Education	0.19*	0.40***	0.35***		0.73***	
Economy	—	0.53***	0.46***	0.73***		
Peer pressure	−0.01					

Note: Model fit statistics: $\chi^2(3) = 4.07$, RMSEA = 0.053, CFI = 0.998, TLI = 0.991, SRMR = 0.016, CD = 0.863. RMSEA = root mean square error of approximation; CFI = comparative fit index; CD = coefficient of determination.

$^*p < .1$; $^{**}p < .05$; $^{***}p < .01$.

and interact. However, democracy was not found to be a key factor in deciding a country's Internet usage level. Hypothesis 3 is not supported by the results.

Hypotheses 4, 5, and 6 are all supported by the study. Higher levels of education lead to higher levels of e-government, democracy, and Internet usage. Singh et al. (2007) argued that human capital can impact e-government development. Countries with better education are more likely to have qualified leaders and employees for governments who are also more innovative and of open enough mindsets to recognize the importance of e-government. Better education makes more citizens able to seek information, demand services, and participate in e-governance.

In addition, education plays an important role in the socialization and cultivation of a citizen's sense of public participation. With better education, citizens have more desire and ability to know the mechanisms of the government, to supervise the government, and to participate in the policy-making process. Thus, the desire to develop a better e-government with improved services and more opportunities to participate will be stronger among such citizens. Lastly, better education enables more citizens to use computers and the Internet.

The results also support Hypothesis 7 in that higher levels of economy lead to higher levels of Internet usage. In countries with higher levels of economy, citizens have more money to buy computers and use the Internet. In addition, the statement of Hypothesis 8 that countries with higher levels of economy would have higher levels of democracy is also supported. The results also support Hypothesis 9: Economy is positively associated with education at the country level. Countries with more wealth are more likely to spend money on education, employ better teachers, and provide better facilities for education. Education can also positively affect the economy by providing more qualified employees, a higher level of technology, and more innovative leaders.

The results do not support the idea in Hypothesis 10, that peer pressure has a significant impact on e-government at the country level. This indicates that countries are relatively independent in developing e-government and are not strongly impacted by the progress of neighboring countries.

3.6 Conclusion

The study had limitations. First, the sample included data only from 2010 and the sample size was 129, which is not very large. More data and a larger sample size would have made for more conclusive results. In addition, because the data used in this analysis were secondhand, the validity of their measurement and data collection impacted the results of this study. For example, the survey conducted by the Economist Intelligence Unit surveyed 167 countries, using 60 questions to measure the state of democracy within them. The questions involved five components: electoral process and pluralism, functioning of government, political participation, democratic political culture, and civil liberties. Both how these five parts were measured and how the survey was conducted impact the validity of how it was applied to this study.

Despite its limitations, this study is one of the few to address whether external environments affect the development of e-government and examine the relationship among these external factors. The study indicated that e-government, at a national level, is impacted by the external environment. A positive environment, including higher levels of economy, education, democracy, and Internet usage, results in higher levels of e-government. The study showed that Internet usage, democracy, education, and economy can directly impact e-government at a national level. In addition, education can also impact e-government through its effects on Internet usage and democracy. Finally, a country's economy serves as foundation for the development of education, democracy, and Internet usage, which indirectly impact e-participation.

Future research is needed to obtain more detailed information about how these factors affect e-government. For example, although democracy impacts e-government at the country level, it would be interesting to investigate how it affects e-government by influencing the internal political environment of the government. In addition, although this study confirmed the statement that external environment—Internet usage, democracy, education, and economy—impacts e-government, one could explore whether the impact of each variable would be the same with regard to different established levels of e-government. For example, it could be discerned whether the impacts of Internet usage on an established low level of e-government would be the same for an established high level of e-government.

Author Biographies

Yueping Zheng is a senior research associate at the E-Governance Institute and is a PhD candidate in the School of Public Affairs and Administration, Rutgers University, Newark. His research interests include e-government, performance management, and citizen participation.

Aroon Manoharan is an assistant professor of public administration at Kent State University. His research focuses on e-governance, performance measurement and reporting, and strategic planning. He holds an MPA from Kansas State University and a PhD from Rutgers University, Newark.

References

Arbuckle, J. L. (2006). *AMOS (Version 7.0)* [Computer software]. Chicago: SPSS.

Banerjee, P., & Chau, P. Y. K. (2004). An evaluative framework for analyzing e-government convergence capability in developing countries. *Electronic Government, 1*(1), 29–49.

Barro, R. (1999). Determinants of democracy. *Journal of Political Economy, 107*(6), 158–183.

Basu, S. (2004). E-government and developing countries: An overview. *International Review of Law, Computers & Technology, 18*(1), 109–132.
Bavec, C., & Vintar, M. (2007). What matters in the development of the e-government in the EU? In M. A. Wimmer, J. Scholl & Å. Grönlund (Eds.), *Electronic government* (pp. 424–435). Berlin, Heidelberg: Springer.
Bwalya, K. J. (2009). Factors affecting adoption of e-government in Zambia. *The Electronic Journal of Information Systems in Developing Countries, 38*(4), 1–13.
Chadwick, A., & May, C. (2003). Interaction between States and citizens in the age of the internet: "E-government" in the United States, Britain, and the European Union. *Governance, 16*(2), 271–300.
Cho, Y. H., & Choi, B. (2004). E-government to combat corruption: The case of Seoul metropolitan government. *International Journal of Public Administration, 27*(10), 719–735.
Clift, S. (2000). *The e-democracy e-book: Democracy is online 2.0.* Retrieved from http://www.publicus.net/ebook/edemebook.html.
Das, J., DiRienzo, C., & Burbridge, J., Jr. (2009). Global e-government and the role of trust: A cross country analysis. *International Journal of Electronic Government Research, 5*(1), 1–18.
Department of Economic and Social Affairs. (2010). *United Nations e-government survey 2010.* New York: UN Publishing Section.
Economist Intelligence Unit. (2011). *Democracy index 2010: Democracy in retreat.* New York: The Economist Group. Retrieved from http://graphics.eiu.com/PDF/Democracy_Index_2010_web.pdf.
Fang, Z. (2002). E-government in digital era: Concept, practice, and development. *International Journal of the Computer, the Internet and Management, 10*(2), 1–22.
Glaeser, E., Ponzetto, G., & Shleifer, A. (2006). *Why does democracy need education?* NBER Working Paper No. 12128. Cambridge, MA: National Bureau of Economic Research.
Guillén, M. F., & Suárez, S. L. (2005). Explaining the global digital divide: Economic, political and sociological drivers of cross-national Internet use. *Social Forces, 84*(2), 681–708.
Ho, A. T.-K. (2002). Reinventing local governments and the e-government initiative. *Public Administration Review, 62*(4), 434–444.
Ho, A. T.-K., & Ni, A. Y. (2004). Explaining the adoption of e-government features: A case study of Iowa county treasurers' offices. *The American Review of Public Administration, 34*(2), 164–180.
Jansen, A. (2005). Assessing e-government progress—Why and what. *NOKOBIT,* 1504–1697.
Ke, W., & Wei, K. K. (2004). Successful e-government in Singapore. *Communications of the ACM, 47*(6), 95–99.
Kibsi, G., Boer, K., Mourshed, M., & Rea, N. (2001). Putting citizens online, not in line. *The McKinsey Quarterly, 2*(1), 65–73.
Kumar, V., Mukerji, B., Butt, I., & Persaud, A. (2007). Factors for successful e-government adoption: A conceptual framework. *Electronic Journal of E-Government, 5*(1), 63–76.
Lee, C. P., Chang, K., & Berry, F. S. (2011). Testing the development and diffusion of e-government and e-democracy: A global perspective. *Public Administration Review, 71(3),* 444–454.
Lipset, S. M. (1959). Some social requisites for democracy: Economic development and political legitimacy. *American Political Science Review, 53*(1), 69–105.
Norris, P. (2001). *Digital divide: Civic engagement, information poverty, and the internet worldwide.* New York: Cambridge University Press.

Organization for Economic Co-operation and Development. (2003). The case of e-government: Experts from the OECD report. The e-government imperative. *OECD Journal on Budgeting, 3*(1), 62–96.

Papaioannou, E., & Siourounis, G. (2005). Economic and social factors driving the third wave of democratization. *Journal of Comparative Economics, 36*, 365–387.

Przeworski, A., & Limongi, F. (1997). Modernization: Theories and facts. *World Politics, 49*(2), 155–183.

Sakowicz, M. (2004). How to evaluate e-government? Different methodologies and methods. *NISPAcee Occasional Papers, 5*(2), 18–26.

Seifert, J., & Peterson, R. E. (2002). The promise of all things E? Expectations and challenges of emergent electronic government. *Perspectives on Global Development and Technology, 1*(2), 193–212.

Silcock, R. (2001). What is e-government. *Parliamentary Affairs, 54*(1), 88–101.

Singh, H., Das, A., & Joseph, D. (2007). Country-level determinants of e-government maturity. *Communications of the Association for Information Systems, 20*(1), 632–648.

Srivastava, S. C., & Teo, T. S. H. (2007). E-government payoffs. *Journal of Global Information Management, 15*(4), 20–40.

Tapscott, D. (1997). The digital media and the reinvention of government. *Canadian Public Administration, 40*(2), 328–345.

Thomas, J. C., & Streib, G. (2003). The new face of government: Citizen-initiated contacts in the era of e-government. *Journal of Public Administration Research and Theory, 13*(1), 83–102.

Tolbert, C. J., & Mossberger, K. (2006). The effects of e-government on trust and confidence in government. *Public Administration Review, 66*(3), 354–369.

U.S. Government. (2002). *The e-government act of 2002*. HR 2458. Retrieved from http://csrc.nist.gov/policies/HR2458-final.pdf.

Van Dyjk, J. (1999). The network society: Social aspects of new media. London: Sage Publications.

Von Haldenwang, C. (2004). Electronic government (e-government) and development. *The European Journal of Development Research, 16*(2), 417–432.

West, D. M. (2004). E-government and the transformation of service delivery and citizen attitudes. *Public Administration Review, 64*(1), 15–27.

Wong, W., & Welch, E. (2004). Does e-government promote accountability? A comparative analysis of website openness and government accountability. *Governance: An International Journal of Policy, Administration and Institutions, 17*(2), 275–297.

Zheng, Y., & Holzer, M. (2013). Explaining e-participation diffusion at the national level: An examination of external environmental influences. *International Journal of eGovernance and Networks, 1*(2), 1–21.

Chapter 4

Web-Based Participatory Democracy: Findings from Italy

Gianluca Sgueo

Contents

Abstract

This chapter focuses on the Italian public administration's use of web-based experiments in participatory democracy designed to address citizens' questions and needs. The chapter aims at investigating why, despite their massive use, such forms of consultation have been largely unsuccessful. A number of causes are identified for this, including the following: (a) lack of access to the web from large parts of the citizenry, (b) miscalculation of the costs from public bodies, and (c) inadequate use of information and communication technology in public administrations. The chapter is divided into three parts. Sections 4.1 through 4.5 provide informational background on the topic of participatory democracy and e-government. To this end, Sections 4.1 through 4.5 begin by focusing on the international scenario; it then turns to describing and classifying a number of Italian experiments of online participatory democracy (at the administrative level). Sections 4.6 through 4.8 introduce and speculate the causes behind the failures of such web-based experiments in participatory democracy. Five reasons are identified: The first consists of the scope of experiments in digital participation; the second involves the target of participants; the third relates with the current state of the Italian digital market; the fourth links to the digital divide in public administrations; and the fifth, and final, argument concerns the budgetary bounds on public bodies. To conclude, Sections 4.9 through 4.11 of this chapter aims at, first, understanding why the advent of the large-scale Internet did not fix the democratic deficit of Italian contemporary politics and public administrations, and second, aim at speculating on possible future evolutions of web-based participatory initiatives in Italy.

Keywords: ICT, democracy, participation, digitalization, online democracy, networking, online polls

4.1 Introductory Remarks

Commentators on civil society's activism have always regarded the spread of new technologies enthusiastically. Already, in 1841, François-René de Chateaubriand wrote that technological advances could be expected to bring about an international society. Few years before, in 1827, Sismondi, in the *Revue Encyclopédique*, celebrated the acceleration of communications that brought the disappearance of distances and speeded up the circulation of thought (Sismondi, 1827). In 1999, Scott Kirsner reported in the article entitled "Nonprofit Motive" in *Wired* magazine: "The new breed of Silicon Valley Philanthropists would make Mother Teresa crunch the numbers" (Kirsner, 1999, p. 22). The article echoed the excitement that surrounded the spread of the upcoming global-scale use of the Internet. The web, argued the enthusiasts, would provide a low-cost and adaptable "platform" where civil society activists could rapidly acquire information, engage in peer-to-peer conversations, share their knowledge, and therefore maximize the results of their efforts. Flowery dot-org fantasies suggested that an epochal shift was about to be realized. At the national level, the Internet seemed to have the capacity to open up the world to users even in shut-in places and could erode dictatorships. At the supranational level, the promise was even greater. The Internet, it was suggested, would enable civil society actors to operate on a global scale, profoundly impacting on the spread of democratic values in transnational policy making.

Undoubtedly, the proliferation of the Internet on a planetary scale has contributed to some of the largest advancements in democracy, social activism, and advocacy. The increased availability of high-speed connections and the expansion of mobile-based services, media-rich, real-time data sharing, and voice-data communications have enhanced the potential of civil society. In the age of "global collaboration," information is disseminated online, awareness and engagement are fostered through social networks, and advocacy relies on heavy usage of web-related tools. When discussing the most visible results of these transcontinental information flows, one might include the Zapatista Movement, the campaign against the Multi-lateral Agreement on Investment, or the campaign for the development of the International Treaty to Ban Landmines. The former began as an almost entirely web-based endeavor; the anti-Multi-lateral Agreement on Investment and the Ban Landmines campaigns provide seminal examples of the usage of web-related technologies in raising awareness and coordinating an ongoing response by a multitude of actors.

Meanwhile, online political organizations and petition platforms attracted millions of members, raised tens of millions of dollars, and campaigned for a vast array of issues. Examples include the United States-based left-leaning groups MoveOn.org and Change.org, the world's large petition platforms, with 70 million users in 196 countries. Also known is the case of Avaaz, an online community involved in campaigning, signing petitions, funding direct actions, e-mailing, calling, and lobbying governments in 15 languages, served by a core team in six continents and thousands of volunteers. Avaaz became internationally recognized after the 2007 climate change summit in Bali, when the delegation of Canada credited it with motivating the delegation's change of position. Finally, smaller initiatives include iPetitions and Petitions Online.

This is not, however, the revolution celebrated by the fanatics of a digital democracy. At the supranational level, the vulgate of a widespread, democratic, decentralized, and virtual network of nonstate actors capable of promoting global values is little more than fable. Supranational activism has not given birth to the nonhierarchical and self-organizing meshwork sketched by Harcourt (2003), nor has it generated the virtual communities described by Howard Rheingold (1993) as "caretakers of electronic public space." At the national level, the spread of the Internet has neither increased the trust in politics nor boosted citizens' engagement in political life. The figures speak for themselves: The turnout in democratic elections across the world has, on average, declined in elections to national parliaments between 1980 and 1984 and again between 2007 and 2013. It has been calculated that, on average, turnout declined by 10 percentage points across both Western and Eastern democracies (Clarke, 2013). Further, the level of trust in political parties has dwindled since 1990. From 1990 to 2006, those who reported having a "great deal" or "quite a lot" of confidence in political parties across the world dropped from 49% to 27% (World Values Survey, 2014). In Europe, between 2002 and 2010, the proportion of Europeans who reported being dissatisfied with politics rose by 12 percentage points, from 31% to 43%. Membership to political parties declined accordingly, especially in Europe. In the last decade, a steep decline has been experienced in almost all major European democracies, including United Kingdom (–70%), Norway (–62%), France (–52%), Germany (–38%), Switzerland (–41%), Belgium (–30%), and, indeed, Italy (–37%) (Van Biezen, Mair, & Poguntke, 2012).

Trends relating to other avenues for political engagement are equally concerning. The case of trade unions is representative of this phenomenon. Since 1980, the proportion of salary earners who are trade union members has dropped in all but two cases across 22 nations surveyed by the Organisation for Economic Co-operation and Development. On average, trade union membership in these countries declined by 14 percentage points (Organisation for Economic Co-operation and Development, 2014). Between 1990 and the late 2000s, decreases were also reported in matters such as the willingness of individuals to engage in activities such as signing a petition or attending a demonstration. Those who reported that they might, or have already, signed a petition dropped by 20 percentage points, from 76% to just over half, at 56%. Over the same period, those who said they had or might participate in a political demonstration dropped from 62% to 51% (World Values Survey, 2014).

In light of these data, a number of problems can be introduced regarding the use of digital democracy from public institutions (both at the political and administrative levels). Two concerns are of greater importance. First, is digital democracy concretely addressing the problem of citizens' engagement in political life? Second, provided that digital participation may increase citizens' awareness, how do we define its usage as "proper usage?" In other words, how do public administrations know that experiments in digital participation are turning into successful outcomes? Is there a reliable way to classify results? As the next section will explain in further detail, this chapter aims at speculating on these issues, providing a few preliminary answers, with specific regard to the Italian scenario.

4.2 Political Parties, Public Administrations, and Digital Inclusiveness in Italy

The same trends described in Section 4.1 have emerged in Italy. This section is therefore focused on the Italian case and specifically on the use of digital democracy by Italian public powers. Let us start with basic information: Over the last 20 years, the economic crisis, the corruption scandals, and the government's inability to make structural reforms have fuelled anger and public demands for a voice in all levels of decision making. According to the *Istituto Cattaneo*, an Italian think-tank devoted to political analysis, the number of political parties' official supporters has halved in the last 50 years. In 1955, there were an estimated 4.2 million political supporters. In 2012, the number of supporters for the four major parties (PdL, PD, Sel, and LegaNord) did not reach 2 million combined. Citizens' alienation from political parties—a trend that increased strongly after the judiciary intervened against political corruption during the *mani pulite* (clean hands) campaign in the 1990s—also negatively affected participation in not-for-profit and social associations. As 14% of the adult population (7 million people) are currently active in volunteering (50% of them at least once a week), their number has registered a constant decline. Almost 1 of 2 (43%) Italians is not socially engaged in any way, whereas another 17% declare no interest in the public sphere (ACLI-IRES, 2006). Even the number of subscribers to consumerism association has decreased by 16% since 2010 (I-Com, 2014).

To counter such phenomena, political parties born in the aftermath of the 2008 economic crisis as well as public administrators have progressively turned communication from a centralized and top-down approach into a bottom-up, and inclusive, approach. The former has relied heavily on "digital inclusiveness" as a means of protest against traditional politics and its apparatus. Public administrations, in turn, have invested in developing digital participation as a form of "social reporting" or as an attempt to empower people to debate policy proposals.

The examples are numerous. The 5 Stars Movement (M5S), a grassroots initiative led by comedian-turned-activist Beppe Grillo, has strongly promoted the idea of a web democracy, in which every citizen gets the chance to play an active part in public decision making. In July 2013, the M5S introduced the "electronic Parliament" platform (named "Five Star Parliament"), which enables citizens to vote, comment, and even write pieces of legislation. These initiatives, however, remain primarily confrontational protest movements. Although they mobilize hundreds of thousands of people for their purpose, in contrast with "public participation," they explicitly distance themselves from the political domain.

The M5S electronic Parliament was launched just weeks after 15 members of the Italian Parliament had unveiled their own interactive platform, *Tu Parlamento*. Few months prior, the government, led by Mario Monti, had attempted to build up a relationship with the public through deliberative strategies like public consultations and the introduction of the "Dialogue with Citizens." The latter was an entirely web-based experiment of participatory democracy aimed at gathering feedback about specific initiatives and designed to address citizens' questions and requests (and feature their voices in decision-making processes). Further, Italian public administrations have increasingly used public online polls to consult citizens about issues of particular importance. More than 16 online consultations were held between January 2012 and January 2014. In April 2014, the government announced

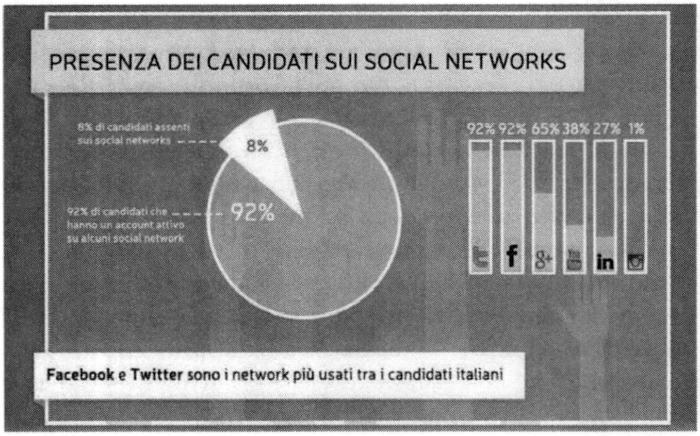

Figure 4.1 Political candidates on social network, 2013. (Courtesy of Augure.)

another broad online public consultation concerning the spending review process. Not accidentally, during the last elections, held in early 2013, the large majority of candidates opened an account on social networks (Figure 4.1) aiming at engaging with discussion and confrontation with citizens.

4.3 Success of Web-Based Participation

However, as is the idea at the basis of this chapter, not every practice of web-based participation has been successful. Public institutions experimenting with online participatory democracy initiatives have faced several problems. In frequent cases, the "public" that has participated in online consultations has represented a very narrow slice of the entire citizenry (e.g., according to the report on the 2013 online consultation on constitutional reforms, participants to the consultation were 78% more likely to have a PhD degree than the average Italian). Not uncommonly, ordinary citizens (prominently from remote areas) have complained because the lack of direct access to the web has left them excluded from online consultations. This could be explained by the fact that, whereas younger citizens show familiarity with use of the Internet, older or less technologically savvy people may be less comfortable with it and thus feel discouraged from participating.

The problem does not stem from just the age gap but more generally from the lack of digital culture within the Italian population. According to *Wired* magazine, 62% of Italians have never interacted online with public administrations. Another 52% still do not know what "e-democracy" or "digital agenda" means. Similar figures are presented in the 2014 Report from Observa Science Society (2014): 3.7 of 10 Italians (37%) have never used the Internet or a personal computer (the European average is 20%).

In other cases, the public institutions have realized too late how costly and time-consuming such experiments of e-democracy can be and eventually abandoned efforts. A recent survey from Confartigianato, an association that represents more than 700,000 businesses and entrepreneurs belonging to 870 business sectors, reported that only 928 of 8000 Italian municipalities interact with the public via the Internet. Not surprisingly, in the European ranking of public digital services, Italy scores poorly, second to last.

Such failures leave unanswered a number of questions regarding the profitable use of digital channels from public institutions (political parties and public administrations) to foster citizens' awareness and participation in public policies. Indeed, as highlighted by the European Institute for Public Participation (2009), pressing issues, such as justice, election laws, and the reform of the central state, would benefit from a more extensive inclusion of citizens. However, there remains a great deal to be done in terms of developing methods aimed at codesigning public policies, improving the quality of digital participative processes, and above all, fostering a culture of participation. The shift from consultation to decisions taken on the basis of a deliberation with and among citizens still seems to be very far from the culture of politicians currently in power.

The following section is divided in two parts. Sections 4.1 through 4.5 aims at describing and classifying a number of Italian experiments in online participatory democracy (at the administrative level). Each case is briefly illustrated in its main characteristics, both quantitatively and qualitatively. In a number of cases, the information was available on the websites of the initiatives under scrutiny. In other cases, however, the data presented are not available to the general public. These data have been harvested chiefly through interviews with key public officials. Sections 4.6 through 4.8 of the chapter turns to speculating the reasons behind the failures of these web-based experiments of participatory democracy. Among these reasons, three are considered of particular relevance. First is the fact that the entire process of "e-democratization" in Italy is still in its infancy; second is the fact that this process, as Section 4.2 explained, born primarily out of discontent and protest; third is the fact that Italian public administrations are still timidly using digital technologies.

4.4 A Look at the Past: A Few Examples of Online Participation in Italy

In the last 5 years, there have been numerous examples in which online participation systems have been utilized in Italy. Among the most memorable cases is *Burocrazia diamoci un taglio!* (BDT). The Ministry of Public Affairs started BDT in 2009. In almost 2 years (the consultation ended in late 2010), 504 citizens—40% from North Italy, 29% from Central Italy, and 34% from the South—posted their opinions and critiques online. Participants were invited to not only post their opinions and critiques but also present policy proposals. The former focused on the taxation system, the construction industry, and the welfare system (it is not by chance that all these topics are closely linked with the professional background of the participants); the latter were almost entirely dedicated to the excessive amount of red tape. The BDT had a specific goal: to translate the most noteworthy ideas into law (which actually happened with two decree-laws approved in 2012: Decree-Law n. 5/2012, named *Semplifica Italia*, and Decree Law n. 179/2012, named *Italia digitale*).

In October 2013, the Italian government launched a second consultation, linked with BDT, entirely dedicated to the top 100 administrative procedures that, according to those who would take parte into it, had to be given priority in the upcoming agenda of administrative simplification. A total of 1953 participants—1428 citizens, mostly professionals in the business sector and civil servants, aged 35–54, and 525 business enterprises, typically (60%) of small or medium size—indicated which administrative burdens were perceived as the most troublesome. Participants pinpointed the Internal Revenue Service, the construction sector, the health care system, and the labor market as the top priorities to be simplified.

Also a remarkable example from the recent past is *Linea Amica* (Friend Line). Described as "the front office of the Italian public administrations," *Linea Amica* was created in 2009 as a spin-off of FormezPA (the Centre for Services, Assistance, Studies and Training for the Modernization of the Public Administration). During 4 years of activity, almost 1.5 million citizens addressed *Linea Amica* to ask for information on public services. Men (48.3%) and women (51.7%), primarily aged 30–60 (74.4%), from Central (30%), Northern (20%), and Southern (15%) regions have contacted *Linea Amica*. The large majority used the telephone (84%). Still, an increasing number of citizens contacted the service through the web (12.8%). According to the official information released by Formez (2013), since 2009, more than 5 million citizens have visited the *Linea Amica* website.

4.4.1 Dialogue with Citizens: The Startup Phase

A noteworthy example of online participatory experiments is the *Dialogo con il Cittadino*, literally "Dialogue with Citizens," introduced by the government led by Mario Monti in early 2012.

The Dialogue was developed for two main reasons. The first involved the search for legitimacy. The government was in urgent need to be perceived as accountable and "democratic" by civil society. The creation of a new web-based channel of participation offered a viable solution to rapidly increase the democratic legitimacy and accountability of the governmental action. In that sense, the government drew inspiration from the European Commission's efforts to promote citizens' participation to address the critiques about its legitimacy and democratic stance. To this extent, the deliberative opinion poll "Tomorrow's Europe" (which polled 3600 European citizens about the future of Europe), the European Citizen's consultations, and multimedia websites such as Radio-Web Europe and the online forum Debate Europe, launched in 2006 and 2005, respectively, were examined.

The second part of the strategy behind the Dialogue aimed at "shielding" the technical government from politics. The idea involved gathering the consensus of the people, to strengthen its choices and overcome traditional political parties' opposition (i.e., the abolition of public funding to political parties).

The third motivation for the Dialogue was that it was the only solution that offered real-time information at low or no cost. In this regard, it might be useful to remember that the Monti Government was appointed during a critical stage for the Italian, and European, economies. The web enabled the government to promptly inform citizens about decisions choosing from a large variety of documents (i.e., in-depth analysis and position papers) at almost no cost (Reda, 2013).

Citizens (as well as media) have responded positively to the Dialogue. During the first quarter of 2012 (let us call it the "startup phase"), 150,537 unique visitors visited the web space that hosted the Dialogue. The trend remained steady in the following two quarters. Overall, at the end of the year, 486,368 unique visitors had accessed the website, a fair result compared with traditional online media. The Dialogue web space could rely on an average of 1332 unique visitors a day, whereas online newspapers in 2012 were visited by 6197 visitors a day (+4.5% compared with 2011).

4.4.2 The Maturity Phase

The Dialogue quickly became the official front office of the government. Although initially established to offer only basic and one-time information, it progressively developed into something very close to a proper conversation. Let us call this second phase "the maturity phase." Once citizens became accustomed to this service, they started to write uninterruptedly. In addition, they began to share complex opinions, share files, ask for detailed information, and eventually, engage the government in a real conversation.

During the 18 months of the project, roughly 90,000 people contributed (approximately 5000 e-mails a month). Citizens wrote mostly from Northern regions (45%, compared with 24% from Central regions, 18% from Southern regions, and 8% from islands). A small percentage, 5% in total, wrote from abroad. Typically, citizens were aged 35–50, with no significant differences in gender. In the first quarter of 2012, 69% of the messages received through the Dialogue were answered within 3 weeks from its reception. Overall, at the end of the year, 95% of the messages received had been answered.

Messages were classified according to a list of topics. Among the most addressed topics were the reform of the welfare and pension systems (addressed by 30% and 18% of the citizens, respectively), the abolition of the new tax on house property (14%), youth policies (10%), the spending review process (7%), and the adoption of a new comprehensive normative system for businesses, startups, and entrepreneurship (4%).

4.5 Online Public Consultations

In concomitance with the Dialogue, the Monti Government organized a number of public online polls to consult citizens about issues of particular importance. Each consultation allowed a defined period during which members of the public could submit comments.

The idea was so successful that consecutive governments have developed the habit of consulting citizens online on a regular basis. Over a 15-month period, the two Prime Ministers who succeeded Mario Monti (Enrico Letta and Matteo Rienzi, respectively) sponsored eight online consultations.

Overall, 11 consultations were held in 2012. At the beginning, the range of variation among the consultations was high. The duration ranged from a minimum of 1 to a maximum of 89 days. Social media (particularly Twitter and Facebook) were used to support the consultation process in nearly half of the cases (40%). In almost 90% of the consultations, citizens sent their opinions through a website. In two cases, the consulting institution also decided to create an e-mail address. In the consultation on the energy strategy, both the website and the e-mail address were available. Interestingly, none of the consultations set a minimum number of citizens to consider before the consultation successfully concluded. This differs from similar initiatives such as the United States-based "We the People" campaign (which, in January 2014, raised the threshold for an official response to 100,000 signatures, from 25,000). Table 4.1 summarizes the key elements of the consultation held by the Monti Government and, specifically, the duration, how citizens' opinions were acquired, and the support of social media.

Of particular interest is the consultation on the spending review process. To date, this has been the consultation with the highest participation rate: 151,536 citizens wrote their opinions. The thematic section of the government website was visited by 550,566 unique visitors (nearly 45% of total access to the government

Table 4.1 Public Consultations (2012–2013)

Consultation	*Duration*	*Participation*	*Social Media*
Legal value of the university degree	30 days	Online	×
Spending review	8 days	Online	×
European digital agenda	35 days	Online	×
Italian digital agenda	35 days	Online	×
Internet principles	44 days	Online	×
Administrative action	59 days	E-mail	–
Guidelines for naval shipping	89 days	E-mail	–

Source: Italian Government website.

website) in the 28 days of the consultation lifespan. The flow of messages increased on given days, but it always kept above 20,000 messages a day. Citizens submitted an array of complaints, from throwing out uneaten hospital food to leaving the heating on during the summer. But the most recurrent theme focused on the outlay required to maintain Italy's political class and their related cost, like chauffeured cars and privileged pension plans. The responses reflected the growing frustration of those who felt overtaxed by elected representatives, who many perceived as having placed their own interests before those of the public good.

The range of variability demonstrates the absence of a common policy on online consultations. Ministries consulted citizens in full independence. It is exactly for this last reason that, in November 2012, the government drafted a series of guidelines to be used for future online public consultations held by central public administrations. The project was interrupted because of the fall of the government in December and the subsequent elections.

As already mentioned, in June 2013, the government led by Enrico Letta launched two large online public consultations. The first was dedicated to constitutional reforms; the second addressed the topic of foreign investments in the Italian market. In both cases, the website hosting the consultations (and the staff that worked on it) was drawn directly from the project elaborated in 2012. As far as the first consultation is concerned, two questionnaires were available. To the first, a shorter version, 131,676 citizens answered. Another 71,385 answered the longer version of the questionnaire. The second consultation was linked to the report "*Destinazione Italia*." This consisted of 50 measures whose goal was to reform a broad range of sectors, from tax to employment and civil justice to research, and to develop investment-focused policies to promote Italy at the international level. The consultation was opened right after the presentation of the report, to give Italian citizens, foreign business communities, business associations, trade unions, and experts at various levels a say on its contents. The government committed to translating these opinions into the report and, thus, into provisions and laws and to begin monitoring their implementation on a weekly basis.

4.6 An Ineffective (Digital) Democracy: Scopes and Target of Web-Based Participatory Experiments

Having described the most relevant experiences of digital participation in Italy, this section is aimed at evaluating the success of such initiatives and expounding the reasons behind their underwhelming results. Five reasons will be considered in evaluating the still ineffective digital democracy in Italy. The first and second reasons involve the scope of experiments in digital participation and the target of participants, respectively. A third reason relates to the current state of the Italian digital market; the fourth reason links to the digital divide in public administrations; the fifth argument concerns the budgetary bounds on public bodies.

Let us start from the scopes and the target of web-based participatory experiments. Although the press has given considerable (and, in general, positive) attention to digital democracy initiatives promoted by the Italian governments, critics have argued that these initiatives have been promoted mostly for political or electoral purposes. In addition, when digital participation has been promoted by public administrations other than the government, the aim, as perceived by many, was that of a "social reporting," rather than a true attempt to empower people's voice within policy-making.

As a consequence (and, at the same time, a substantiation) of the partial success of experiments in digital democracy, it might be relevant to consider the target of participants. According to the trend registered in many of the Western democracies, the number of people connected to the web is in constant growth, and their willingness to participate in policy-making is thus assumed. As shown by the Deloitte Digital Democracy Survey (2014), for instance, the so-called digital omnivores in the United States—those consumers who own a trio of tablets, smartphones, and laptops—have continued to grow in number, driven by the proliferation of new platforms and increased device adoption. Deloitte classifies one third (37%) of U.S. consumers as digital omnivores, a 42% growth over the previous year. The report indicates a growth, which has been driven by continued tablet adoption (33% increase from 2012 to 2013) and, to a lesser extent, smartphone ownership (18% increase).

Italy is no exception to this trend. Almost 62% of Italians are connected to the web, with an average increase of 9 percentage points per year. Increasingly, Italians are connected through smartphone (+10% from 2011 to 2012) and tablets. Nevertheless, Italy is still suffering from a significant gap between the proportion of the population that is connected and actively engaged with the web and the still considerable number of citizens who either are not connected to the Internet or do not engage in any form of online activism, including forms of digital democracy. As already mentioned, the 2014 report from the Observa Science Society notes that 3.7 of 10 Italians have never used Internet or a personal computer. According to another think-tank devoted to research on politics and society, Demos&PI, 6 of 10 Italians have an Internet connection at home (they doubled in 10 years: from 23% in 2000 to 58% in 2013). However, less than half of those citizens (roughly 40%) are actively engaged online. Also, the level of engagement varies. Demos&Pi distinguishes between the "cives.net" (25% overall), those who consider the web as an arena for discussion and political confrontation, and the "infonauts" (15% in total), who prefer to use the web mostly for getting information.

4.7 A Market in Backsliding and the Digital Divide in Local Administrations

Let us move to the third and fourth reasons. The third reason that might explain the limits of web-based participatory initiatives in Italy relates to the Italian digital market. In 2013, the Italian digital market has shown a considerable backsliding, with a loss of 4.4 percentage points compared with the previous year. According to the 2014 report from Assinform—the national association of information and communication technology (ICT) companies operating in Italy—the Italian digital market is now worth a total of €65.2 billion, having registered a reduction in growth of 1.8% per year since 2009. Over the same period, ICT worldwide has an annual average growth of 3.8% (3.5% in North America, 6.6% in Asia, and 5.8% in Latina America). Investment on ICT in relation to gross domestic product in Italy is 4.8%. The European average is 6.5%, with Germany at 6.8%, France at 7.0%, and United Kingdom at 9.6%. In essence, Italy invests €25 billion per year less than Europe does on ICT.

A fourth factor to consider when evaluating attempts by Italian public administration to engage citizens online relates to the digital divide in the public sector, especially in local administrations.

The 2014 Smart Culture & Travel Report examines the development of ICT in 116 municipalities. The report highlights the gap in a number of online services to the public, such as tourism. Although all the municipalities analyzed in the report have a website, the report notes that less than 50% have a page dedicated to tourism, and only 17% have a page dedicated to culture. Most importantly, only 14% of the municipalities provide booking services for hotels, whereas in less than 3% of the cases, it is possible to buy museum tickets online. The report also analyzes the use of social media by municipalities. Of the 10 bigger cities, 6 have at least a Facebook, Twitter, or YouTube account. However, only 29% have activated a profile on all those social media. In fact, whereas 54% have a Facebook profile, 445 and 41% of the municipalities use Twitter and YouTube, respectively.

As reported by the E-Democracy Centre of the Faculty of Law of the University of Geneva, in 2005, 16.5% of regional public administration bodies in Italy had an ICT office, which often deals also with e-democracy initiatives. The highest concentration was found in the regions of Emilia Romagna and Tuscany. More specifically about the presence of digital democracy tool from Italian municipalities, the latest data available (Saviano & Iorio, 2010) reveal that only 52.6% provide at least one digital democracy tool. However, 50% of these tools are just very simple and informative (e.g., information contents on the websites or newsletters). Deliberative tools (e.g., forums with public administrators or blogs) are almost nonexistent, as they do not even number 1% of the total.

4.8 A Matter of Costs

Ultimately, the current stagnation of web democracy in Italy is connected with the budgetary restraints in the public sector. Even when public bodies have admittedly relied on the web to engage civil society as a way to maximize the outcomes at almost no cost, as in the case of the Dialogue with Citizens experimented by the Monti Government, budgetary restraints have impeded attempts to view web participation as a long-term investment. Although initially fulfilled in the presence of a low budget and with a small staff, the Dialogue soon led to additional costs due to the need to manage a constant increase in interactions and to deal with emergencies. As a consequence, the staff dedicated to the Dialogue almost doubled in 1 year. Additional expenses come from the workflow delay. As already argued, a substantial and systematic increase in citizen comments might lead public powers to strive to satisfy those who file comments instead of selecting the policy option that best fulfills the statutory mandate or public interest (Coglianese, 2005).

What is particularly striking is the fact that, as argued by many scholars, public administrations could not noticeably reduce their costs through the web (Wang & Bryer, 2009). With specific reference to the Italian case, in 2012, the School of Management of the Polytechnic Institute of Milano estimated that the introduction of a system of electronic payment, the full implementation of e-procurement, and the digitalization of the management and conservation of administrative acts would allow Italian public administrations to save €20 billion in 3 years.

There is not a single successful strategy to limit the expenses. Public institutions interested in e-democracy can decide only on a case-to-case basis. In 2012, the Italian government relied on the usage of the web, as well as on what the Aspen Institute has defined "soft power," namely, the use of communication built on persuasion trough transparency and information, to secure public support of interests, values, and policies (Bollier, 2003).

4.9 Conclusions: Dilemmas of E-Participation in Italy

Sections 4.1 through 4.5 of this chapter has touched on the intense debates dating from the early phases of widespread Internet use. As Section 4.1 has explained, this debate involved academics, media outlets, and politicians who believed that the web would herald a new age of democratic participation. Many, at that time, pointed to the web's capacity to host deliberative dialogues among large groups of individuals. Others focused on the capacity of the web to host rapid real-time interactions and described it as a beneficial mechanism for citizens and their political leaders to interact more regularly and openly. The effects, they implied, would be threefold. First, citizens would be better able to communicate their needs and preferences to their political representatives. Second, having received this input, these representatives would make better decisions that more accurately reflected the will of the people, and citizens, in turn, would develop a greater sense of trust in their political system. Third, drawing on the concept of "digital natives"—a term that some have applied to the millennial generation, given the ubiquitous role that ICTs tend to play in the lives of those born from the early 1980s onward—authors argued that the Internet would prove particularly potent as a means of engaging youth and thus curb their withdrawal from politics. Lastly, early commentators also welcomed the massive use of technology as an encouragement for the rise in a range of web-based media outlets, thanks to which information has become widely accessible to the world population.

The data presented in Sections 4.1 through 4.5 (and with specific regard to the Italian case in Sections 4.6 through 4.8) of this chapter suggest that the advent of the large-scale Internet did not fix the democratic deficit of contemporary politics and public administration. To begin with, the rise in access to the Internet has influenced neither the voting rates nor party membership. Similarly, disengaged youth did not become interested in politics because it moved online (Barnard, 2009). In addition, the extent to which opportunities for online citizen engagement are enjoyed equally by different segments of the population has been questioned. Those who engage online, it has been noted, are often the same as those who engage offline (Di Gennaro & Dutton, 2006). The Internet, in other words, seems to attract less those citizens who Lance Bennett (2008) would describe as "actualizing citizens"—people who are distrustful of traditional forms of authority and are inclined to adopt more privatized responses to changing social circumstances—and more of the "dutiful citizens"—to whom involvement in civic life is an obligation to be fulfilled through conventional activities, such as voting. According to this interpretation, the Internet might paradoxically broaden the gap between those who are politically active (and would therefore benefit from new opportunities to participate in their democracies via the web) and those who are not.

The basic purpose of this section of this chapter is to summarize such dilemmas with specific reference to the Italian scenario. More specifically, from the analysis conducted in this chapter, three motives of reflection arise: (1) The first settles on the search for the maximization of profits and the minimization of costs from Italian public institutions engaged in experiments of e-participation. How can a public institution avoid incurring excessive costs while increasing its transparency and accessibility through online participation? (2) The second motive of reflection considers citizens' reaction to web participation. Arguably, the more the citizens are given the opportunity to engage in policy-making, the less they will be prone to accept delays or blackouts in communication from the institution. (3) Finally, the third motive of reflection links with the future of e-participation in Italy: Is this one of expansion or regression?

4.10 Maximization of Profits and Citizens' Expectations

As in the cases of many other European and extra-European democracies, also in Italy, the use of web-related tools has demonstrated its high potential to address large communities. Nearly unrestricted access, ample space of expression, and the possibility of participating anonymously represented strong incentives to citizens for participating. Thus far, however, the Italian public administrations did not attempt to face the issue of costs rationally. In some cases, announcements of new online consultations have followed long periods of silence. In March 2014, for instance, the government published a call for experts to be hired in a task force being set up to manage all future initiatives of participatory democracy. The deadline for applications expired in April 2014, but the government did not give any sign to proceed on the creation of the task force. In other cases, the Italian administrations have simply decided to postpone any initiatives of web democracy, especially if the estimated costs were too high. Exemplary is the case of the Ministry of Public Affairs, who announced in April 2014 a broad consultation on the reform of Italian public administration. The consultation lasted 30 days. A brief summary was then published by the ministry to inform about the consultation, and contextually, the phase of confrontation with stakeholders was postponed. Given the absence of a unitary approach from Italian administrations on the topic of web democracy, it might be only assumed that this is currently hinged to the will of each administration. When the spending review process approved by the Italian Parliament in 2013 will enter into force, a progressive harmonization of public administrations' budgets and, possibly, a less varying usage of web participation at the administrative level might be expected.

Moving to the second issue, one should note that, even if discontinuous, the spread of channels of web democracy contributed to the increase in the expectations of Italian citizens of good administrative behavior from public institutions. Initiatives from civil society aimed at supporting public administrations in experimenting with e-democracy have increased over the last 2–3 years. The scope is to fill the gap created by public administrations' inactivity in web democracy and the increasing citizens' expectations. Unfortunately, the self-financed nature of these grassroots projects thus far impelled a greater experimentation. Examples include Airesis, a social network created with the specific aim of promoting e-democracy in the public sector, and Openpolis. The latter is a not-for-profit association founded in 2006 by citizen-activists that develops and implements projects to enable free access to public information on political candidates, elected representatives, and legislative activity. The data are extracted from Italian public administrations' websites and then made available for free to citizens and the media. Openpolis has rapidly turned into a civic observatory on Italian politics, allowing experts and ordinary people to shape their own view. The project currently monitors more than 225,285 politicians and includes more than 17,348 official declarations. Almost 19,000 users access and share the information available on its website on a regular basis. The goal is not only to increase the opportunities for dialogue between citizens and their representatives in central and local governments but also to contribute to constantly and effectively pressure local governments and individual politicians. It is for this reason that the municipalities that adhere to the initiative are invited to let citizens post their opinions and comments online.

4.11 Future of Online Participation in Italy

Having discussed the issue of the costs of web democracy and the problem of citizens' expectations, the last dilemma to be addressed relates to the future developments of e-democracy in Italy. The findings of this chapter seem to indicate that Italian public administrations have a common perception of the basics of online democratic participation. They also share similar problems in dealing with e-democracy (namely, the problem of costs and that of citizens' expectations). As a further, and major, consequence, the experiments on online participation that have been discussed in this chapter may have good chances of being replicated in very different contexts across the country. This could possibly lead to a common model for online participatory rights. On the one hand, this model draws from the incorporation of shared core fundamental principles into domestic procedural rules/practices of administrative law. Despite its relative adolescence, web-based experiments in participatory democracy have a robust collection of projects in general and can count on a relatively high level of cooperation and coordination among and between the different levels of government. Although at the local level, there are no explicitly defined government bodies that are directly responsible for e-government or e-democracy actions, the fact that, at the central level, all electronic initiatives are still monitored by a single, central, administrative body, the National Centre for Public Administration and Informatics, can be considered as progress. The role of the National Centre for Public Administration and Informatics is to stimulate, support, and promote ICT usage within national public bodies.

On the other hand, it seems that the success of this process toward uniformity increasingly depends on some sort of higher-level supervision—either from the European Union or from international bodies. Not casually, in the Italian case, the government has replicated the European model, which has been adopted also by a number of international bodies. In 2012, the Italian Council of Ministers introduced the "Community Initiative," enabling citizens to take part in the legislative activities of the European Parliament through the European Commission. This instrument has been issued in response to the member countries' wish to consolidate the principles governing democracy, by bringing the citizens closer to the institutions and making them "frontline activists" in initiatives and lawmaking.

Author Biography

Gianluca Sgueo is a postdoctoral researcher in Democracy and Innovation at the Center for Social Studies of the University of Coimbra (Portugal). He currently holds the positions of adjunct professor in European Law at the University of Viterbo (Italy) and lecturer in Lobbying & Democracy at the Bocconi University (Italy). He graduated with degrees in law (University of Rome, La Sapienza, 2004) and political sciences (University of Viterbo, 2006). He holds an MA in Administrative &

Public Law (Scuola Superiore della Pubblica Amministrazione e degli Enti Locali, Rome, 2005), an MA in European Law (European Public Law Academy, Athens, 2009), and a PhD in Public Law (University of Salento, Italy, 2010). Between 2011 and 2013, Gianluca Sgueo was appointed head of the Press Office and coordinator of the "Dialogue with Citizens" of the Italian government.

References

ACLI-IRES. (2006). *Gli anticorpi della società civile – nono rapporto sull'associazionismo sociale.* Retrieved from http://www.ideeinrete.coop/aggiunte/rapporto_associazionismo.pdf.

Barnard, R. (2009). *Coding youth engagement.* Presentation delivered at the Library of Parliament's Youth and Democracy Dialogue Session, Ottawa, Canada.

Bennett, L. W. (Ed.). (2008). *Civic life online: Learning how digital media can engage youth.* Cambridge, MA: Cambridge University Press.

Bollier, D. (2003). *The rise of netpolitik: How the internet is changing international politics and diplomacy.* Washington, DC: The Aspen Institute.

Clarke, A. (2013). *Exploiting the web as a tool for democracy: New ways forward in the study and practice of digital democracy.* Strasbourg: World Forum for Democracy.

Coglianese, C. (2005). The internet and citizen participation in rulemaking. *Journal of Law and Policy, 1*(1), 33–57.

Deloitte Digital Democracy Survey. (2014). Retrieved from http://www.deloitte.com/view/en_US/us/Industries/media-entertainment/digital-democracy/index.htm.

Di Gennaro, C., & Dutton, W. (2006). The internet and the public: Online and offline political participation in the United Kingdom. *Parliamentary Affairs, 59*(2), 299–313.

European Institute for Public Participation. (2009). *Public participation in Europe. An international perspective.* Retrieved from http://www.participationinstitute.org/wp-content/uploads/2009/06/pp_in_e_report_03_06.pdf.

Harcourt, W. (2003). The impact of transnational discourses on local community organizing. *Development, 46,* 74–79.

I-Com, Rapporto 2014 sui consumatori. (2014). Retrieved from http://www.i-com.it/en/2014/04/24/documento-di-output-su-fast-track-linnovazione-al-servizio-dei-consumatori/.

Kirsner, S. (1999). Nonprofit motive. *Wired Magazine, 7*(9), 1–4.

Observa Science Society (OSS). (2014). Retrieved from http://www.observa.it/scienza-societa-dieci-anni-tendenze-nel-nuovo-annuario-scienza-tecnologia-societa-observa/?lang=it.

Organisation for Economic Co-operation and Development Economic Surveys. (2014). Retrieved from http://www.oecd.org/eco/surveys/.

Reda, V. (2013). Managing communication in the Prime Minister's Office, 1988–2012. *Contemporary Italian Politics, 5*(3), 277–291.

Rheingold, H. (1993). *The virtual community: Homesteading on the electronic frontier.* Cambridge, MA: MIT Press.

Saviano, M., & Iorio, G. (2010). How far from participatory governance? A survey on e-democracy in Italian municipalities. *Prayaa Journal of Management, 1*(2), 1–18.

Sismondi, C. L. (1827). *L'Amérique, Revue Encyclopédique*, 22.

Van Biezen, I., Mair, P., & Poguntke, T. (2012). Going, going…gone? The decline of party membership in contemporary Europe. *European Journal of Contemporary Research, 51*(1), 24–56.

Wang, X., & Bryer, T. A. (2009). Assessing the costs of public participation. A case study of two online participation mechanisms. *The American Review of Public Administration, 43*(2), 179–199.

World Values Survey. (2014). Retrieved from http://www.worldvaluessurvey.org.

Chapter 5

Comparative Scientometric Analysis in Social Media: What Can We Learn? And What Is Next?

Manuel Pedro Rodríguez Bolívar and
Laura Alcaide Muñoz

Contents

Abstract

In this chapter, we seek to analyze the research into the field of social media implementation in public administrations, examining the theories that support the use of social media, the main use of this technology for public administrations, and the trends and research innovations in this area in the future. In addition, a comparative study will be performed to identify differences of research, research gaps, and interest on the different domains of this topic into different contexts. To achieve the aims, the scientometric methodology will be applied to a sample of papers published in journals listed in the fields of Public Administration, Information Sciences, and Communication. In this sense, the research about social media has not been the same in all areas. Knowledge gaps and research opportunities are identified from these observations, which reveal changes in the research methods applied, reinforcing the development of a theoretical framework so that the application of social media may efficiently contribute to improving management in the public sector.

Keywords: social media, scientometric study, research opportunities, United States, Europe, Asia

5.1 Introduction

E-strategies have been the key elements for governments to perform investment planning on information communication and technologies and to manage social change (Anthopoulos & Fitdilis, 2014). In this regard, e-government development has been characterized by a three-stage process (Taylor, 2012). The first one is called the era of "direct government" and is characterized by the pronouncements of a set of guiding principles to underpin development; the main ones were choice, confidence, and accessibility. Under this stage, governments offered only static, limited, and basic information, such as links to ministries/departments, archived information, and regional/local government services. In addition, governments usually implement unidirectional government-to-citizen information flows such as online policies, laws and regulation, reports, newsletters, and downloadable databases, among other things (United Nations, 2012).

In the second stage, called "orthodox government," a new wave of investment by the government in information communication and technology applications was expected. Services tailored to individual needs, more joined-up government services, and opportunities for a "mixed economy" of service provision that could include organizations from private and voluntary sectors were promised in this stage. Under this stage of development, Brown (2010) proposed the strategy called

Mygov in the United Kingdom, which marked the end of the one-size-fits-all and implemented the need for a more interactive and personalized service. In addition, "connected government" initiatives, in terms of cross-agency e-services, are promoted in this stage (United Nations, 2012). In fact, two-way financial and nonfinancial transactional services and advanced technical features such as mobile apps are present in this stage of development (citizen-to-government interactions). It includes options for paying taxes and applying for ID cards, birth certificates/passports, license renewals, and other similar citizen-to-government interactions by allowing citizens to submit these online 24/7.

Nonetheless, e-government initiatives over the past decade have been based mainly on first-generation web-based resources: HTML, a relatively primitive, static page mark-up technology that simply outlines what a page should look like onscreen. Finally, from 2005 to the present time, a shift in strategic emphasis has occurred whereby the separate designation and practice of e-government have been removed largely in favor of a whole of government approach, carrying the title "transformational government." In this stage, the government encourages participatory deliberative decision making and is willing and able to involve the society in a two-way, open dialogue. An initiative performed under this stage of development is the so-called Open Government Initiative, which prioritizes transparency, participation, and collaboration and promotes interactive features such as the web comment form and innovative online consultation mechanisms to actively solicit citizens' views on public policy, law making, and democratic participatory decision making (Commonwealth Network of Information Technology for Development Foundation, 2002; Peedu, 2011).

In this regard, under this third stage, governments must strengthen their capacity to assess the needs of users (both private and commercial) and involve user groups through the use of second-generation web technologies (Web 2.0) to listen, to engage users in the design of services and in the production of policies, and to forge collective initiatives and interaction (Organisation for Economic Co-operation and Development, 2010). Indeed, the implementation of Web 2.0 technologies has favored reforms and modernization within public administrations, changing the nature of political and public dialogue (Osimo, 2008) and encouraging participation by citizens, allowing them greater involvement in public affairs and enabling public managers to create more affordable, participatory, and transparent models of public sector management (McMillan, Medd, & Hughes, 2008).

Various popular Web 2.0 technologies facilitate interactive information sharing, interoperability, and collaboration (United Nations, 2010) and can promote open, user-driven governance (Bertot, Jaeger, & Hansen, 2010a, 2010b; Bertot, Jaeger, Munson, & Glaisyer, 2010; Millard, 2009). Social-media-related tools/applications include collaborative projects (such as wikis), blogs, content communities, social networking sites (such as Facebook and Twitter), virtual game worlds, and virtual social worlds (Kaplan & Haenlein, 2010). Social media, particularly Internet social networking, has profound effects on all facets of social life and has fundamentally altered how we interact and maintain our social relations in a society (Kaplan & Haenlein, 2010). In addition, the growing participation of citizens in social networking sites is altering the nature of social relations (Christofides, Muise, & Desmarais, 2009; Li, 2011), changing the nature of political and public dialogue (Osimo, 2008). Therefore, this chapter is focused on analyzing social media as a new means of communication between public administrations and the citizenry.

Continuing interest in the question is reflected in the large number of studies published in this respect in the last years. It draws upon various reference disciplines, including Public Administration, Information Science, and Communication. Since its appearance, social media researchers in public administration have analyzed mainly the usefulness of social media for different purposes such as political campaigns (Aparaschivei, 2011; Williams & Gulati, 2013), disclosure of a greater volume of information to a wider range of citizens (Bonsón, Torres, Royo, & Flores, 2012), and citizen coproduction initiatives (Linders, 2012). The first two aspects concern the transparency and visibility of local government actions, whereas the second, in addition, favors more participative management (Kovac & Decman, 2009).

Other researchers have explored the past, present, and future development of social media (Boyd & Ellison, 2007) to create a theoretical support for the discipline (Sidorova, Evangelopoulos, Valacich, & Ramakrishnan, 2008). Nevertheless, a systematic, unifying approach has yet to be achieved, and without this, no new theoretical arguments can be created nor concepts and categories provided to help us understand the social media phenomenon in public administrations.

To understand the construction of theoretical support underpinning the question of social media, it is of critical importance to explore its intellectual core, by analyzing the cumulative body of knowledge rather than looking at individual works (Holsapple, 2008). Therefore, comprehensive reviews are needed to integrate contributions and to provide a critical outlook on work in this field, to improve our understanding of e-government, and to gain a broad view of the current situation and of possibilities for future research. For this purpose, we propose a descriptive scientometric methodology (Neufeld, Fang, & Huff, 2007) that has been comprehensively tested in the field of information science (Cocosila, Serenko, & Turel, 2011; Waltman, Yan, & Van Eck, 2011). This specific approach has been used, among other reasons, as an attempt to understand the identity of a scientific discipline in a particular academic area (Straub, 2006). In this chapter, the underlying assumption of our approach is that the theoretical framework of a scholarly discipline is built upon the high-quality body of knowledge published in the leading channels established for this purpose. Nevertheless, to date, prior research has not analyzed this item in the social media field of knowledge.

To fill this gap, the objective of this chapter is to analyze the research into the field of social media implementation in public administrations, examining the theories that support the use of social media, the main use of this technology for public administrations, and the trends and research innovations in this area in the future. In addition, a comparative study will be performed to identify differences in research, research gaps, and interest on the different domains of this topic into different contexts. The ultimate objective of this chapter is to assist researchers in the development and direction of future analysis in the implementation of social media in public administrations into different contexts. To achieve the aims proposed in this chapter, the scientometric methodology will be applied to a sample of papers published in high-quality journals in the fields of Public Administration, Information Sciences, and Communication.

The rest of this chapter is organized as follows. Section 5.2 presents the literature review and sets out the research questions proposed. Then, Section 5.3 describes the sample selection process and the analytical methodology used and then presents the results obtained from this approach. Finally, we highlight the findings obtained and the main conclusions drawn and identify possibilities for the development of further analysis in the area of social media.

5.2 Background

There are multiple channels used by citizens to be in contact with governments. Although e-government has promoted the use of digital means for citizens to initiate contact with the government (Kanat & Ozkan, 2009), prior research has demonstrated that where citizens have multiple choices to contact government, they can use the channel that best suits their needs (Reddick & Turner, 2012). In this milieu, governments need to ensure consistency of information and service response across channels and to create a positive experience with e-government if they want to promote digital channels of communication (Reddick & Turner, 2012).

In the last years, social media has become an alternative channel for citizens to access their governments. The advent of social media using Web 2.0 technologies has opened up unprecedented new possibilities of engaging the public in government work and has changed the public's expectations about how government work should be done (Chun, Shulman, Sandoval-Almazan, & Hovy, 2010; Lathrop & Ruma, 2010; McDermott, 2010). Indeed, social media applications provide channels not just for mass dissemination but also for mass production and collaboration (Benkler, 2006) and have become acceptable information and communication channels in governments (Mergel, 2013; Mossberger, Wu, & Crawford, 2013), playing an important role in implementing open government and in rendering online public services (Noveck, 2009).

In this regard, driven by rising citizen expectations and the need for government innovation, social media has become a central component of e-government in a very short period (Bertot, Jaeger, & Hansen, 2012). Social media differs from first-generation web-based resources in at least three significant ways: It is participatory, it is pervasive, and it is integrated (Mintz, 2008). Web 2.0 has the potential to transform public administration services, enabling the development of better policies and eliminating data silos (Klein, 2008; see Table 5.1).

Therefore, Web 2.0 technologies (Web 2.0) have the potential to change the way the government delivers services and its relationship with the public, enabling more effective citizens' engagement and collaboration with the community; more personalized, faster, easier-to-use, and deliverable services; effective collaboration and teamwork; and higher productivity than the Web alone can provide (Accenture, 2009). In this regard, Government 2.0 is more than simply the adoption of Web 2.0 tools by the government. Government 2.0 is also a philosophy and a culture that reflect the society's new way of interacting and communicating, and one that governments must accept if they wish to keep up (Goldsmith & Eggers, 2004).

In any case, with the aim of analyzing the contribution of social media to the sphere of public administrations, it would be necessary to undertake scientometric analyses that seek to help organize the information available and to highlight useful areas for future research. In addition, this analysis could help to know the theoretical underpinnings of the use of social media in the public sector environment. Nonetheless, to date, scientometric projects in social media in public

Table 5.1 Differences between Government 1.0 and Government 2.0

	Direct and Orthodox Government	*Transformational Government*
Dimension	*Government 1.0*	*Government 2.0*
Operating model	Hierarchical	Networked
	Rigid	Collaborative
		Flexible
New models of service delivery	One-size-fits-all	Personalized
	Monopoly	Choice based
	Single channel	Multichannel
Performance	Input oriented	Outcome driven
	Closed	Transparent
Decision making	Spectator	Participative

Source: Based on Deloitte (2008). *Change your world or the world will change you. The future of collaborative government and Web 2.0.* Quebec: Deloitte & Touche LLP and affiliated entities; and Taylor, J. A. (2012). *Information Polity, 17*, 227–237.

administrations have not been examined. This absence of comprehensive scientometric studies could mean that an interesting aspect of social media research has remained unexamined, despite the fact that social media, in terms of the use of information technology, has been studied from diverse standpoints. Therefore, it would be interesting to analyze the research published in the main fields of knowledge and thus acquire different perspectives of social media research. Taking into account that the majority of published research in e-government corresponds to the field of Public Administration, Information Science, and Communication, our first research question is as follows:

RQ1: How many social media articles have been published in Journal Citation Report (JCR) journals in the fields of Public Administration, Information Science and Library Science, and Communication?

As noted above, scientometric projects seek to identify a discipline, structure the information available, and highlight potential areas for future research. In this regard, the analysis of the most commonly published questions in the field of social media could inform researchers of the state of the art and highlight research gaps. In other words, this analysis could be valuable as a tool guiding social media research. Therefore, the following research question is proposed:

RQ2: What aspects of social media are most commonly addressed in published research in the fields of Public Administration, Information Science and Library Science, and Communication?

The existence and identity of a scientific domain are defined, to some degree, by a set of academic journals (Paul, 2004). Discipline-specific outlets lay the foundation, impact the evolution, and form directions for the development of a scholarly domain (Serenko & Bontis, 2009). Journals allow researchers to directly communicate their ideas to a wider audience, become aware of recent developments, learn about seminal works, accumulate references, and preserve the scientific body of knowledge for the future generations of scholars and practitioners (Serenko & Jiao, 2012). Therefore, in the field of e-government research, it could be interesting to analyze the leading high-quality journals that include social media as one of their areas of interest. Accordingly, the third research question addressed is as follows:

RQ3: Which journals publish such papers most frequently?

It could be of interest to determine whether research aim influences the methodology used to analyze the subject. If such an analysis were made, it could be of assistance to researchers, informing them of the most appropriate methodology to be adopted for the purposes of their research, thus avoiding possible methodological mistakes (Lan & Anders, 2000; McCurdy & Cleary, 1984; Plümber & Radaelli, 2004). The following research question is therefore derived:

RQ4: What research methodologies are employed in analyzing social media?

Finally, a study of different social media experiences in the United States, Europe, and Asia could enhance our understanding of the instruments used to promote the social media implementation in public affairs and thus facilitate a resolution of democratic deficits (Nabatchi, 2010), improving the efficiency, acceptance, and legitimacy of political processes. In addition, the different interests of researchers in this field of knowledge could be the result of the universities to which they belong and could be a reflection of the concerns of the country in which social media is analyzed by the universities. The following research question is therefore derived:

RQ5: Are there differences in social media research among American, European, and Asian countries?

To address these research questions, we conducted a descriptive scientometric study (Neufeld et al., 2007) of social media articles published in JCR-listed international journals in the areas of Information Science and Library Science, Communication, and Public Administration, as described in the methodology section of this chapter (Figure 5.1).

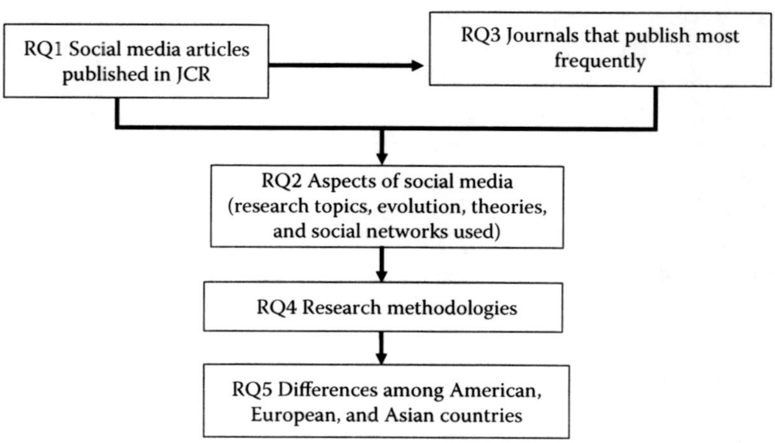

Figure 5.1 Research framework. JCR = Journal Citation Report.

5.3 Research Methodology

5.3.1 Sample Selection

5.3.1.1 Journals Analyzed

This study is focused on analyzing journal publications, in the view that they constitute a resource that is often used by academics as a source of new knowledge and as a medium for its disclosure (Nord & Nord, 1995) and at the same time as an indicator of scientific productivity (Legge & Devore, 1987). Symposia, summaries of communications, letters to the editor, articles of a professional nature, and book reviews were excluded from this analysis because, in our opinion, they offer a limited view of the subject. However, we did take into account articles included in special issues of journals, considering that these reflect a greater interest in the study of a particular issue and in the need to examine it further (Rodríguez, Alcaide, & López, 2010).

Institute for Scientific Information-listed journals in the areas of Communication, Information Science and Library Science, and Public Administration for the year 2012 were selected for analysis. In selecting the journals to be analyzed, we used objective indicators (Forrester & Watson, 1994), seeking references that provided useful and reasonably valid statements in terms of research consumption (Garfield, 1972). For this reason, and taking into account the findings of previous studies (Plümber & Radaelli, 2004), we excluded listed journals of marginal importance, that is, those with an impact factor of less than 0.25 or with fewer than 50 total citations for the year 2012.

For the analysis of these journals, we could have used the descriptions offered by the journals in their aims and scopes as indicators to determine whether they publish articles about social media. However, to avoid the possible exclusion of articles published in multidisciplinary journals with general descriptions, we analyzed all journals and articles that met the exclusion/inclusion criteria. In consequence, our sample was composed of 27 journals listed in the area of Communication, 45 journals listed in the area of Public Administration, and 69 journals listed in the area of Information Science and Library Science, indexed in the Institute for Scientific Information Web of Knowledge for the year 2012. However, not all of these journals have published articles on social media. Therefore, Table 5.2 lists only the journals that, within the sample, have published articles on social media.

5.3.1.2 Articles Selected

In determining the articles to be included in the sample, we analyzed all the articles published in journals that met the above-described criteria for inclusion. To do this, we first examined the title and the keywords of each one (Lan & Anders, 2000; Plümber & Radaelli, 2004). These descriptors provided an initial idea of the topic

Table 5.2 Articles about Social Media and Journals That Are Published (Chronological Evolution)

Journals[a]	*Year*						*Total*[b]
	2008	*2009*	*2010*	*2011*	*2012*	*2013*	
Communication							
Journal of Communication	—	—	—	—	4	1	5
New Media and Society	—	1	1	4	11	1	18
Journal of Computer-Mediated Communication	—	—	—	—	—	1	1
Telecommunication Policy	—	—	—	—	1	1	2
International Journal of Press-Politics	1	—	—	—	—	1	2
International Journal of Public Opinion Research	—	—	—	—	1	—	1
Public Relations Review	—	—	—	—	1	4	5
Information, Communication, and Society	—	—	—	2	—	2	4
European Journal of Communication	—	—	—	1	1	2	4
Total	1	1	1	7	19	13	42 (38.18%)
Information Science							
Scientometrics	—	—	—	—	1	—	1
Government Information Quarterly	—	—	2	2	10	16	30
Journal of Computer-Mediated Communication	—	—	—	—	—	1	1

Information and Management	1	–	–	–	–	–	1
Telecommunication Policy	–	–	–	–	1	1	2
Social Science Computer Review	–	–	–	8	1	8	17
Online Information Review	–	–	–	–	2	1	3
Aslib Proceedings	–	1	–	1	–	1	3
Total	1	1	2	11	15	28	58 (52.73%)
Public Administration							
Journal of Public Administration Research and Theory	–	–	–	–	–	1	1
Policy Studies Journal	–	–	–	1	–	–	1
Public Administration Review	–	–	–	1	–	1	2
International Review of Administrative Science	–	–	–	–	2	–	2
Australian Journal of Public Administration	–	–	1	–	1	–	2
Administration and Society	–	–	–	1	–	–	1
Transylvania Review of Administrative Sciences	–	1	–	–	–	–	1
Total	–	1	1	3	3	2	10 (9.09%)
Total	2	3	4	21	37	43	110

Source: Own Elaboration.

[a] This table shows only those journals that have published articles about social media.

[b] There are two journals—*Journal of Computer-Mediated Communication and Telecommunication Policy*—that are classified in Communication and Information Science, so they appear in both areas.

addressed in the article. If the keywords offered were generic and did not provide a clear idea about the research topic discussed, we then read the abstract to obtain a better view of the article. If doubts remained, we then read the introduction to identify the research goals and to determine the main factors analyzed. Finally, in the very few cases when these discriminant criteria were insufficient, we read the whole article. As a result, we obtained a database composed of 107 articles published during the period 2000–2013, although the first article that we found was published in 2008 (see Table 5.2).

5.3.2 Research Methodology

To achieve our research goals, each of the articles included in our database was manually examined and catalogued, using MS Excel software, by the year of publication, the journal title, the country in which the study was carried out, the main topic addressed, the principal methodology used, the level of analysis of these data, the theories used by the author to support the knowledge, and the social network analyzed.

To determine the research topics analyzed in each study, we carried out an exploratory content analysis of each of the articles in the sample (Krippendorff, 1980). Research topics are directly expressed in the text or are derived from them through analysis. In the coding process, researchers using content analysis create or develop a coding scheme to guide coders to make decisions in the analysis of content. A coding scheme is a translation device that organizes date into categories (Poole & Folger, 1981). A coding scheme includes the process and rule of data analysis that are systematic, logical, and scientific. Keywords are derived from review of literature. During this phase of the study, QSR NVivo v.8 software was used to automate the coding of the articles (Fraser, 2000). This coding was conducted using the random tags option of the software, which enabled us to obtain a hierarchical concept structure to group and adapt this published research on social media.

In this encoding phase, the researchers held several meetings to decide the labels to be assigned and the topics to be included (see Table 5.3). Subsequently, each of the articles incorporated in the study sample was encoded separately (Lan & Anders, 2000), and any disagreements concerning the definition of the categories to be analyzed were discussed and resolved.

Once the specific areas of study concerning social media had been catalogued and systematized, in accordance with the criteria set out in Sections 5.1 and 5.2, the methodologies employed in each line of research were analyzed to identify the research trends present. When an article incorporated various methodologies, we avoided double counting by focusing on the main research goal and on the methodology used to achieve it. For this reason, it was essential to identify the principal intention of the scientific document.

Table 5.3 Main Social Media Topics Addressed in the Leading Fields of Study

Research Topics	*Communication*	*Information Science*	*Public Administration*
1. Disclosure information and transparency	9.52%	17.24%	10.00%
2. Delivery public services	9.52%	15.52%	40.00%
3. Legal and regulatory aspects	2.38%	3.45%	10.00%
4. Social media in general terms	2.38%	12.07%	10.00%
5. Organizational theory and reforms	—	5.17%	—
6. Electronic participation	76.19%	46.55%	30.00%
a. Use of social media by politicians in election campaigns (e-Campaign)	28.57%	20.69%	—
b. Social media use for reducing inequalities, enhancing democracy, and increasing participation	47.62%	25.86%	30.00%

Source: Own elaboration.

5.3.3 Analysis of the Results

RQ1: How many social media articles have been published in JCR journals in the fields of Public Administration, Information Science and Library Science, and Communication?

RQ3: Which journals publish such papers most frequently?

Public administrations have adopted different Web 2.0 tools, such as blogs, microblogging, wikis, social networking, multimedia sharing, mash-up applications, tagging, and virtual worlds, among others, which have attracted the interest of researchers and scholars, and this has been reflected in a gradual increase in the number of studies published in international journals since 2011 (see Table 5.2). In addition, we can see that this emerging research on social media within the e-government community began in 2008.

In this sense, most of these articles were published in Information Science and Library Science journals (52.73%), followed closely by Communication journals (38.18%), and the rest (9.09%) are set out in Public Administration journals. In this regard, 81.03% (n = 47) of the articles are published in two journals in Information Science and Library Science—*Government Information Quarterly* (51.72%) and *Social Science Computer Review* (29.31%). Meanwhile, in the case of Communication journals, Table 5.1 shows that 42.86% of the articles published correspond to a single journal, *New Media and Society*, which accounts for 18 of the 42 articles included in this knowledge. Finally, in the case of Public Administration, there is no clear preference for published articles about social media.

RQ2: What aspects of social media are most commonly addressed in published research in the fields of Public Administration, Information Science and Library Science, and Communication?

Table 5.3 shows that, recently, it has been suggested that the web has moved into a new phase, which is characterized by social networking activities, online collaboration, and user sharing, that is, a shift from a publishing medium to a platform for social participation and interaction. This is due to the way that Internet brought with it various visions and scholarly discussions regarding it impact on politics and election campaigns. Hence, the main social media topics are as follows: (a) how the politicians use social media tools in their elections campaign: Communication, 28.57%, and Information Science and Library Science, 20.69%; and (b) how social media reduces inequalities, enhances democracy, and increases participation; Communication, 47.62%; Information Science and Library Science, 25.86%; and Public Administration 30.00%. In addition, Table 5.4 shows that these research topics have increased interest among scholars and academic researchers since 2011.

These research topics are followed distantly by how public administrations are using social media to promote information disclosure, transparency, and accountability; that is, there are studies that reflect on how social media technologies may help public sector professionals to get involved in the process of distribution and communication and the real-time information sharing and responsiveness challenge in government: Communication, 9.52%; Information Science and Library Science, 17.24%; and Public Administration, 10.00%. In addition, it is interesting that researchers analyze how social media facilitate and involve the public services delivery: Communication, 9.52%; Information Science and Library Science, 15.52%; and Public Administration, 40.00%.

Meanwhile, there are few studies that have analyzed the problem of adaption to the existing organizational culture and institutional structure of public sector organizations. Although it is very important to analyze whether social

Table 5.4 Chronological Evolution of the Main Social Media Topics

	Years						
Research Topics	*2008*	*2009*	*2010*	*2011*	*2012*	*2013*	*Total*
1. Disclosure information and transparency	—	1	1	6	4	3	15
2. Delivery public services	—	—	1	1	6	8	16
3. Legal and regulatory aspects	—	1	—	—	2	—	3
4. Social media in general terms	—	—	—	1	2	6	9
5. Organizational theory and reforms	1	—	1	1	—	—	3
6. Electronic participation	1	1	1	12	22	24	61
a. Use of social media by politicians in election campaigns (e-Campaign)	1	—	—	3	8	11	23
b. Social media use for reducing inequalities, enhancing democracy, and increasing participation	—	1	1	9	14	13	38

Source: Own elaboration.

media tools have the potential to improve interactions with citizens through dialogue, citizen engagement, openness, transparency, good governance, and trust (Charalabidis & Loukis, 2012), it is equally important examine what organizational and contextual factors are also essential to understand their results in public bureaucracies.

In this context, there are 66.36% published articles that do not use theories to support their findings; only 33.64% of the articles use theories, and these theories used to analyze the same topic are varied (see Table 5.5). Similarly, the same theory is used to analyze different research topics.

Table 5.5 Research Topics and Theories Used

Research Topics	*Theories*
1. Disclosure information and transparency	Commodity theory
	Communication privacy management theory
	Gatekeeping theory
	Habermas theory
	Post-Habermasian theories
	Principal-agent theory
	Diffusion innovation theory
2. Delivery public services	Software platforms and ecosystems theory
	Wicked problems theory
	Innovation diffusion theory
	New public services theory
	Communities of public services support
	Galbraith's information-processing theory
	Transactions cost theory
	Interorganizational networks theory
3. Legal and regulatory aspects	Merkl-Kelsen's theory
4. Social media in general terms	Media affordances theory
	Social-technical systems theory
5. Organizational theory and reforms	Strategic business alignment approach
6. Electronic participation	
a. Use of social media by politicians in election campaigns (e-Campaign)	Exemplification theory
	Equalization and normalization theory
	Social information processing theory
	Priming theory
	Framing theory

Table 5.5 (Continued) Research Topics and Theories Used

Research Topics	*Theories*
	Agenda setting theory
	Mediatization theory
	Diffusion of innovation theory
	Cognitive dissonance theory
b. Social media use for reducing inequalities, enhancing democracy, and increasing participation	Cognitive dissonance theory
	Democratic theory
	New media theory
	Social network theory
	Domestication theory
	Convergence culture
	Inglehart's postmaterialist theory
	Political communication theory
	Habermas theory
	Two step flow of communication
	Incidental byproduct model of learning theory
	Political engagement gateway theory
	Public sphere and representation theories
	Lasswell's policy sciences

Source: Own elaboration.

The main social media used are social networking, Facebook (37.84% country level, 31.46% central government level, and 33.33% local government level; see Table 5.6), followed closely by microblogging, Twitter (31.08% country level, 35.96% central government level, and 25.00% local government level). In addition, state and local public administrations use multimedia sharing YouTube, and Flickr is used by municipalities. In general, Facebook, Twitter, YouTube, Flicker, MySpace, and LinkedIn are the most widespread social media technologies in bureaucracies.

RQ4: What research methodologies are employed in analyzing social media?

Table 5.6 Social Networks Analyzed in the Different Levels of Governments

Social Network	*Country Level*	*Central Government Level*	*Local Government Level*	*Total*
Facebook	37.84%	31.46%	33.33%	34.12%
Twitter	31.08%	35.96%	25.00%	31.75%
YouTube	5.41%	15.73%	20.83%	13.27%
Flickr	6.76%	3.37%	10.42%	6.16%
MySpace	9.46%	2.25%	2.08%	4.74%
LinkedIn	2.70%	3.37%	8.33%	3.79%
Google +	2.70%	2.25%	—	1.90%
Others	4.05%[a]	6.74%[b]	—	4.27%

Source: Own elaboration.

[a] Dodgeball, Bebo, and Cyworld.
[b] Me2day, Hyves, Foursquare, Tumblr, and Daum.

Focusing on the methodology used in these articles addressing social media, we observed a preference for the use of empirical research methods (93.46%, 100/107; see Table 5.7). The remaining 6.54% (7/107) of the articles concern nonempirical studies, most of which are traditional literature reviews, followed by critical reviews and discussions.

Among the empirical methods used, qualitative tools (51.40%, 55/107) are more common than quantitative ones (42.06%, 45/107; Table 5.7). The most frequently used qualitative instruments are case studies (47.27%, 26/55), followed distantly by other methods such as content analysis (20.00%, 11/55) and social network analysis (7.27%, 4/55). In addition, many articles described the use of bibliometrics tools, webmetrics–hyperlinks, and meta-analysis (Gayo-Avello, 2013; Meraz, 2013; Tkach-Kawasaki, 2011; Vergeer, 2012). These tools allow data to be updated and clearer conclusions to be drawn and highlight any conceptual and theoretical weaknesses present in previous studies. Moreover, the actual progress of research on a particular topic can be assessed. Methodological tools such as meta-synthesis and meta-analysis are used to integrate diverse studies with common goals, thus providing objective evidence and enabling reliable conclusions to be drawn, benefiting present and future research.

Regression analysis is the quantitative methodology most commonly employed (62.22%, 28/45), followed by logistic regression analysis (11.11%, 5/45; see Table 5.7). Regression analyses are widely used in all these fields of e-government research, but especially when the researcher examines the question of citizens' participation in public policy making or seeks to identify the factors that enhance the disclosure of economic and financial information or those acting to reduce the incidence of corruption.

Table 5.7 Methodologies Used in Analyzing Social Media

Methodologies	*United States*	*Asia*	*Europe*	*Total*
Nonempirical	—	—	—	6.54%
Qualitative Methodologies	50.00%	58.33%	55.17%	51.40%
Action research approach	—	—	6.25%	1.82%
Bibliometric study	—	—	—	1.82%
Case study	50.00%	42.86%	43.75%	47.27%
Content analysis	22.73%	14.29%	25.00%	20.00%
Delphi method	—	—	—	1.82%
Ethnographic analysis	4.55%	—	—	1.82%
Exploratory study	9.09%	—	—	5.45%
Heuristic approach	—	—	—	1.82%
Hyperlink analysis—webometric	—	14.29%	—	1.82%
Interpretivist method	4.55%	—	—	1.82%
Legal statutory analysis	4.55%	—	12.50%	5.45%
Meta-analysis	—	—	6.25%	1.82%
Social network analysis	—	28.57%	6.25%	7.27%
Quantitative Methodologies	50.00%	41.67%	44.83%	42.06%
Content analysis	4.55%	20.00%	7.49%	6.67%
Difference test	—	20.00%	—	2.22%
Exploratory analysis	—	—	7.49%	2.22%
Logistic regression analysis	13.64%	—	15.38%	11.11%
MANCOVA-ANOVA	9.09%	—	—	6.67%
Regression analysis	68.18%	20.00%	69.23%	62.22%
Social network analysis	—	20.00%	—	2.22%
Structured equation method	4.55%	20.00%	—	4.44%
Webmetric approach	4.55%	—	—	2.22%

Source: Own elaboration.

Note: There are two journals—*Journal of Computer-Mediated Communication* and *Telecommunication Policy*—that are classified in Communication and Information Science, so they appear in both areas (110 – 3 = 107). ANOVA = analysis of variance; MANCOVA = multivariate analysis of covariance.

RQ5: Are there differences in social media research among American, European, and Asian countries?

Regarding the methodologies used in different countries, there are not many differences (see Table 5.8). Qualitative tools are more common than quantitative ones in all cases, although these ones are used more often when researchers analyzed Asian countries. In this sense, the most frequently used qualitative tools are case studies in all cases, followed distantly by content analysis. In the case of European countries analysis, research used legal statutory approach when examining laws that regulate disclosure of information, review content online, person responsible for answering certain questions, and so on.

In the case of quantitative methodologies, results are very similar to those above. Regression analysis is the quantitative methodology most commonly employed in all cases. Structure equation model is used more often when scholars and researchers analyzed Asian countries and determinants that promote online participation (see Table 5.7).

The main social media topics are how social media improve interactions with citizens through dialogue, citizen engagement, and online participation in all cases, although they are higher in European countries. In the case of the United States, academics are very interested in examining how social media improved public services delivery (see Table 5.8).

Table 5.8 Main Social Media Topics in Different Countries

Research Topics	*United States*	*Asia*	*Europe*
1. Disclosure information and transparency	15.91%	16.67%	10.34%
2. Delivery public services	20.45%	—	10.34%
3. Legal and regulatory aspects	4.55%	—	3.45%
4. Social media in general terms	11.36%	—	—
5. Organizational theory and reforms	—	16.67%	3.45%
6. Electronic participation	47.72%	66.66%	72.42%
a. Use of social media by politicians in election campaigns (e-Campaign)	20.45%	33.33%	24.14%
b. Social media use for reducing inequalities, enhancing democracy, and increasing participation	27.27%	33.33%	48.28%

Source: Own elaboration.

5.4 Conclusions and Discussion

According to our results, there are an increasing number of papers published in JCR journals about the use of social media in public administrations, and the number is growing in the last years (RQ1). In fact, social media in public administrations has mainly attracted the attention of researchers in the last 3 years. It is a reflection of the increasing incorporation of these technologies in the social life of people and the need of public administrations to implement these technologies in public affairs (Table 5.9).

In addition, our results indicate that many papers have analyzed the social media applications used by governments and the use that public administrations have made

Table 5.9 Research Findings

Research Questions	*Main Conclusions*
RQ1. Social media articles published in JCR journals	Social media has attracted the attention of researchers in the last 3 years.
RQ2. Aspects of social media	Many papers have been addressed to analyze the social media applications used by governments, showing that public administrations have implemented the most widely used social media applications: Facebook, Twitter, YouTube, Flickr, and LinkedIn-. The use has been homogeneous by all level of public administrations. E-participation has been the main research topic published in Communication and Information Science journals. Delivery of public services is the key research topic in Public Administration journals. Several different theories have been called to be applied in the application of social media.
RQ3. Journals publish most frequently	Most of papers about the use of social media in governments have been published in Information Science and Library Science journals, followed closely by Communication journals. *Government Information Quarterly* and *Social Science Computer Review* are the main outlets for papers about the use of Web 2.0 applications in governments.
RQ4. Research methodologies	Qualitative methodologies are more frequently used.

(*Continued*)

Table 5.9 (Continued) Research Findings

Research Questions	*Main Conclusions*
RQ5. Differences among American, European, and Asian countries	Case study and content analysis are the most widely qualitative methodology used in the United States and Europe, and regression analysis as quantitative methodology. Case study and social network analysis are the most frequently methodologies used in Asian countries, and clear tendency in the use of a quantitative methodology is not shown. Public services, disclosure information, and electronic participation are the main research topics analyzed in the United States and Europe. Electronic participation is the main research topic analyzed in Asian countries.

Source: Own elaboration.

of these applications (RQ2). According to these papers, public administrations have implemented the most widely used social media applications, such as Facebook, Twitter, YouTube, Flickr, and LinkedIn, mainly for political purposes and for reducing inequalities, enhancing democracy, and increasing participation (RQ2). In addition, the use of these social media applications by all levels of public administrations has been homogeneous, which means that governments have tried to reach a wider range of stakeholders because these social media have been indicated as those with a higher number of users (Meijer, Grimmellikhuijsen, & Brandsma 2012).

Nonetheless, the research about social media implementation and, similarly, their use in public administrations has not been the same in all fields of knowledge (RQ1) or in the journals in which they are published (RQ3). Indeed, whereas electronic participation has been the main research theme published in the Communication and Information Science fields of knowledge, mainly in the journals *Government Information Quarterly* and *Social Science Computer Review*, the delivery of public services is increasingly being the key theme in the public administration area (no preference for publishing in a particular journal in this field exists). This result highlights the different perspectives of the fields of knowledge analyzed in this chapter and reflects the concerns of social media in its implementation in public administrations. In fact, results indicate that electronic participation has been the first concern of public administrations in implementing these new technologies. It seems that governments have tried to take advantage of Web 2.0 technologies as other channels for a wider representation of the government actions or the elected political party into the Internet. Therefore, public administrations have embraced social media as a means of representation in both political activities and public actions (RQ2).

In addition, the application of social media to public services is a new concern in the last 2 years, mainly in the field of Public Administration, whereas the disclosure of information and transparency issues has been decreasing gradually. It could indicate that after a first stage of studies on implementing social media as communication channels for governments, public administrations are now analyzing how to use these new technologies to improve the delivery of public services (RQ2). Does it mean a next step in the way to the maturity of this field of research? In any case, it means that future research could be addressed to analyze the implementation of social media for public sector services in order to capture experiences, to be useful for public managers to introduce these new technologies into governments.

On the other hand, this new field of research needs grounded theories to support social media application into public administrations (RQ2). Our review indicates that several different theories have been called to be applied in the application of social media in public administrations. But which one of these theories prevails? Many of them are embedded from other areas of study. In this regard, is social media a field of knowledge that needs a deep analysis of theories or can it embed theories of other fields of knowledge? Why? All these questions currently do not have appropriate answers. Therefore, future research could undertake theoretical studies on this issue.

Regarding the methodologies used to analyze the phenomenon of Web 2.0 technologies (RQ4), results indicate that qualitative methodologies are more frequently used, mainly in Asia and Europe. In any case, the use of methodologies of research seems to be different in the different continents analyzed. In this regard, whereas case study and content analysis are the most widely used qualitative methodologies in the research about social media implementation in public administrations in the United States and Europe, the analyses of social media implementations in governments in Asian countries have usually used the case study and social network analysis techniques. In addition, the research in European countries and in the United States has been undertaken using regression analysis when the research was built on quantitative methodologies of research. In contrast, research on social media implementations in Asian countries does not show a clear tendency in the use of a particular quantitative methodology. Therefore, these results for RQ4 could indicate the need to take into account the use of different research methodologies according to the countries or areas in which the research is going to be undertaken and the journals to which the research is going to be submitted for possible publication. In both areas, as noted previously (United States, Europe, and Asia), have some gaps in research methodologies that could be filled in future research according to the aim of the book chapter and to the area in which the social media is being analyzed.

Finally, this tendency in research methodology is also shown in the topics analyzed in the different countries (RQ5). This way, American and European countries are focused on implementing social media for disclosing information, for rendering public services, and mainly, for electronic participation. In addition, although social media in Asian countries is also analyzed for electronic participation, these countries have not analyzed social media implementations for public services or for general terms. In this regard, future research could analyze the reason for these differences between countries, and it could be addressed to fill these gaps into the Asian context.

To conclude, the review presented in this chapter provides a comprehensive summary of the research into social media within the fields of Communication, Information Science, and Public Administration, highlighting the main research topics and methodologies used. Explanations and clarifications are given whenever possible. Knowledge gaps and research opportunities are identified from these observations, which reveal changes in the research methods applied, reinforcing the development of a theoretical framework so that the application of social media may efficiently contribute to improving management in the public sector.

Author Biographies

Prof. Manuel Pedro Rodríguez Bolívar is a professor at the University of Granada. He has authored numerous articles in international journals, including *Public Money & Management, Government Information Quarterly, Public Administration and Development, Online Information Review, International Review of Administrative Sciences, American Review of Public Administration, ABACUS*, and *Administration and Society*, among others. He is also the author and editor of international journals and several books and book chapters in Kluwer Academic Publishers, Springer, Taylor & Francis, and IGI Global.

Dr. Laura Alcaide Muñoz is an assistant professor in accounting and finance in the Department of Finance Economy and Accounting, University of Granada. She is interested in how E-government has favored the process of reform and modernization of public administrations, giving rise to greater accessibility to public information and services and information transparency, this latter aspect being a key factor in the accountability in the public sector. Also, she is interested in financial sustainability in local governments. She was the author of some articles published in the *Administration & Society, Public Money and Management, International Public Management Journal*, and *Information Technology for Development*, among others, and has written book chapters in prestigious editorials like IGI Global, Springer, and Routledge-Taylor & Francis.

References

Accenture. (2009). *Web 2.0 and the next generation of public service: Driving high performance through more engaging, accountable and citizen-focused service*. London: Accenture. Available at http://www.epractice.eu/files/Web%202.0%20and%20the%20Next%20Generation%20of%20Public%20Service%20-%20Driving%20high%20performance%20through%20more%20engaging,%20accountable%20and%20citizen-focused%20service.pdf.

Anthopoulos, L., & Fitsilis, P. (2014). Trends in e-strategic management: How do governments transform their policies? *International Journal of Public Administration in the Digital Age, 1*(1), 15–38.
Aparaschivei, P. (2011). The use of new media in electoral campaigns: Analysis on the use of blogs, Facebook, Twitter and YouTube in the 2009 Romanian presidential campaign. *Journal of Media Research, 4*(2), 39–60.
Benkler, Y. (2006). *The wealth of networks: How social production transforms markets and freedom*. New Haven, CT: Yale University Press.
Bertot, J. C., Jaeger, P. T., & Grimes, J. M. (2010a). Crowd-sourcing transparency: ICTs, social media, and government transparency initiatives. *Proceedings of the 11th Annual International Conference on Digital Government Research*, Puebla, Mexico, May 17–20.
Bertot, J. C., Jaeger, P. T., & Grimes, J. M. (2010b). Using ICTs to create a culture of transparency: E-government and social media as openness and anti-corruption tools for societies. *Government Information Quarterly, 27*(3), 264–271.
Bertot, J. C., Jaeger, P. T., & Hansen, D. (2012). The impact of policies on government social media usage: Issues, challenges, and recommendations. *Government Information Quarterly, 29*(1), 30–40.
Bertot, J. C., Jaeger, P. T., Munson, S., & Glaisyer, T. (2010). Social media technology and government transparency. *Computer, 43*(11), 53–59.
Bonsón, E., Torres, L., Royo, S., & Flores, F. (2012). Local e-government 2.0: Social media and corporate transparency in municipalities. *Government Information Quarterly, 29*(2), 123–132.
Boyd, D. M., & Ellison, N. B. (2007). Social network sites: Definition, history, and scholarship. *Journal of Computer-Mediated Communication, 13*(1), 210–223.
Brown, G. (2010). Every citizen to get their own MyGov. Retrieved, January 2011 from http://www.publicservice.co.uk/news_story.asp?id=12491.
Charalabidis, U., & Loukis, E. (2012). Participative public policy making through multiple social media platforms utilization. *International Journal of Electronic Government Research, 8*(3), 78–97.
Christofides, E., Muise, A., & Desmarais, S. (2009). Information disclosure and control on Facebook: Are they two sides of the same coin or two different processes? *CyberPsychology and Behavior, 12*(3), 3411–3450.
Chun, S., Shulman, S., Sandoval-Almazan, R., & Hovy, E. (2010). Government 2.0: Marking connections between citizens, data and government. *Information Polity, 15*, 1–9.
Cocosila, M., Serenko, A., & Turel, O. (2011). Exploring the management information systems discipline: A scientometric study of ICIS, PACIS, and ASAC. *Scientometrics, 87*(1), 1–16.
Commonwealth Network of Information Technology for Development Foundation. (2002). *Country profiles of E-governance*. Paris: UNESCO, Available at http://unesdoc.unesco.org/images/0012/001276/127601e.pdf.
Deloitte. (2008). *Change your world or the world will change you. The future of collaborative government and Web 2.0*. Quebec: Deloitte & Touche LLP and affiliated entities, Available at http://www.deloitte.com/assets/Dcom-Sweden/Local%20Assets/Documents/se_Change_your_world_230508.pdf.
Forrester, J. P., & Watson, S. S. (1994). An assessment of public administration journals: The perspective of editors and editorial board members. *Public Administration Review, 54*(5), 474–482.

Fraser, D. (2000). *QSR Nvivo. NUDIST Vivo. Reference guide*. Melbourne: QSR International Pty. Ltd.

Garfield, E. (1972). Citation analysis as a tool in journal evaluation. *Science, 178*(60), 471–479.

Gayo-Avello, D. (2013). A meta-analysis of state-of-the-art electoral prediction from Twitter data. *Social Science Computer Review, 31*(6), 649–679.

Goldsmith, S., & Eggers, W. D. (2004). *Governing by network: The new shape of the public sector*. Washington, DC: Brookings Institution Press.

Holsapple, C. W. (2008). The pulse of multiparticipant systems. *Journal of Organizational Computing and Electronic Commerce, 18*(4), 333–343.

Kanat, I. E., & Ozkan, S. (2009). Exploring citizens' perception of government to citizen services: A model based on theory and planned behavior (TBP). *Transforming Government: People, Process and Policy, 3*(4), 406–419.

Kaplan, A. M., & Haenlein, M. (2010). Users of the world, unite! The challenges and opportunities of social media. *Business Horizons, 53*(1), 59–68.

Klein, P. (2008). Web 2.0: Reinventing democracy. *CIO Insight*, 30–43.

Kovac, P., & Decman, M. (2009). Implementation and change of processual administrative legislation through an innovative Web 2.0 solution. *Transylvanian Review of Administrative Sciences, 28E*, 65–86.

Krippendorff, K. (1980). *Content analysis: An introduction to its methodology*. Thousand Oaks, CA: Sage Publications.

Lan, Z., & Anders, K. K. (2000). A paradigmatic view of contemporary public administration research: An empirical test. *Administration and Society, 32*(2), 138–165.

Lathrop, D., & Ruma, L. (2010). *Open government: Collaboration, transparency, and participation in practice*. Sebastopol, CA: O'Reilly.

Legge, J. S., Jr., & Devore, J. (1987). Measuring productivity in U.S. public administration and public affairs programs 1981–1985. *Administration and Society, 19*(2), 147–156.

Li, D. (2011). Online social network acceptance: A social perspective. *Internet Research, 21*(5), 562–580.

Linders, D. (2012). From e-government to we-government: Defining a typology for citizen coproduction in the age of social media. *Government Information Quarterly, 29*(4), 446–454.

McCurdy, H. E., & Cleary, R. E. (1984). A call for appropriate methods. *Public Administration Review, 44*(6), 49–55.

McDermott, P. (2010). Building open government. *Government Information Quarterly, 27*(4), 401–413.

McMillan, P., Medd. A., & Hughes. P. (2008). *Change the world or the world will change you: The future of collaborative government and Web 2.0*. Deloitte Touche Tohmatsu. Retrieved from http://www.deloitte.com.

Meijer, A., Grimmellikhuijsen, S., & Brandsma, G. J. (2012). Communities of public service support: Citizens engage in social learning in peer-to-peer networks. *Government Information Quarterly, 29*(1), 21–29.

Meraz, S. (2013). The democratic contribution of weakly tied political networks: Moderate political blogs as bridges to heterogeneous information pools. *Social Science Computer Review, 31*(2), 191–207.

Mergel, I. (2013). Social media adoption and resulting tactics in the U.S. federal government. *Government Information Quarterly, 30*(2), 123–130.

Millard, J. (2009). Government 1.5: Is the bottle half full or half empty? *European Journal of ePractice, 9*(1), 35–50.

Mintz, D. (2008). Government 2.0—Fact or fiction? *The Public Manager, 36*(4), 21–24.
Mossberger, K., Wu, Y., & Crawford, J. (2013). *Connecting citizens and local governments? Social media and interactivity in major U.S. cities.* Paper presented at Public Management Research Conference, Madison, WI.
Nabatchi, T. (2010). Addressing the citizenship and democratic deficits: The potential of deliberative democracy for public administration. *American Review of Public Administration, 40*(4), 376–399.
Neufeld, D., Fang, Y., & Huff, S. (2007). The IS identity crisis. *Communications of the Association for Information Systems, 19*, 447–464.
Nord, J. H., & Nord, G. D. (1995). MIS research: Journal status and analysis. *Information and Management, 29*(1), 29–42.
Noveck, B. (2009). *Wiki government: How technology can make government better, democracy stronger, and citizens more powerful.* Washington, DC: Brookings Institution Press.
Organisation for Economic Co-operation and Development. (2010). *Denmark: Efficient e-government for smarter service delivery.* Paris: OECD Publishing.
Osimo, D. (2008). *Web 2.0 in government: Why? and how?* Institute for Prospective Technological Studies, Joint Research Centre, European Commission. Luxembourg: Office for Official Publications of the European Communities. Available at http://ftp.jrc.es/EURdoc/JRC45269.pdf.
Paul, K. (2004). Business and society and business ethics journals: A citation and impact analysis. *Journal of Scholarly Publishing, 35*(2), 103–117.
Peedu, G. (2011). *Enhancing public service user experience in information society* (Master's thesis). Tallinn University, Estonia.
Plümber, T., & Radaelli, C. M. (2004). Publish or perish? Publications and citations of Italian political scientists in international political science journals, 1990–2002. *Journal of European Public Policy, 11*(6), 1112–1127.
Poole, M. S., & Folger, J. P. (1981). Modes of observation and the validation of interaction analysis schemes. *Small Group Behavior, 12*, 477–493.
Reddick, C. G., & Turner, M. (2012). Channel choice and public service delivery in Canada: Comparing e-government to traditional service delivery. *Government Information Quarterly, 29*(1), 1–11.
Rodríguez, M. P., Alcaide, L., & López, A. M. (2010). Trends of e-government research. Contextualization and research opportunities. *The International Journal of Digital Accounting Research, 10*, 87–111.
Serenko, A., & Bontis, N. (2009). A citation-based ranking of the top business ethics scholarly journals. *International Journal of Business Governance and Ethics, 4*(4), 390–399.
Serenko, A., & Jiao, C. (2012). Investigating information systems research in Canada. *Canadian Journal of Administrative Sciences, 29*(1), 3–24.
Sidorova, A., Evangelopoulos, N., Valacich, J. S., & Ramakrishnan, T. (2008). Uncovering the intellectual core of the information systems discipline. *MIS Quarterly, 32*(3), 467–482.
Straub, D. (2006). The value of scientometric studies: An introduction to a debate on IS as a reference discipline. *Journal of the Association for Information Systems, 7*(5), 241–245.
Taylor, J. A. (2012). The Information Polity: Towards a two speed future? *Information Polity: The International Journal of Government & Democracy in the Information Age, 17*(3/4), 227–237.

Tkach-Kawasaki, L. (2011). Mapping political connections in Japan: The functions of hyperlinks on Japanese diet member websites. *Social Science Computer Review, 29*(3), 300–312.

United Nations. (2010). *E-Government Survey 2010. Leveraging e-government at a time of financial and economic crisis*. New York: United Nations. Available at http://unpan3.un.org/egovkb/Portals/egovkb/Documents/un/2010-Survey/Cover.pdf

United Nations. (2012). *E-Government Survey 2012. E-Government for the People*. New York: United Nations. Available at http://unpan3.un.org/egovkb/Portals/egovkb/Documents/un/2012-Survey/unpan048065.pdf

Vergeer, M. (2012). Politics, elections and online campaigning: Past, present... and a peek into the future. *New Media & Society, 15*(1), 9–17.

Waltman, L., Yan, E., & Van Eck, N. J. (2011). A recursive field-normalized bibliometric performance indicator: An application to the field of library and information science. *Scientometrics, 89*(1), 301–314.

Williams, C. B., & Gulati, G. J. (2013). Social networks in political campaigns: Facebook and the congressional elections of 2006 and 2008. *New Media & Society, 15*(1), 52–71.

Chapter 6

E-Government as an Innovative Product: Theories and Case Study

Leonidas Anthopoulos, Christopher G. Reddick, Irene Giannakidou, and Nikolaos Mavridis

Contents

Abstract

E-government was first introduced more than 20 years ago as an ambitious "vehicle," which would result in a better, more efficient and effective, transparent, and accountable government. However, after all these years and tremendous funding, scholars argue over the effectiveness of e-government, and various types of failures are being discussed: project failures, adoption, online bureaucratic processes, delays in objectives, and so forth. Although all agree on the potential of e-government, the performance of the existing efforts appears to lag far behind what was expected. To this end, e-government is starting to be viewed more as an alternative channel for citizens to access their government, rather than an innovative product for organizational success. The aim of this chapter is to consider e-government as an "innovative product" for the public sector and to suggest e-government deployment with the means of innovation management. Respective theories are discussed and speculations are made for such an e-government alignment. Moreover, data from an analysis of Twitter feeds, after the failed roll out of the Healthcare.gov website, or the Obamacare website, in the United States are used to examine this theory.

Keywords: e-government, innovation management, social media analysis, performance

6.1 Introduction

E-government was first introduced in the early 1990s as a means for governments to utilize information and communications technologies (ICTs) to become more effective and efficient in delivering information and services to the public; more accountable and transparent regarding its internal processes, procurement, and auctioning; more open with regard to citizen engagement in decision and policy making; even more friendly; and able to deliver customized and modern services (Heeks & Bailur, 2007; Kim, Pan, & Pan, 2007).

However, after all these years, e-government outcomes are being questioned, and various scholars have argued about its potential. Failures, which vary from not establishing project success, to missing citizen expectations of adoption, even to preferences in turning back to traditional channel selection (i.e., face-to-face visits and voice phone calls), are being illustrated in the literature (Heeks & Bailur, 2007; Reddick & Turner, 2012).

On the other hand, scholars try to reach solutions that can enhance e-government development, and various proposals have been given so far, some of them worth mentioning, including the following: management frameworks that emphasize e-government, legal framework adjustments that release e-government potentials (i.e., digital signatures), key driver definition (i.e., eID), political declarations that drive e-government vision (i.e., European Union Malmo Declaration on e-Government and U.S. Paperless Action, etc.; Anthopoulos & Fitsilis, 2014), and international monitoring and measurement (i.e., by United Nations, the WorldBank, and the Organisation for Economic Co-operation and Development) (Anthopoulos & Fitsilis, 2014).

Results from international measurements have illustrated some of the factors that impact e-government development and could be defined as barriers and drivers. For instance, digital literacy and Internet penetration are drivers for e-government growth, whereas digital divide and trust are some of the documented barriers (Anthopoulos & Fitsilis, 2014; Heeks & Bailur, 2007; Reddick & Turner, 2012).

The purpose of this chapter is to focus on the problem of e-government's slow implementation and corresponding failures through both theory and a case study. Our study differs from existing research because although the combination of "innovation and e-Government" appears to attract some scholarly research, no works can be located that address the approach addressed here. Moreover, findings from the case study are novel, recent, and result in important outcomes. To this end, our study aims to consider e-government as an "innovative product" for government and verify whether governments have approached e-government with respective means. The authors will attempt to justify this speculation and to present how its implementation would be managed. Findings from this analysis will be compared with existing e-government implementation approaches, and respective conclusions will be presented. In this regard, the authors aim to answer the following research questions:

1. Can e-government be considered an "innovative product"? Although this question appears to be trivial or cliché, such a speculation might change the ways e-government is defined and managed.
2. Do governments approach e-government as an innovation? This question follows up the results from the first question and compares theories from innovation management with findings from corresponding international e-strategies.
3. What would change from e-government consideration as an innovation? This question will demonstrate how the e-government implementation process will be updated and the benefits that this consideration has for governments.

The answers to these questions aim to justify that e-government failures could be avoided if governments approached it as an innovation. Then, the previously mentioned initially grounded and rather ambitious e-government objectives could be established. Otherwise, it would perhaps be more appropriate for e-government to focus on the challenge to become another channel for public service delivery

(Reddick & Turner, 2012). To this end, the remainder of this chapter is organized as follows: In Section 6.2, a corresponding literature review is performed. Section 6.3 presents findings from international e-strategies and a case study and compares them with innovation management perspectives. Section 6.4 contains the conclusions and some future thoughts.

6.2 Background

This section outlines theoretical aspects, which concern innovation, innovation management techniques, e-government and innovation, and e-government development. Although the corresponding literature review is not complete, findings show the interconnection between innovation and e-government.

6.2.1 Innovation

Innovation can be defined as the first attempt to carry out an invention into practice (Fagerberg, 2004), whereas invention is limited to describing the generation of the idea of a new product or process (Howells, 2005). To this end, innovation concerns the practical refinement and development of an original invention into a usable technique or product or a process, in which creativity is applied to every facet of an organization's value chain, from beginning to end, to develop new and better ways of creating value for customers (Maital & Seshadri, 2007).

According to the Organisation for Economic Co-operation and Development (2005) and its Oslo Manual, there are various types of innovation, which vary from products to processes, marketing techniques, or organizational methods. Moreover, technological innovation differs from social innovation because the first is based on specific discoveries, whereas the second concerns extensive changes (i.e., mail, telephone, etc.; Maital & Seshadri, 2007). Additionally, innovation is distinguished to sustaining and disruptive proportionally on customer demand. According to the degree of the invention's novelty, innovations vary from incremental (when they are continuous) to radical (when a product series change) to systemic (when the entire production technique changes). Finally, innovation can be open when it is developed outsourced or closed when it is developed insourced.

Innovation comes to solve particularly defined problems (Maital & Seshadri, 2007), where the idea is the smallest part of the innovation process. Its success is supported—but not guaranteed—by appropriate management, funding, and marketing efforts, while it has to balance properly between technology push and market pull. Porter and Schwab (2008) are only some of the active scholars in the innovation domain who realize the potential that innovation has for economic growth. A comparison between various countries (Porter & Schwab, 2008) justifies that innovation supports economic growth and competition because it enhances the added value of the offered services and/or products. This study results in impulses that empower national innovation:

1. A strong technological and scientific base
2. Investments (both private and public) on innovation
3. Close interconnection between enterprises and research centers
4. A high-level education system
5. Political transparency
6. The establishment of an ecosystem that encourages entrepreneurship and risk-taking

These impulses show the contribution of government acts in innovation because almost all of the previously mentioned drivers can be supported or hindered by political decisions, whereas transparency aligns to the e-government top priorities.

Moreover, government strategic thinking influences innovation. Anthopoulos and Siozos (forthcoming) used data from European strategic planning and compared Lisbon and Horizon 2020 strategies to justify whether governments realized these potentials. This study concluded that governments who invest more on education and innovation create the premises that can deal with fiscal crisis, whereas countries capitalize innovation in the international competition.

6.2.2 Innovation Management

Innovation management can be seen as the process of creating and implementing a business, with the goal of capitalizing an innovation (Fagerberg, 2004; Kamarck, 2003), while various techniques can be found that support innovation management. For the purposes of this chapter, emphasis is given to process innovation management models because the authors aimed to compare the requirements that each process step places for the innovative idea and compare these requirements with the respective ones that seem to be followed in e-government deployment.

Process models either recognize different stages in the innovation life cycle or ignore these stages in the early beginning (fuzzy front end) and emphasize on a project plan later (when the estimated product has been specified: new product development). The stage gate process defines six phases (initial success, second screen, business case decision, post-development, pre-commercialization, and review) and applies various filters in the beginning of each for the manager to decide whether the process should proceed. On the other hand, the interdisciplinary view process consists of five phases (concept development, system-level design, detail design, testing and refinement, and production ramp-up), where various alternative sciences collaborate in each step to realize the innovation's potentials.

6.2.3 E-Government and Innovation

The combination of the keywords *innovation* and *e-government* returns limited results from the scientific community. SCOPUS search returns 443 related articles, whereas Google Scholar returns 121 related works. Most of these articles (e.g., Arduini et al., 2010; Kamarck, 2003; Lee, Tan, & Trimi, 2005; Moon, 2002) focus on ICT innovations that have been utilized for e-government. Furthermore, an analysis of various government innovations (Lee et al., 2005) results in the conclusion that "all innovations in government deal with 'quality' government." Quality is understood as public service improvements, which are not expected to return economic benefits but to enhance citizens' trust in their government.

Moreover, some researchers realize the positive role that innovation in public administration plays in e-government development. For instance, Ho (2002) recognizes that public seniors shift from emphasizing producer concerns, such as cost-efficiency, to focusing on user satisfaction and control flexibility in service delivery. This approach stresses innovation, organizational change, and entrepreneurship within public administration, which leads to government re-invention.

The role of innovation in e-government was also seen by Moon (2002), who considers aspects from technological innovation to understand how e-government can succeed in social diffusion and adoption. In Moon's study, this author discovered a relationship between organization size and probability of adopting innovation. His findings were later investigated for local governments by Moon and Norris (2005). In their analysis of numerous American cities, they conclude on determinants that affect e-government adoption and they justify that innovation management's orientation concerns a respective key enabler. More specifically, this study recognizes the organization's ICT capacity as critical for e-government adoption, while it concludes that large municipal governments with high innovation orientations tend to be more receptive to new approaches and more likely to be early adopters of e-government.

Finally, Kim et al. (2007) recognize innovation as a part of the overall e-government development process and, more specifically, during the initiation process (technology push), as well as a requirement for continuous public service improvement.

6.2.4 E-Government Development

Information and communications technology has contributed much to the public sector's modernization. An analysis of government e-strategies from eight national, one regional, and two supranational cases from the early 1990s to today (Anthopoulos & Fitsilis, 2014) shows that governments around the world share common e-government visions, which have evolved from a citizen-centered public administration to an accountable, inclusive, and accessible government and to a connected government (Table 6.1). This study returned findings that interrelate

Table 6.1 Government Strategic Visions

Strategic Vision	*Year of First Appearance*
Information society, citizen-centered public administration	2001
Inclusive and innovative government	2006
Economic growth, leadership	2007
Open government (transparency, accountability, participation)	2009
Connected government (cross-agency/cross-border e-services)	2010
Digital economy	2010

Source: Anthopoulos, L., & Fitsilis, P. (2014). *International Journal of Public Administration in the Digital Age, 1*(1), 15–38.

governments' willingness for e-government with their aims for economic growth and for international or supranational leadership in the ICT industry.

The corresponding e-government progress is being measured with various, mainly quantitative, models and indexes, which vary from the World Bank's online availability, implications, investments, and satisfaction; to the United Nations' readiness, e-service stages, e-participation, progress, and performance; to the European Committee's e-service availability and Internet indicators; and to American Customer Satisfaction Index (ACSI) citizen satisfaction (Fitsilis, Anthopoulos, & Gerogiannis, 2010). Most of these measures emphasize established government's efficiency and effectiveness: Efficiency returns public sector performance or productivity rates and is associated mainly with public spending effects to socioeconomic indicators (Hauner & Kyobe, 2008), and effectiveness deals with the quality of public services, the degree of its independence from political pressures, the quality of policy formulation and implementation, and the credibility of the government's commitment to such policies (The World Bank, 2012).

With regard to the e-government progress, various scholars emphasize on adoption. Indicative works are from Carter and Weerakkody (2008), who defined an adoption measurement model with the comparison among cultural variables in the United States and United Kingdom, and from Reddick (2009), who identified and measured various determinants that influence adoption, whereas Tung and Rieck (2005) used Rogers' Innovation Diffusion Theory to ground their adoption framework, and they realized that an innovative environment strengthens adoption potentials.

However, there is an increasing argument with regard to e-government success. Some known failures concern project failures (Heeks & Bailur, 2007), utopian vision statement (Tung & Rieck, 2005), losing citizen expectations and keeping traditional service channels (Bekkers & Homburg, 2007), or evolved types of bureaucracy (Bovens & Zouridis, 2002).

6.3 Research Methodology

This section uses alternative methods to answer this paper's initially grounded research questions.

6.3.1 Case of the Obamacare Website

The first research method is a case study of the Healthcare.gov website, which is more commonly referred to as the Obamacare website and was rolled out on October 1, 2013. This website allows uninsured Americans to enroll in healthcare insurance in a private exchange marketplace run through the Healthcare.gov website. After the website went live on October 1, there were numerous reports of glitches, error messages, and crashes on the website. It took almost a month for the website to get fixed and fully functional, with problems still being reported after the deadline.

This case study provides a good example of a very public e-government project failure. Moreover, it can be considered an innovative product for the U.S. government because it uses novel methods as a means of dealing with the real social problem concerning uninsured Americans. It is also a recent and representative e-government case and demonstrates what can go wrong when innovation is not put front and center with e-government development. To this end, the results from this case study's analysis can illustrate whether the government recognized the e-government project as an innovation and performed respective management practices.

The analysis that follows is from a mixed-methods approach, using NVivo 10 software, analyzing Twitter feeds with the search term *Obamacare website* from October 1, 2013, to January 20, 2014. Authors collected the top 320 tweets from this period and did a word search for their most common occurrences using NVivo. These top tweets included those on Twitter's top tweets feed and those that were retweeted more than 10 times.

Examining the word frequency summary, some of the most common words found in top tweets were predominantly negative referring to the Obamacare website, with issues of security (2.68%), crashes (0.91%), and problems (0.71%).

These findings depict that this project was launched before its completion and proper testing. Although this conclusion sounds obvious, findings from the tweet analysis reflect significant technical issues, which could have been prevented with simple project management processes. The government's delayed reactions to the technical problems indicate a lack in precrisis risk analysis and preparation for post-crisis management. Moreover, in this website rollout, there was much less attention devoted toward the website acting as an important vehicle to implement the new healthcare law. The overall dissatisfaction of the website rollout as seen through the tweets illustrates that the government did not pay close attention to their customers and, to this end, to product success, which is a prerequisite for innovation product success (Fagerberg, 2004). All these outcomes validate that this extremely innovative project was not managed with respective best practices.

This case study shows how e-government project failures can occur and have widespread impacts beyond just failed management practices. This Healthcare.gov website was a significant "embarrassment" for President Obama and fuelled anger against the new healthcare law providing coverage to millions of uninsured Americans. Moreover, this case study shows how politics can intersect with project failures and exacerbate issues and differences among stakeholders. Normally, for e-government projects, failures are seen as isolated incidents with very little media attention. The failure of the Obamacare website created a very visible event, which demonstrates what can go wrong if there is no proper project planning and implementation. Finally, this case study depicts what can happen when e-government is not viewed as an innovative product.

6.3.2 E-Government as an Innovation

The first research question is answered with the findings from the corresponding literature review: In case the public sector is seen as a unique organization, then its mission concerns the delivery of goods (i.e., public infrastructure) and services to citizens and enterprises, the definition of policies, and the legislation of rules that ground these policies. Then, the "e" stands for the utilization of ICT to streamline the above processes (Ho, 2002). This definition of e-government leads to the conclusion that "e" updates the original government processes in a rather incremental manner, meaning that "e" concerns an evolving substitute of the original product. It is difficult to see e-government as radical or systemic because traditional processes (i.e., paper-based and face-to-face contacts and negotiations) are still performed. Thus, e-government can be considered an innovation for the public domain, which could be classified as "incremental," "process" and/or "organizational," and "technological" because it is based on emerging ICT.

With regard to the second research question, literature findings illustrate that governments approach e-government with e-strategic planning, which is specialized with program planning and transformed to deliverables with project planning (Anthopoulos & Fitsilis, 2014). This process is iterative and can be repeated according to e-strategic evolution. This approach does not concern an innovation management process. Moreover, the outcomes from the case study validate that representative e-government projects with high social impacts and extreme, even political, implications are not analyzed and treated carefully, even with simple project management techniques; this will not lead to project success. Both the theoretical and the case studies justify that past and existing e-government implementations have the following characteristics. They:

1. Are rather politically oriented, lacking the appropriate capacity that the public sector must have.
2. Aim to "boost" public administration's efficiency, but without calculating the required cost, neither the added values that will be created to citizens, enterprises, and the public sector.

3. Are driven by technology push, without balancing it with market pull, and, to this end, fail to meet end users' expectations.
4. Are not loyal to the initially grounded vision and mission statements, while performance measurement does not compare outcomes with regard to the initially stated objectives.
5. National economic growth lies behind all government e-government strategies (Anthopoulos & Fitsilis, 2014), which is the outcome of innovation.

The above characteristics do not meet two of the conditions for innovate concerns (Fagerberg, 2004): hard, concentrated, and thorough work; and market focus or market pull (what citizens and enterprises really need). The third condition for innovation seems to be complied with (Fagerberg, 2004), meaning that e-government appears to be important for governments because extensive public funding has been and is being invested in this discipline.

Furthermore, existing e-government approaches depict the following contradiction: Governments feel the change that is demanded for the public sector to enhance national economic growth, they name it e-government, and they invest hugely in this direction; however, the corresponding implementation process appears to be fragmentary and weak because it results in various types of failures. Of course, it is not easy for a government to perform as good management as an enterprise does, and government failures cannot be seen only in an e-government arena: Poverty, economic recession, and wars are also the results of ineffective governance.

If e-government is seen as an innovative government product, then the process-based innovation management methods (Figure 6.1) suggest the required changes to the existing implementation approaches. For the purposes of this chapter, the five-phase interdisciplinary process for innovation management method was followed because it involves contributions from various sciences, something that occurs in the e-government domain (i.e., social and political sciences, ICT, management and finance, etc.). This process is combined with the one proposed by Kim et al. (2007), which consists of initiation, adoption, adaptation, acceptance, routinization, and infusion. The combination (Kim et al., 2007) was performed because it follows a strong theoretical basis, it is quite recent, and it has been tested using case studies, while helping to develop better understanding for both ICT and organizational issues during the e-government implementation process.

Phase 0: A fuzzy-front-end period will determine the corresponding innovative product (i.e., how a digital public service would be. How will it be delivered? etc.).

Phase 1: Concept development: This phase deals with the context of the initiation phase (Kim et al., 2007), where requirements evaluation has to be combined with the clarification of how the expected added value (i.e., efficiency improvement, but how much? Cost reduction, but how much?) would be clearly calculated.

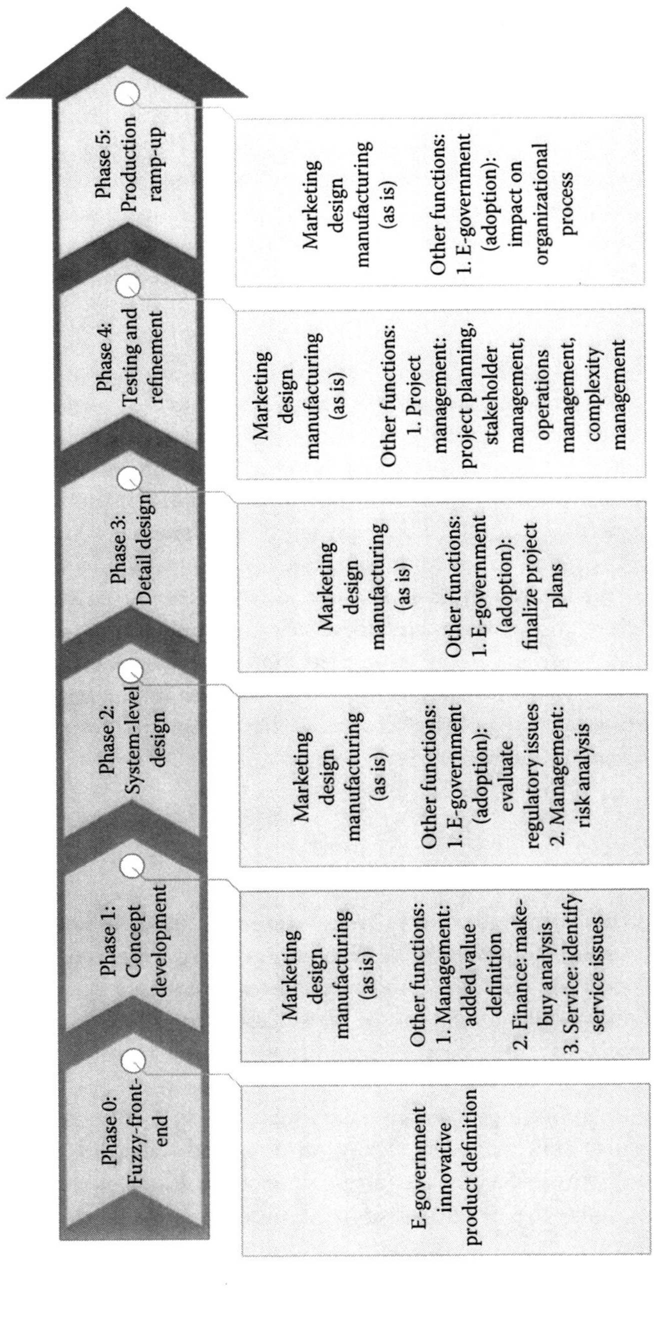

Figure 6.1 Proposed innovation management process for e-government.

Phase 2: System-level design: Beyond existing calculations, this stage relates with the regulatory evaluation issues from the adoption phase of Kim et al. (2007). Moreover, the undertaken risks that are not defined (i.e., What might be wrong with regard to process reengineering?) would be clearly estimated.
Phase 3: Detail design: The context of this phase is combined with the project plans' finalization from the adoption phase (Kim et al., 2007).
Phase 4: Testing and refinement: This phase is extended with the project management demands from the adaptation phase (Kim et al., 2007).
Phase 5: Production ramp-up: This phase concerns e-government deliverable production with the combination of the impacts' management from the acceptance phase (Kim et al., 2007).

This process gives the answer to the first part of the third research question: The e-government implementation process will be updated from e-strategic implementation to an interdisciplinary process-based innovation management. This suggested process incorporates strong commitments regarding focus and management, which are expected to bring in the two identified missing conditions for innovation (strong and thorough work; market pull) to the e-government implementation process.

Moreover, recognizing e-government as an innovation would make governments emphasize the innovation impulses and justify prerequisites that have to be accomplished before an e-government initiative's launch. A strong scientific base, for instance, both inside the public administration and the society, will enhance e-government development education. Moreover, all these impulses can secure the technological, financial, and political capacities that Moon and Norris (2005) recognize that positively influence e-government adoption.

6.4 Conclusions

This chapter presents a different concept for e-government development: It addresses existing corresponding failures with the recognition of e-government as an "innovative product." To this end, the authors stated three research questions, which investigated whether e-government can be considered innovation, whether existing development views e-government as an innovation, and what would change from such a consideration.

Literature review findings demonstrate that e-government can concern an "innovative product" and, more specifically, an improved substitute to the offered services. Moreover, past and recent e-strategic development follows the usual strategic management approach—where strategic definition is followed by program and project management—which do not meet innovation prerequisites. Moreover, the case study of the launch of the Healthcare.gov website shows what happens when novel e-government projects do not follow simple project management methods, which could prevent these technical failures, and innovation management, which is estimated to secure project success. Instead, this case study shows what happens when governments do not consider the innovativeness of e-government and a strategic tool in public service delivery.

To this end, authors introduce an innovation management process for e-government, where fuzzy-front-end is combined with an interdisciplinary process and with the e-government implementation process defined by Kim et al. (2007). This proposed process "closes the gaps" with innovation prerequisites and remains to be tested. The overall concept can benefit both governments and e-government progress because innovation asserts economic growth, which appears to be one of the primary e-government objectives, and a properly adjusted innovation environment will secure public administration and society with the appropriate capacities to adopt e-government. Otherwise, e-government seems to evolve to another service delivery channel, alternative to existing traditional ones.

Author Biographies

Dr. Leonidas Anthopoulos is an associate professor at the Business School of the TEI of Thessaly. At his previous job positions, he was responsible for planning and managing the development of multiple IT systems for the Greek government and for various public organizations. His research interests concern, among others, smart city, e-government, enterprise architecture, and strategic management.

Christopher G. Reddick is a professor and the chair of the Department of Public Administration at the University of Texas–San Antonio. Dr. Reddick's research and teaching interests are in information technology and public sector organizations. Some of his publications can be found in *Public Administration Review*, *Government Information Quarterly*, *Public Performance*, and *Management Review*.

Irene Giannakidou received her PhD, MSc, and BSc degrees in information technology from the Aristotle University of Thessaloniki. She is currently serving as a data analyst with the Historical Archives of Thessaloniki, and her research interests include data mining in social networks, with applications in e-government and smart cities.

Nikolaos Mavridis received his PhD degree from the Massachusetts Institute of Technology, after receiving his MSc degree from UCLA and MEng degree from the Aristotle University of Thessaloniki. He has served as a faculty member at New York University and UAEU, and is a researcher at NCSR Demokritos. He is the vice-chair of the Hellenic Artificial Intelligence Society. His research interests include technologies for participatory governance and e-government, human–robot interaction, and artificial intelligence.

References

Anthopoulos, L., & Fitsilis, P. (2014). Trends in e-strategic management: How do governments transform their policies? *International Journal of Public Administration in the Digital Age, 1*(1), 15–38.

Anthopoulos, L., & Siozos, P. (forthcoming). Can IT innovation become a tool against fiscal crisis? Findings from Europe. *International Journal of Public Administration in the Digital Age*, Special Issue on Fiscal Crisis, IT, and Government Reform.

Arduini, D., Belotti, F., Denni, M., Giungato, G., & Zanfei, A. (2010). Technology adoption and innovation in public services: The case of e-government in Italy. *Information Economics and Policy, 22*(3), 257–275.

Bekkers, V., & Homburg, V. (2007). The myths of e-government: Looking beyond the assumptions of a new and better government. *The Information Society, 23*(5), 373–382.

Bovens, M., & Zouridis, S. (2002). From street-level to system-level bureaucracies: How information and communication technology is transforming administrative discretion and constitutional control. *Public Administration Review, 62*(2), 174–184.

Carter, L., & Weerakkody, V. (2008). E-government adoption: A cultural comparison. *Information Systems Frontiers, 10*, 473–482.

Fagerberg, J. (2004). Innovation: A guide to the literature. In J. Fagerberg, D. C. Mowery & R. R. Nelson (Eds.), *The Oxford handbook of innovation*. Oxford: Oxford University Press.

Fitsilis, P., Anthopoulos, L., & Gerogiannis, V. (2010). An evaluation framework for e-government projects. In C. Reddick (Ed.), *Citizens and e-government: Evaluating policy and management*. Hershey, PA: IGI Global.

Hauner, D., & Kyobe, A. (2008). *Determinants of government efficiency*. International Monetary Fund (IMF) Working Paper. Retrieved December 2013, from http://www.imf.org/external/pubs/ft/wp/2008/wp08228.pdf.

Heeks, R., & Bailur, S. (2007). Analyzing e-government research: Perspectives, philosophies, theories, methods, and practice. *Government Information Quarterly, 24*(2), 243–265.

Ho, A. T.-K. (2002). Reinventing local governments and the e-government initiative. *Public Administration Review, 62*(4), 434–444.

Howells, J. (2005). *The management of innovation and technology: The shaping of technology and institutions of the market economy*. London: Sage Publications.

Kamarck, E. (2003). *Government innovation around the World*. Paper developed for the Ash Institute for Democratic Governance and Innovation John F. Kennedy School of Government, Harvard University.

Kim, H. J., Pan, G., & Pan, S. L. (2007). Managing IT-enabled transformation in the public sector: A case study on e-government in South Korea. *Government Information Quarterly, 24*, 338–352.

Lee, S. M., Tan, X., & Trimi, S. (2005). Current practices of leading e-government countries. *Communications of the ACM, 48*(10), 99–104.

Maital, S., & Seshadri, D. V. R. (2007). *Innovation management: Strategies, concepts and tools for growth and profit*. New Delhi, India: Response Books, Sage Publications.

Moon, M. J. (2002). The evolution of e-government among municipalities: Rhetoric or reality? *Public Administration Review, 62*(4), 424–433.

Moon, M. J., & Norris, F. D. (2005). Does managerial orientation matter? The adoption of reinventing government and e-government at the municipal level. *Information Systems Journal, 15*, 43–60.

Organisation for Economic Co-operation and Development. (2005). *Oslo manual: Guidelines for collecting and interpreting innovation data* (3rd ed.). Paris: OECD Publishing.

Porter, M., & Schwab, K. (2008). *Global competitiveness report 2008–2009*. Geneva: World Economic Forum.

Reddick, C. G. (2009). The adoption of centralized customer service systems: A survey of local governments. *Government Information Quarterly, 26*(1), 219–226.

Reddick, C. G., & Turner, M. (2012). Channel choice and public service delivery in Canada: Comparing e-government to traditional service delivery. *Government Information Quarterly, 29*(1), 1–11.

The World Bank. (2012). *The Worldwide Governance Indicators (WGI)* [online]. Retrieved December 2013, from http://info.worldbank.org/governance/wgi/index.aspx#home.

Tung, L. L., & Rieck, O. (2005). Adoption of electronic government services among business organizations in Singapore. *Journal of Strategic Information Systems, 14*, 417–440.

Chapter 7

Aligning Strategy and Information and Communication Technology in Public Organizations: A Critical Management Challenge

Eamonn Caffrey and Joe McDonagh

Contents

Abstract

The challenge in aligning strategy and information and communication technologies (ICTs), commonly referred to as strategic alignment, has persisted as a key issue for ICT management for many years now. The purpose of this chapter is to provide insights into the strategic alignment construct as it relates to public organizations based on an assessment of empirical works. Over time, strategic alignment has evolved, and greater understanding about the intricate nature of the concept has become known. Impediments to strategic alignment efficacy that serve to obstruct the maximization of value are discussed. A perspective on strategy process as a useful mechanism by which to align strategy and ICT highlights some key activities for consideration. While much has been discovered in relation to strategic alignment, a good deal continues to elude us, particularly in relation to public organizations, and some potential opportunities to advance further inquiry into this important management domain are presented.

Keywords: alignment, e-government, ICT, ICT management, ICT strategy, ICT strategy-making, public organization, strategic alignment, strategy, strategy process

7.1 Introduction

This chapter presents an overview of the topic on aligning strategy and information and communication technologies (ICTs) in public organizations, the aim of which is to provide a reasonably high-level view of the strategic alignment construct and offer some perspectives on its relevance to the public sector. This chapter begins with a perspective on the evolving nature of aligning strategy and ICT in organizations. For some 50 years now, aligning strategy and ICT has been a topic of great interest to the fields of both organization and management and ICT management. Since 1983, numerous researchers have been interested to learn about the ICT management challenges that organizations face on an almost annual basis. In almost every study since, the topic of strategic alignment has persistently ranked highly. A review of the dynamic features of strategic alignment is presented, which lays the foundation for the definitional meaning of the topic along with some of the related challenges and benefits. This is followed by a perspective on the role of strategic alignment in public organizations. The next section offers some empirical-based insight into the nature of public-based strategic alignment research and performance implications for the sector. Some of the known impediments to strategic alignment in public organizations are then reviewed, highlighting distinct challenges for the sector. A perspective is then presented that considers strategic alignment from a process-orientated point of view. This section elaborates on some of the strategic activities considered necessary when aligning strategy and ICT. A number of potential opportunities for further research are presented, followed by concluding comments.

7.2 The Evolving Nature of Strategic Alignment

The internal organizational elements were of particular interest to Chandler (1962), who especially focused on the strategy-structure relationship and concluded that "strategy drives structure." Chandler highlighted that there must be a fit between strategy and structure so that organizational change can be managed effectively. Four internal organizational elements were identified by Leavitt (1965), who emphasized the important relationship between all four. These elements consisted of (a) strategy, (b) structure, (c) systems, and (d) individuals. According to Leavitt, organizations are a "dynamic equilibrium" between all four elements, and each element must be considered in coordination with the other three. No single element is more dominant than any other given that effectiveness is determined by how each affects, and is affected by, the others.

In studying organizational anatomy, Ackoff (1971) described organizations as purposeful systems consisting of elements with a common purpose. Behavior is deeply embedded in all functions and subsystems; "its functionally distinct subsets can respond to each other's behavior through observation or communication"

(Ackoff, 1999, p. 550), and hence the need to orientate study to incorporate the linkage between strategy, structure, and system behavior. During the 1970s, the requirement to locate a "fit" between ICT development and an organization's strategy led to the creation of proprietary ICT planning methodologies by management consulting houses, such as Coopers & Lybrand and Andersen Consulting (Lederer & Gardiner, 1992) and Business Systems Planning by International Business Machines Corporation (1981), as well as a range of modeling techniques, such as information engineering (Martin, 1989), enterprise-wide information management (Parker & Benson, 1989), and total information systems management (Osterle, Brenner, & Hilbers, 1993). This form of mechanistic planning paved the way for the strategic alignment construct to emerge in rigid fashion based on the analysis of structured data schemes. These types of traditional planning methods lacked the involvement of organizational users (e.g., lines of business) and consequently failed to deliver functional ICT solutions.

By the 1980s, significant advances in ICT were having a profound impact on organizational performance, and researchers set about to investigate the effects of ICT-enabled change. The Massachusetts Institute of Technology Sloan School of Management undertook a major 7-year research initiative led by a group of ICT researchers to identify the key organizational elements associated with the development of ICT. The research initiative commenced in 1984, and the findings were revealed 7 years later and became better known as "Management in the 1990s" (Scott-Morton, 1991). The study highlighted the importance of an alignment among organizational strategy, ICT strategy, organizational infrastructure, and ICT infrastructure. It was shown that an alignment among these four components was essential to the success of modern organizations. The conceptual model underpinning the study had identified five components of organizational functionality that must be kept in check while undertaking major change initiatives: (a) strategy, (b) structure, (c) ICT, (d) management processes, and (e) individuals and roles.

The Management in the 1990s study had shown that strategy and ICT are essentially the change stimuli, whereas structure, management processes, and individuals and roles formed the totality of organizational structure. Essentially, the study highlighted the high level of interdependence between organizational elements when effecting ICT-enabled change. "IT [information technology] is a key enabler of strategic direction and that an important problem is to find the link between strategic ideas and the application of IT" (Scott-Morton, 1991, p. 91). From this point on, scholarly interest in strategic ICT practice and the role of ICT in support of organizational performance advanced in a direction that would seek to gain a more detailed, holistic, and integrated understanding of the strategic alignment construct. Recognition of the alignment challenge continued to grow, and this led to the development of more modern planning approaches (Silvius, 2007). The models that emerged from the 1990s onward in an attempt to bridge the alignment gap were less formal in procedure but more practical. Chan and Reich (2007a) listed over 150 articles that revealed more practical elements that comprised of the following features: focus on business strategy, pragmatic versus methodological planning approaches, project prioritization, greater business involvement in ICT development, and improved planning of ICT infrastructure and capabilities.

7.3 The Dynamic Features of Strategic Alignment

For more than three decades now, strategic alignment has persistently shown to be highly ranked as a key issue for ICT management (Caffrey & McDonagh, 2009). Accordingly, it is considered to be an important contemporary ICT management issue. The strategic alignment dilemma for business leaders and ICT executives worldwide has proven to be a "long-standing pervasive conundrum" (Luftman & Ben-Zvi, 2010a, p. 265). A good deal of empirical inquiry has studied the strategic alignment construct in order to gain a better understanding of its meaning, the embedded constraints, and performance-related benefits. Each of these aspects is discussed more fully next.

7.3.1 Definition

Over time, the concept of strategic alignment has been variously defined. According to the Massachusetts Institute of Technology, strategic alignment is based on "the premise that the inability of organizations to realize value from IS (information systems)/IT investments is, in part, due to lack of alignment between business and IS/IT strategies" (Ward & Peppard, 2002, p. 44). An adequate definition for strategic alignment is often vague (Maes, Rijsenbrij, Truijens, & Goedvolk, 2000). It was noted by Silvius (2009) that in over 1 million Google search hits, a sound definition is challenging to find. Nevertheless, over the last two decades, meaningful attempts have been made to define the construct. These have included creating a strategic fit between "business strategy, information technology strategy, organizational infrastructure and processes, and information technology infrastructure and processes" (Henderson & Venkatraman, 1993, p. 4). It is concerned with "the degree to which the IT mission, objectives, and plans support and are supported by the business mission, objectives and plans" (Reich & Benbasat, 1996, p. 56). It was shown to be the task of "applying IT in an appropriate and timely way and in harmony with business strategies, goals, and needs" (Luftman & Brier, 1999, p. 109). Strategic alignment "is the business and IT working together to reach a common goal" (Campbell, Kay & Avison, 2005, p. 662). It is also considered to be "the degree to which the IT application portfolio converges with business strategies" (Oh & Pinsonneault, 2007, p. 244). Each of the above is, in one way or another, based on the perspective posited by Edwards (1994, p. 51) insofar as strategic alignment generally refers to "congruence, match, agreement, or similarity between two conceptually distinct constructs." This concept was clearly rooted in the work of Nadler and Tushman (1980), whereby "the degree to which the needs, demands, goals, objectives, and/or structure of one component are consistent with the needs, demands, goals, objectives, and/or structure of another component" (p. 40). Hence, by embellishing the essence of the viewpoint of Nadler and Tushman and distilling essential themes made apparent by earlier definitions shown above, we define strategic alignment to mean "the extent to which the ICT strategy is consistent with the organizational strategy by way of supporting major business priorities and advancing the long term organizational direction based on the various business and ICT management levels working congruently to generate real business value."

7.3.2 Challenges

The scholarly work heretofore has revealed many reasons for why strategic alignment persistently challenges organizations and management to generate real business value from ICT investments. By examining some of these studies (Luftman & Ben-Zvi, 2010a, 2010b; Luftman & Brier, 1999; Luftman & Kempaiah, 2007, 2008; Luftman, Kempaiah, & Rigoni, 2009), it was possible to distil the high-level reasons to explain the challenges embedded in strategic alignment, and these included the following: the inability of business and ICT executives to work together by way of cocreation, the lack of an interdependent enterprise-wide view, the lack of effective strategic planning that sets out a vision for the role of ICT, an underperforming role enacted by senior executives in support of ICT-enabling possibilities, the failure to position ICT as a business investment, and insufficient and inadequate levels of ICT management capability and resource capacity. These issues, among others, are discussed more fully throughout this chapter.

7.3.3 Benefits

Many different types of benefits can stem from strategic alignment efficacy that have the potential to impact on organizational performance, and these include service improvement (Sabherwal & Chan, 2001; Wang & Tai, 2003); adaptation to potential risk situations (Chague, 1996; Gupta, Karimi, & Somers, 1997); leverage efficiency, stability, and economies of scale (Gupta et al., 1997; Oh & Pinsonneault, 2007; Sabherwal & Chan, 2001); ability to overcome environmental uncertainties (Croteau & Bergeron, 2001; Huber, 1984); support of the control and delegation of decision authority (Tavakolian, 1989); greater coordination and integration of complex goals, functions, and relationships by facilitating a higher degree of formality and specialization (Leifer, 1988); satisfying agency operations and individual user needs more robustly and completely (Huang & Hu, 2007); improvement of cooperation among system partners (Tallon, 2007–2008; Wang & Tai, 2003); driving innovation (Raskino, 2011; Silvius, 2010); providing of unique administration competencies (Chan, Huff, Barclay, & Copeland, 1997); and facilitation of greater knowledge exchange (Hsu, Hu, & Chen, 2006). On the basis of empirical findings, no more than 15% of organizations can claim to fully maximize the benefits from ICT investments (Breyne, Cumps, & Viaene, 2005; Luftman & Kempaiah, 2007; Shpilberg, Berez, Puryear, & Shah, 2007). The opportunity for public organizations to leverage these types of benefits via ICT investment is rather significant given that there is broad-based scope for more effective strategic public management (Bryson, Berry, & Yang, 2010).

7.4 Strategic Alignment in Public Organizations

In recent years, government and federal agencies have made "significant investments in IT applications aimed at reducing costs and improving operational effectiveness while exchanging information across organizational boundaries, functional areas, or jurisdictions" (Fedorowicz, Gelinas, Gogan, & Williams, 2009, p. 53). A variety of contextual factors can impact on strategic ICT effectiveness in public organizations, which includes politics, term limits, goal ambiguities, conflicts of interest, resistance to change, measurement issues, and legal concerns (Bryson et al., 2010). Therefore, alignment must take into account an organization's strategic focus, content, visioning, and implementation. "ICTs may be used to achieve greater government efficiency, better service quality, and more democratic participation" (Bryson et al., 2010, p. 510). Based on leading public chief information officer (CIO) insights, ICT is regarded to be a critical enabler in support of agency operations, but extracting the real business value resides in ICT being effective, cost-efficient, secure, and well supported (Tech America, 2013). The high-level challenges center on the need to adapt to rapidly evolving and continuously changing technology, with lower ICT budgets, reduced workforces, and ineffective acquisition support, and all the while facing the exponential challenge offered up by privacy, control, and security threats. However, for ICT to be effective, such issues must be considered from an interconnected perspective that goes beyond tactical ICT, which so often seldom happens.

Many public organizations today are faced with highly complex and fragmented ICT environments. Such environments are made up of silo structures operating large numbers of individual hardware and software systems with little or no integration and minimal heterogeneity and costs running excessively high. Project teams are often isolated and capabilities are thinly spread, with consultants and suppliers mantling the latest wave of ICT. "Most IT organizations are amazingly complex and have individualized initiatives that are like independent countries, each with its own business applications, technologies, culture, data definitions, and orientation" (Feld & Stoddard, 2004, p. 76). The challenge herewith is to develop and implement appropriate ICT solutions aligned with priorities that penetrate multiple interconnections by linking concepts with solutions that provide deeply embedded integrity.

In orientating an alignment between strategy and ICT in public organizations, a review of current concerns from a practical point of view can be learned by examining the context within which it holds relevance and meaning to the U.S.

Government Accountability Office (GAO). The key areas of management concern in aligning strategy and ICT highlighted the following areas for improvement: needing better management in order to achieve significant savings, establishing and implementing incremental development policies, refining ICT investment management processes to improve reliability and transparency, anticipating and meeting accountability challenges, overcoming limitations in management controls and human capital planning to support modernization efforts, overcoming information exchange challenges resulting from the lack of specific prioritized actions, leveraging best practices to ensure the success of major acquisitions, improving project management practices, developing strategic capabilities to improve interactive services, and needing strengthened oversight to improve overall performance and control. These issues offer only a flavor of the types of challenges that public organizations face in aligning strategy and ICT today. Further insight into these issues and more can be explored by directly accessing the GAO website (http://gao.gov).

Nearly all state and local governments now formulate strategic ICT plans. However, such plans lack transparency and responsiveness (Yang & Melitski, 2007). According to Sommer (2002), many governmental ICT projects fail. Furthermore, ICT projects classified as e-government initiatives are considered to be unsophisticated and nontransactional, and only few governments can attribute major transformational change to the successful execution of ICT (Coursey & Norris, 2008). To better understand ICT failure in public organizations, one way of doing this is to reflect on the outcomes associated with strategic ICT initiatives (Bloch, Blumberg, & Laartz, 2012; Nelson, 2007). The level of total success for ICT initiatives based on planned outcomes is posited to be as low as 10% (McDonagh & Coghlan, 2006, 2010). A further 20% of initiatives are delivered in full but fall far outside the range of planned investment cost, functionality, and scheduled delivery time (McDonagh & Coghlan, 2006, 2010). An additional 20% of initiatives fall short in that fulfillment results in only 50% of the functionality, at twice the planned level of investment and scheduled delivery time (McDonagh & Coghlan, 2006, 2010). Finally, the remaining 50% fail to deliver anything of value and accomplish nothing more than consuming scarce public resources in pursuit of failed ICT programs and projects (Loonam, McDonagh, Kumar, & O'Regan, 2014; McDonagh & Coghlan, 2006, 2010).

7.5 Empirical Inquiry into Strategic Alignment

The study of the strategic alignment construct has been dominated by research that examined mostly its relevance and meaning to private sector organizations. The distinction between public and private sectors can be described as the constrained generalizability of findings. However, a small number of empirical studies (approximately 10%) have, in contrast, taken a particular interest in exploring the relevance and performance of strategic alignment in the public sector (Caffrey,

2013). Such studies examined strategic alignment as it related to e-government (Fedorowicz et al., 2009; Hsu et al., 2006; Tarafdar & Qrunfleh, 2009), healthcare (Bush, Lederer, Li, Palmisano, & Rao, 2009; Sauer, Southon, & Dampney, 1997; Wijnhoven, Spil, Stegwee, & Fa, 2006), public administration (Burn, 1996), academia (Sabherwal & Chan, 2001; Sabherwal & Kirs, 1994; Silva, Figueroa, & Gonzalez-Reinhart, 2007), and defense (Hartung, Reich, & Benbasat, 2000). Furthermore, following a series of high-level government reviews of poorly performing large-scale ICT initiatives, a range of pathologies that contributed to both underperformance and/or failure of ICT-enabled change initiatives in public organizations such as the Government of Ireland, Government of United Kingdom and Northern Ireland, and the United Nations System are highlighted by McDonagh (in press). A high-level summary of the reasons that fermented the seeds of failure in relation to ICT-enabled change initiatives as noted by McDonagh included the following: inadequate strategic capabilities to guide resource allocation, inadequate strategic capabilities to craft coordination and integration of technological and organizational change, inadequate strategic capabilities to guide the formation and execution of programmed change, poor understanding of the enabling potential of ICT to modernize and reform, ICT investment considered from a highly technocentric perspective, and inadequate transformational leadership capabilities to deliver radical organizational change.

The transformation pathologies embody multilevel impacts. The level of impact can be traced to each of the following areas: institution, organization, and individuals and groups. Aligning strategy and ICT is a multilevel activity, and each of the above levels has distinct roles to play. Consequently, given the comprehensive lack of understanding for ICT-enabled transformation from a multilevel perspective in public organizations, the outcome from significant ICT initiatives traditionally results in greater fragmentation rather than any kind of tighter integration or cohesiveness. Therefore, ICT environments typically become only more burdensome via an abundance of over complexity (McDonagh, in press). Some of these issues are discussed next in terms of impeding alignment efficacy based on empirical observation.

7.6 Impediments to Strategic Alignment

Major impediments to aligning strategy and ICT in public organizations were found to comprise of tactical versus strategic orientation, lack of top management support and understanding, organizational structure, investment governance, and resistance to change.

7.6.1 Tactical versus Strategic Orientation

The perception of strategic alignment in public organizations is one where the organization and ICT structure "walk hand-in-hand," but it has also been found that strategic expectations merely extend to providing ICT services as needed (Silva et al., 2007). In a similar vein, "the IT department was perceived as a support department, its primary function being to implement the chosen applications at the lowest possible cost" (Tarafdar & Qrunfleh, 2009, p. 341). Strategic ICT is treated as a tactical response rather than taking into account the wider needs and consideration for the interdependencies among organizations. Comprehensive and effective strategic ICT planning is rare (Bryson et al., 2010). Such plans are focused on operational matters rather than on setting out any long-term joined up approach to leverage ICT from a strategic perspective (Holley, Dufner, & Reed, 2004). In terms of planning and delivering ICT projects at the federal level, "agencies generally have goals associated with IT, these goals are not always linked to specific performance measures...many agencies do not monitor actual-versus-expected performance against enterprise-wide IT performance measures in their IRM [information resource management] plans" (GAO, 2004, p. 3). Essentially, strategy content is directed to consider a variety of factors distinct to public organizations (Bryson et al., 2010), but it does not necessarily support the exploitation of strategic opportunities via ICT enablement. An example would point to the high levels of ICT fragmentation in many public organizations. Information and communication technology solutions are designed and implemented for the sole purpose of providing capability to a local division or business area but done so without due regard for the wider need to make information accessible and transferable, that is, lack of an integrative and joined-up view at the functional, business, strategic, and system levels.

7.6.2 Lack of Top Management Support and Understanding

The role of senior managers was described as one of expectancy rather than participatory (Pyburn, 1983). A decade later, Earl (1993) argued that the most important enabling condition to effectively align strategy and ICT is top management involvement and support. According to Broadbent and Weill (1993, p. 176), "extensive and on-going senior management responsibility for long-term IS developments" is essential. "Senior managers need to ensure that IT spending and initiatives are aligned with and further the company's strategy and goals" (Ross & Weill, 2002, p. 91). According to Gartner research, the average organization squanders away 20% of its ICT capital budget. This puts the annual cost of global ICT mal-investment at approximately US$500 billion (Feld & Stoddard, 2004). Such waste "is a direct result of the fact that IT has so far operated without the constructive involvement of the senior management team" (Feld & Stoddard, 2004, p. 74). Strategic alignment requires a common understanding; however, senior management commonly complains that, too often, the performance returns on the high cost of ICT investment are substantially below expectation. Meanwhile, the same senior managers observe a continuous rise in seemingly necessary ICT capabilities and expenditure. Arguably, many organizations are not generating the value from ICT investments that they could or perhaps should. The distinguishing factor for organizations that demonstrate highly positive returns has shown that "senior managers take a leadership role in a handful of key IT decisions" (Ross & Weill, 2002, p. 85). In contrast, senior management abdication of executive responsibility for ICT investment decisions usually results in wasted investment and missed opportunities. Some examples of this would include the following: Senior managers fail to fully interpret the transformational significance and inadequately prepare an organization accordingly; conservatism is present among senior managers, whereby there is limited appreciation for the potential of ICT enablement; and decision making is confined to the senior management level, thereby failing to capture the innovative potential located in other areas such as the middle management level.

7.6.3 Organization Structure

Organization structures are usually more rigid in comparison with the private sector, that is, red tape (public) versus flexibility (private) (Silva & Hirschheim, 2007). Public organizations tend toward higher internal complexity, supported by stable hierarchies with centralized and formalized rules (Rainey, 2009). The role of ICT aims to support the attainment of political and public service goals, with deployment undertaken in relatively rigid organizational structures (Dufner, Holley, & Reed, 2003). The bureaucratic attributes and political contexts of public administration impede and reduce the pace of ICT development. Strategy is treated from a command perspective led from the top down and reduces the opportunity to leverage ICT by way of fermenting a strategic approach that cultivates development through social learning.

7.6.4 Stakeholder Engagement and Management

The ICT structure cannot, in isolation, articulate a vision for ICT and selectively set out the major priorities to tackle. "Championing technology is as much about collaboration and persuasion as it is about setting strategy" (Harvey Nash, 2013, p. 6). Arguably, collaboration and persuasion are key components of strategy. The process of stakeholder engagement is a critical impediment for many public organizations to overcome and features strongly in determining whether organizations find themselves lowly or highly aligned. Stakeholder engagement involves building support with public executives for crafting the ICT vision. This form of stakeholder management is necessary at multiple levels, such as between the CIO and senior management team, IT executives, and functional leaders, and is particularly relevant to the public context in terms of gaining support for critical initiatives from government leaders, policy decision makers, and high-level institutional executives. Stakeholder engagement can be classified in terms of its strategic significance via the impact at functional, business, strategic, or network level (De Wit & Meyer, 2002). Technological success is underpinned by collaborative working among ICT and public executives and should be recognized as the necessary need to partner in forming and executing strategic ICT ambitions.

7.6.5 Investment Governance

Governance is an important feature of ICT management. At a federal level, however, "most CIOs lack the direct control over the bulk of IT spending" (Tech America, 2013, p. 6). Subsequently, federal CIOs argue that they cannot be responsible for how public agencies invest in ICT when given little budgetary control and powers. The spread of ICT investment control cuts across three institutional structures: department CIO, component/bureau CIO, and program office. Investment governance is a major issue for public organizations when department CIOs have little or no control over vast proportions of ICT budgets, and consequently, the value of ICT investments is negated and thus greatly impacting on the degree of alignment. The main reason for why most organizations (85%) are neither highly aligned nor highly effective is that core and support functions unilaterally develop ICT with complete disregard for the bigger picture, resulting in an overlap of ICT systems. This unnecessarily creates substantial layers of complexity that support short-term purpose but gradually corrode strategic performance (Shpilberg et al., 2007).

7.6.6 Resistance to Change

Within public organizations, change is often viewed with suspicion. The transition to ICT-dependency mode requires "significant culture change which takes time to develop both from an IS and business perspective" (Burn, 1996, p. 8). Although ICT stimulates transformation, the process requires continual adjustment and improvisation based on broad-based understanding, resulting in large-scale differences made to the profile of work enacted and changes in knowledge, level and style of interaction among organizational members, work allocation, accountability, responsibility and control, and coordination activities (Orlikowski, 1996). As pointed out by McDonagh (in press), ICT-enabled change, especially as it relates to the public sector, is strongly contingent on an effective organizational diagnosis to craft and execute a strategic response that will guide a series of focused change interventions. Such interventions must fully consider important elements such as organizational practices (Floyd, Cornelissen, & Delios, 2011; Hendry & Seidl, 2003), organizational processes (Bizzi & Langley, 2012; Feldman, 2004), and organizational routines (Polites & Karahanna, 2013). A good example of this particular impediment would point to the concept of "strategy blindness" and the impact of "cognitive entrenchment" (Arvidsson, Holmstrom, & Lyytinen, 2014).

7.7 Crafting a Process Approach to Strategic Alignment

The manner in which strategies emerge and develop is referred to as the strategy process. Strategy process considers the "how, who and when of strategy" (De Wit &

Meyer, 2002, p. 5). Process is concerned with how strategy is, and should be, developed, formulated, implemented, and changed; the identity of the people involved; and when relevant process activities take place. Four enabling conditions are considered necessary to carry out an effective strategy process: (a) identifying, (b) diagnosing, (c) conceiving, and (d) realizing (De Wit & Meyer, 2002). Each condition forms a key feature of the strategic activities.

7.7.1 Strategic Activities

A central component of strategic alignment is the organizational ability to build and sustain partnerships. This form of partnership arrangement takes the shape of taskforces, multi-tiered and cross-functional teams, and collaborative planning networks. Organizational learning is a central facet of strategic alignment in that it facilitates knowledge exchange (Henderson & Sifonis, 1988). "Active involvement in firm-wide strategy formation processes, which enhanced knowledge of the content of strategic directions, built relationships, and increased the dialogue between different business and functional managers" (Broadbent & Weill, 1993, p. 176) brought about higher levels of alignment efficacy. This style of active involvement illustrates what can be described as the social and intellectual dimensions. An implicit feature of these two dimensions is the fundamental tenet of linkage (Reich & Benbasat, 1996). Linkage has been shown to require the involvement of multiple sources of knowledge and expertise to build an integrated understanding of the internal and external organizational complexities.

7.7.2 Issue Identification

Strategic issue identification refers to the activities performed by an organization that help to provide a better understanding of what can be viewed as problematic (De Wit & Meyer, 2002). The purpose of which is to identify the issues that constitute an important opportunity or threat to the organization. If an organization intends to maintain its core purpose then such issues must be carefully discussed and explored in-depth by members of the organization's executive ranks. The social dimension aims to unlock executive and ICT intellect to identify key issues. The strategy process centers on the problems and issues that matter most (Shpilberg et al., 2007). Such an approach recognizes a tight link between executive priorities and ICT development aligned to improving performance outcomes. Alignment is dependent on an awareness of the underlying issues, and these must be well known and understood. Interaction between the various strands of strategy and the connection to ICT is crucial. Public executives must have strong knowledge of the contextual issues that will impact on future direction and the connection to ICT development must be made clear.

7.7.3 Issue Diagnosis

Most strategy issues are not simple and straightforward. To come to terms with an issue, executives must fully understand its structure and underlying causes to learn about what is really going on. Effectively aligning strategy and ICT then building greater understanding is crucial to fully comprehend how public affairs and ICT interrelate. Fostering the social dimension among "users, functional managers, corporate managers, and outside parties" (Reich & Kaarst-Brown, 2003, p. 92) enables greater understanding through the combination of competence and the commitment to apply one's abilities (Ulrich, 1998). The diagnostic activity centers on the knowledge capability of an organization to act with intent. This is contingent on ICT executives having a good understanding of organizational processes, routines, and technologies. The diagnostic activity requires ICT executives to develop sufficient levels of shared cognition with public executives based on assumptions, knowledge, and expectations. Strategic alignment can be made possible only when there is an alignment of core beliefs and assumptions about the underlying issues that need to be overcome.

7.7.4 Option Conception and Selection

This activity involves the conceiving of ideas that contribute to determining the course of action to be pursued. Organizations face challenges in the form of complex problems and threats. The aim here is to identify appropriate ICT solutions, and this requires logical and generative thinking to formulate strategic options in response. The activity requires public executives to question assumptions, challenge ideas, rally support for strategic agenda items, encourage learning, champion new initiatives, steer change, and build political support (De Wit & Meyer, 2005). Knowledge sharing is a key feature of the activity that aims to reduce knowledge gaps leading to improved outcomes (Ghosh & Scott, 2009). Knowledge sharing is supported by a series of activities that include bonding, linking, and bridging among organizational members. These types of socially based activities help to identify issues, conceive options, and select appropriate solutions. Option conception is dependent on four enabling conditions: (a) identify the information needs of each business unit; (b) assess and visualize; (c) detect gaps between the business and its information; and (d) identify ICT solution strategies, projects, and systems to support priorities (Peak, Guynes, & Kroon, 2005). This requires linking ICT planning to executive planning processes. Option selection must be factually based using rigorous analysis to derive fact-driven conclusions about the selection of ICTs. Investments on ICT must focus on building capability and avoiding the labyrinth of ICT complexity.

7.7.5 Realization Activities

Realization of activities is concerned with the intended actions necessary to implement ICT. It concerns the practical actions performed by the organization. This relates to all hands-on activities to execute the strategy. Despite organizations being able to appropriately identify strategic ICT solutions, implementation was found to hinder the process because organizations lack top management support and suitable control mechanisms. Insufficient planning produces problems as a consequence of low implementation detail (Newkirk, Lederer, & Srinivasan, 2003). Conversely, too much implementation planning can complicate, delay, or otherwise impede implementation. Overall, "the data strongly supported the simplistic view that more SISP [Strategic Information Systems Planning] is better" (Newkirk et al., 2003, p. 221) in that there is a strong correlation between the degree of planning comprehensiveness and realized ICT-enabled success. Comprehensive attention to planning the implementation must consider the tasks of defining the change management approach, defining the action plan, evaluating it, and defining the follow-up and control procedures. Most intended strategies are not fully implemented due to execution difficulties encountered at the implementation phase, and consequently, only a small number of planned ICT strategies are ever realized (Wang & Tai, 2003).

7.8 Toward an Agenda for Further Inquiry

This section sets out a number of high-level opportunities considered suitable for further inquiry to gain new and relevant insights that can sharpen our understanding of the strategic alignment construct in a public sector context.

7.8.1 Toward a Role-Based Perspective on Strategic Alignment

Strategy stimulates organizational structure, management, and process (Chandler, 1962). Later, Scott-Morton (1991) concluded that strategic ICT had evolved to become the stimulating force behind driving structural, role, and process transformation. Arguably, however, organizations do not make strategy; people do. Subsequently then, management actors who make decisions about the selection of ICT investments are responsible for stimulating organizational transformation (albeit quite often unbeknownst to them, as this chapter has revealed). Therefore, greater understanding about management roles and routines as they relate to aligning strategy and ICT could highlight some of the embedded constraints that hinder ICT strategy-making.

7.8.2 Toward a Networks-Based Perspective on Strategic Alignment

The level of interdependency among public organizations is rising. This chapter revealed the importance of leveraging ICT from a high-level system perspective that trickles down to the network of interrelated organizations. Thus, any strategic ICT initiative must reflect the bigger picture. It could be useful to gain insight into how process activities focus on building interorganization alliances at the local, state, and federal levels to steer large-scale ICT-enabled change.

It is the responsibility of executives to "recognize that the value of ICTs is not only about efficiency and economy but also about engaging citizens in policy discourse, managing knowledge in public organizations, and integrating organizational cultures" (Bryson et al., 2010, p. 511). Now that public organizations are becoming more interconnected and a reliance on interdependency is essential for many, managing knowledge in and between networks of organizations is crucial. Fostering a shared cultural philosophy among networks rather than individual organizations about the strategic agenda to instill belief and gather support for mission-critical advances is a key feature of strategy-led transformational initiatives. However, as it relates to setting the strategic ICT agenda, how are public executives working together to visualize a future of highly interconnected systems? What form of management routines are embedded into the process of acting strategically and how do public executives respond to and overcome major constraining factors such as politics, power, and timing when planning multiyear ICT initiatives that extend far beyond any single government's remit? More research around such themes could be rewarding.

7.8.3 Toward an Interdisciplinary Perspective on Strategic Alignment

Much of the scholarly literature as it relates to aligning strategy and ICT can be described as overly mechanistic and lacks a strong process orientation (Chan & Reich, 2007b). The argument for an interdisciplinary multimethod approach to inquiry into major ICT-enabled transformation whereby public executives hold within themselves the necessary knowledge to cocreate the strategic ICT agenda with management peers and executive leaders has been well articulated (McDonagh, in press). The approach calls for studying ICT-enabled change by investigating the process directly. Such an approach looks beyond traditional ICT management views and instead recognizes that strategy is a multifaceted domain with roots embedded in the fields of strategic management, organization and change management, and program and project management. Such an approach could be based on multiple fields of scholarly interest tied to fostering organizational change and development as it relates to ICT-enabled initiatives. Tracing the sequence of events over time, first-hand, offers the potential to glean new insight into the complexities attached to strategic alignment in public organizations and thus critically illuminate the embedded nature of interdisciplinary components.

7.8.4 Toward an Ambidextrous Capability Perspective on Strategic Alignment

Little is known in terms of how public organizations explore knowledge to identify new opportunities to leverage ICT to drive efficiency, effectiveness, and interactivity and to exploit knowledge to maximize the value of core strengths and capabilities by way of transferring and embedding best practice to other public sector areas. It could be useful to learn more about the role of ICT in "achieving an ambidexterity capability" (Huang, Newell, Huang, & Pan, 2014, p. 29) as it might potentially relate to public organizations.

7.9 Conclusion

This chapter described the evolving nature of aligning strategy and ICT over the last 50 years. Here, it was shown that strategic alignment is concerned primarily with maximizing value from ICT investments by way of raising organizational performance. Strategic alignment was shown to be a dynamic construct, reflected by its various features that can hinder or heighten performance. However, essentially, it requires the collective knowledge capacity of ICT managers and public executives to map the strategic trajectory by astutely selecting and executing ICT initiatives that support and challenge organizational performance.

During the last decade or two, public organizations have generally raised the level of ICT investment to support mission-critical aims toward improved service delivery, greater efficiency, and higher interactivity. However, as was pointed out, only a small number of these ICT initiatives deliver on what was originally intended. For the most part, a great many ICT plans in the public sector struggle to realize value or make a difference to the running and performance of organizations. A number of reasons were highlighted to demonstrate why this is the case. Arguably, the more things change, the more they really remain the same insofar as the public sector struggles to derive value from major ICT-enabled transformation initiatives.

Some of the known impediments to realizing high strategic alignment efficacy were presented, and these pointed to themes such as inadequate strategic capabilities to orientate ICT-enabled change, lack of senior executive support for ICT, stifling and complex organization structures, low participatory engagement, misaligned investment controls, and resistance to change. In terms of the last point, it is not so much that public organizations resist change as it is that they fail to fully interpret the intertwined strategic, structural, and socially systemic complexities tied to leading ICT-enabled transformation.

In aligning strategy and ICT, a comprehensive approach to ICT strategy-making is essential. This chapter set out at a high level only some of the process activities to select, develop, and execute appropriately aligned ICT initiatives. Any such strategizing activity must identify the issue, thoroughly diagnose the underlying causes, conceive ideas to overcome challenges, make astute selection choices, and then go about executing the plan to deliver in full that which was intended.

This chapter highlighted that strategic alignment in public organizations has received only sporadic scholarly interest in comparison to studies of the construct in a corporate context. To further expand knowledge within a public context, some suggestions were made that could potentially lead to new insights into this particular domain of interest.

Key Terms and Definitions

Alignment: process of conducting relevant activities that continuously adjust core organizational elements to realize the most optimum position, for example, ICT-enabled transformation.

E-government: use by government organizations of ICT with the enabling capability to transform interaction with citizens, businesses, and other government-related agencies.

ICT: ICTs associated with the development and management of the computer and networking aspects of information systems.

ICT management: efficient and effective development and management of ICT resources in support of the major organizational needs and core priorities.

ICT strategy: major ICT plans and activities enabling transformation deemed supportive and challenging to an organization's core mission and purpose.

ICT strategy-making: deliberate and emergent processes enacted during the planning and execution of high-level organizational ICT initiatives intended to realize performance benefits.

Public organization: state-run organizations owned or controlled by a government and paid for by revenues collected from taxation.

Strategic alignment: congruent organizational action to manipulate ICT toward high performance.

Strategy: organizational planning and execution of the necessary actions to achieve high-level aims aligned to its core purpose and long-term direction.

Strategy process: multilevel tasks and activities to formulate and implement plans with the intended purpose and aim to achieve a favorable outcome.

Author Biographies

Dr. Eamonn Caffrey holds BSc and MA degrees in business and information technology (IT). He was awarded a PhD degree by the University of Dublin in 2013. His doctoral thesis focused on the roles of middle managers in aligning business and IT strategies in public organizations. He has extensive experience working with multinational corporations and small-medium enterprises, both in Ireland and internationally. Current interests include research, teaching, consulting, writing, and publishing. Research foci are concerned with strategic IT alignment and business performance; corporate-level issues in relation to strategy and technology; and facilitating management teams to develop and execute IT strategy. This includes maximizing business value from IT investments, building an IT resource capability, and strengthening managerial strategic agency. He currently teaches in the subject areas of strategic management, and strategy and leadership, at the Dublin Institute of Technology.

Joe McDonagh is an associate professor of business at the School of Business, Trinity College Dublin, where he also serves as the director of doctoral studies.

He holds a BSc degree in computer science and MA degree from the University of Dublin, an MBS degree in strategic management and information systems from the National University of Ireland, and a PhD degree in organization development and change from the University of Warwick. He also holds FBCS and CITP qualifications.

His research, teaching, and advisory work focus on the process of leading large-scale change across complex systems of organizations, particularly change enabled by modern technological systems. He has extensive international experience and advises executive and technology leaders in governments and large corporations on the effective integration of strategic, organizational, and technological change. Some recent government assignments include Ireland, United Kingdom, United States of America, and the United Nations, while corporate assignments include large private sector organizations.

He is a regular conference speaker and contributor to the media while also publishing widely on the dynamics of large-scale change. Research output has appeared in a wide range of conference papers, book chapters, journal articles, and government reports. Specific journals include *Administration, Communications of the AIS, Irish Journal of Management, Organization Development Journal, Public Administration Quarterly*, and *Strategic Change.*

Prior to joining the Trinity College, he spent almost two decades in industry in a number of management and consulting roles with Continental, Imperial Chemical Industries, Philips, and Price Waterhouse.

In between researching, teaching, and consulting, he devotes himself to hiking, biking, and fundraising for Dublin's homeless.

He lives in Dublin with his wife Majella and two sons Colin and Sean.

References

Ackoff, R. L. (1971). Towards a system of systems concepts. *Management Science, 17*(11), 1–15.

Ackoff, R. L. (1999). *Re-creating the corporation: A design of organizations for the 21st century.* New York: Oxford University Press.

Arvidsson, V., Holmstrom, J., & Lyytinen, K. (2014). Information systems use as strategy practice: A multi-dimensional view of strategic information system implementation and use. *Journal of Strategic Information Systems, 23*, 45–61.

Bizzi, L., & Langley, A. (2012). Studying processes in and around networks. *Industrial Marketing Management, 41*, 224–234.

Bloch, M., Blumberg, S., & Laartz, J. (2012). *Delivering large-scale IT projects on time, on budget, and on value*. New York: McKinsey & Company, October 1–6. Retrieved July 30, 2014, from http://www.mckinsey.com/insights/business_technology/delivering_large-scale_it_projects_on_time_on_budget_and_on_value.

Breyne, P., Cumps, B., & Viaene, S. (2005). *European survey on ICT value management: Strategic alignment alone is not enough*. Belgium: PriceWaterhouseCoopers.

Broadbent, M., & Weill, P. (1993). Improving business and information strategy alignment: Learning from the banking industry. *IBM Systems Journal, 32*(1), 162–179.

Bryson, J. M., Berry, F. S., & Yang, K. (2010). The state of strategic management research: A selective literature review and set of future directions. *The American Review of Public Administration, 40*(5), 495–521.

Burn, J. M. (1996). IS innovation and organizational alignment—A professional juggling act. *Journal of Information Technology, 11*, 3–12.

Bush, M., Lederer, A. L., Li, X., Palmisano, J., & Rao, S. (2009). The alignment of information systems with organizational objectives and strategies in health care. *International Journal of Medical Informatics, 78*(7), 446–456.

Caffrey, E. (2013). *Exploring the roles of ICT middle managers in aligning business and ICT strategies—The case of information and communication technology in public service organizations* (Unpublished Doctoral Thesis). Trinity College Dublin, Dublin.

Caffrey, E., & McDonagh, J. (2009). *A longitudinal perspective on critical issues in the management of information systems*. Paper presented at the Proceedings of the 1st International Conference on Business Innovation and Information Technology, Dublin, Ireland.

Campbell, B., Kay, R., & Avison, D. (2005). Strategic alignment: A practitioner's perspective. *Journal of Enterprise Information Management, 18*(5/6), 653–664.

Chague, V. (1996). Gérer la Technologie dans les PME. *Direction et Gestion des Entreprises, 157*, 13–21.

Chan, Y. E., Huff, S. L., Barclay, D. W., & Copeland, D. G. (1997). Business strategy orientation, information systems orientation and strategic alignment. *Information Systems Research, 8*(2), 125–150.

Chan, Y. E., & Reich, B. H. (2007a). IT alignment: An annotated bibliography. *Journal of Information Technology, 22*, 316–396.

Chan, Y. E., & Reich, B. H. (2007b). IT alignment: What have we learned? *Journal of Information Technology, 22*, 297–315.

Chandler, A. D. (1962). *Strategy and structure: Chapters in the history of the American industrial enterprise*. Cambridge, MA: MIT Press.

Coursey, D., & Norris, D. (2008). Models of e-government: Are they correct? An empirical assessment. *Public Administration Review, 68*, 523–536.

Croteau, A., & Bergeron, F. (2001). An information technology trilogy: Business strategy, technological deployment and organizational performance. *Journal of Strategic Information Systems, 10*(2), 77–99.

De Wit, B., & Meyer, R. (2002). *Strategy: Process, content, context—An international perspective* (2nd ed.). London: Thomson Learning.

De Wit, B., & Meyer, R. (2005). *Strategy synthesis: Resolving strategy paradoxes to create competitive advantage* (2nd ed.). London: South-Western Cengage Learning.

Dufner, D., Holley, L. M., & Reed, B. J. (2003). Strategic information systems planning and U.S. county government. *Communications of the Association for Information Systems, 11*, 219–243.

Earl, M. J. (1993). Experiences in strategic information systems planning. *MIS Quarterly, 17*(1), 1–24.
Edwards, J. (1994). The study of congruence in organizational behavior research: Critique and a proposed alternative. *Organizational Behavior and Human Decision Processes, 58*(1), 51–100.
Fedorowicz, J., Gelinas, U. J., Jr., Gogan, J. L., & Williams, C. B. (2009). Strategic alignment of participant motivations in e-government collaborations: The internet payment platform pilot. *Government Information Quarterly, 26*(1), 51–59.
Feld, C. S., & Stoddard, D. B. (2004). Getting IT right. *Harvard Business Review, 82*(2), 72–79.
Feldman, M. S. (2004). Resources in emerging structures and processes of change. *Organization Science, 15*(3), 295–309.
Floyd, S. W., Cornelissen, J. P., Wright, M., & Delios, A. (2011). Processes and practices of strategizing and organizing: Review, development, and the role of bridging and umbrella constructs. *Journal of Management Studies, 48*(5), 933–952.
Ghosh, B., & Scott, J. E. (2009). Relational alignment in offshore IS outsourcing. *MIS Quarterly Executive, 8*(1), 19–29.
Government Accountability Office. (2004). *Information technology management (GAO-04-49)*. Washington, DC: Government Accountability Office.
Gupta, Y. P., Karimi, J., & Somers, T. M. (1997). Alignment of a firm's competitive strategy and information technology management sophistication: The missing link. *IEEE Transactions on Engineering Management, 44*(4), 399–413.
Hartung, S., Reich, B. H., & Benbasat, I. (2000). Information technology alignment in the Canadian forces. *Canadian Journal of Administrative Sciences, 17*(4), 285–302.
Harvey Nash. (2013). CIO survey. *Harvey Nash CIO Survey 2013*, 1–55.
Henderson, J. C., & Sifonis, J. G. (1988). The value of strategic IS planning: Understanding consistency, validity, and IS markets. *MIS Quarterly, 12*(2), 187–200.
Henderson, J. C., & Venkatraman, N. (1993). Strategic alignment: Leveraging information technology for transforming organizations. *IBM Systems Journal, 32*(1), 4–16.
Hendry, J., & Seidl, D. (2003). The structure and significance of strategic episodes: Social systems theory and the routine practices of strategic change. *Journal of Management Studies, 40*(1), 175–196.
Holley, L. M., Dufner, D., & Reed, B. J. (2004). Strategic information systems planning in U.S. county governments: Will the real SISP model please stand up? *Public Performance & Management Review, 27*, 102–126.
Hsu, F., Hu, P. J., & Chen, H. (2006). *Examining the business-technology alignment in government agencies: A study of electronic record management systems in Taiwan*. Paper presented at the Proceedings of the Tenth Pacific Asia Conference on Information Systems, Kuala Lumpur, Malaysia.
Huang, C. D., & Hu, Q. (2007). Achieving IT-business strategic alignment via enterprise-wide implementation of balanced scorecards. *Information Systems Management, 24*, 173–184.
Huang, J., Newell, S., Huang, J., & Pan, S. (2014). Site-shifting as the source of ambidexterity: Empirical insights from the field of ticketing. *Journal of Strategic Information Systems, 23*, 29–44.
Huber, G. P. (1984). The nature and design of the post-industrial organization. *Management Science, 30*(8), 928–951.

International Business Machines Corporation. (1981). *Business systems planning. Information systems planning guide*. Application Manual GE20-0527-2. Armonk, NY: IBM Corporation.
Leavitt, H. J. (1965). Applied organizational change: Structural, technological, and humanistic approaches. In J. March (Ed.), *Handbook of organizations*. Chicago: Rand McNally Publishing.
Lederer, A. L., & Gardiner, V. (1992). Strategic information systems planning: The Method/I approach. *Information Systems Management, 9*, 13–20.
Leifer, R. (1988). Matching computer-based information systems with organizational structures. *MIS Quarterly, 12*(1), 63–73.
Loonam, J., McDonagh, J., Kumar, V., & O'Regan, N. (2014). Top managers and information systems: Crossing the Rubicon. *Strategic Change, 23*, 205–224.
Luftman, J., & Ben-Zvi, T. (2010a). Key issues for IT executives 2010: Judicious IT investments continue post-recession. *MIS Quarterly Executive, 9*(4), 263–273.
Luftman, J., & Ben-Zvi, T. (2010b). Key issues for IT executives 2009: Difficult economy's impact on IT. *MIS Quarterly Executive, 9*(1), 49–59.
Luftman, J., & Brier, T. (1999). Achieving and sustaining business-IT alignment. *California Management Review, 42*(1), 109–122.
Luftman, J., & Kempaiah, R. (2007). An update on business-IT alignment: "A line" has been drawn. *MIS Quarterly Executive, 6*(3), 165–177.
Luftman, J., & Kempaiah, R. (2008). Key issues for IT executives 2007. *MIS Quarterly Executive, 7*(2), 99–112.
Luftman, J., Kempaiah, R., & Rigoni, E. H. (2009). Key issues for IT executives 2008. *MIS Quarterly Executive, 8*(3), 151–159.
Maes, R., Rijsenbrij, D., Truijens, D., & Goedvolk, H. (2000). *Redefining business-IT alignment through a unified framework*. Amsterdam: University of Amsterdam.
Martin, J. (1989). *Information engineering: Book 1—Introduction*. Englewood Cliffs, NJ: Prentice-Hall.
McDonagh, J. (in press). Why change programmes don't produce change: The case of IT-enabled change in public service organizations. In S. Gao & L. Rusu (Eds.), *Modern techniques for successful IT project management*. Hershey, PA: IGI Global.
McDonagh, J., & Coghlan, D. (2006). Information technology and the lure of integrated change: A neglected role for organization development? *Public Administration Quarterly, 30*(1), 22–55.
McDonagh, J., & Coghlan, D. (2010). Information technology and the lure of integrated change: A neglected role for organization development. In D. Coghlan & A. B. Shani (Eds.), *Fundamentals of organization development, volume IV, organization development—Emerging issues and challenges* (pp. 383–401). London: Sage Publications.
Nadler, D. A., & Tushman, M. L. (1980). A model for diagnosing organizational behavior. *Organizational Dynamics, 9*(2), 35–51.
Nelson, R. R. (2007). IT project management: Infamous failures, classic mistakes, and best practice. *MIS Quarterly Executive, 5*(2), 67–78.
Newkirk, H. E., Lederer, A. L., & Srinivasan, C. (2003). Strategic information systems planning: Too little or too much? *Journal of Strategic Information Systems, 12*, 201–228.
Oh, W., & Pinsonneault, A. (2007). On the assessment of the strategic value of information technologies: Conceptual and analytical approaches. *MIS Quarterly, 31*(2), 239–265.
Orlikowski, W. J. (1996). Improvising organizational transformation over time: A situated change perspective. *Information Systems Research, 7*(1), 63–92.

Osterle, H., Brenner, W., & Hilbers, K. (1993). *Total information systems management: A European approach*. Chichester: John Wiley & Sons.

Parker, M. M., & Benson, R. J. (1989). Enterprise wide information management: State-of-the-art strategic planning. *Journal of Information Systems Management, 6*, 14–24.

Peak, D., Guynes, C. S., & Kroon, V. (2005). Information technology alignment planning—A case study. *Information & Management, 42*, 635–649.

Polites, G. L., & Karahanna, E. (2013). The embeddedness of information systems habits in organizational and individual level routines: Development and disruption. *MIS Quarterly, 37*(1), 221–245.

Pyburn, P. J. (1983). Linking the MIS plan with corporate strategy: An exploratory study. *MIS Quarterly, 7*(2), 1–14.

Rainey, H. G. (2009). *Understanding and managing public organizations* (4th ed.). San Francisco: Jossey-Bass.

Raskino, R. (2011). *Overview: Gartner's CEO and board survey research 2011*. Gartner Report No. G00211941. Stamford, CT: Gartner.

Reich, B. H., & Benbasat, I. (1996). Measuring the linkage between business and information technology objectives. *MIS Quarterly, 20*(1), 55–81.

Reich, B. H., & Kaarst-Brown, M. L. (2003). Creating social and intellectual capital through IT career transitions. *Journal of Strategic Information Systems, 12*(2), 91–109.

Ross, J. W., & Weill, P. (2002). Six IT decisions your IT people shouldn't make. *Harvard Business Review, 80*(11), 84–91.

Sabherwal, R., & Chan, Y. E. (2001). Alignment between business and IS strategies: A study of prospectors, analyzers, and defenders. *Information Systems Research, 12*(1), 11–33.

Sabherwal, R., & Kirs, P. (1994). The alignment between organizational critical success factors and information technology capability in academic institutions. *Decision Sciences, 25*(2), 301–330.

Sauer, C., Southon, G., & Dampney, C. N. G. (1997). *Fit, failure, and the house of horrors: Toward a configurational theory of IS project failure*. Paper presented at the International Conference on Information Systems, Proceedings of the Eighteenth International Conference on Information systems, Atlanta, GA.

Scott-Morton, M. S. (1991). *The corporation of the 1990s: Information technology and organizational transformation*. New York: Oxford University Press.

Shpilberg, D., Berez, S., Puryear, R., & Shah, S. (2007). Avoiding the alignment trap in information technology. *Sloan Management Review, 49*(1), 51–58.

Silva, L., Figueroa, E., & Gonzalez-Reinhart, J. (2007). Interpreting IS alignment: A multiple case study in professional organizations. *Information and Organizations, 17*, 232–265.

Silva, L., & Hirschheim, R. (2007). Fighting against windmills: Strategic information systems and organizational deep structures. *MIS Quarterly, 31*(2), 327–354.

Silvius, A. J. G. (2007). *Aligning IT strategy to business strategy in a multi business company*. Paper presented at the America's Conference on Information Systems (AMCIS), Denver, CO.

Silvius, A. J. G. (2009). *Business and IT alignment: What we know and what we don't know*. Paper presented at the International Conference on Information Management and Engineering, Utrecht, Netherlands.

Silvius, A. J. G. (2010). A conceptual model for aligning IT valuation methods. *International Journal of IT/Business Alignment and Governance, 1*(3), 36–54.

Sommer, M. (2002). Quo Vadis, e-government? *Journal of Political Marketing, 1*, 233–236.

Tallon, P. (2007–2008). A process-orientated perspective on the alignment of information technology and business strategy. *Journal of Management Information Systems, 24*(3), 227–268.

Tarafdar, M., & Qrunfleh, S. (2009). IT-business alignment: A two-level analysis. *Information Systems Management, 26*(4), 338–349.

Tavakolian, H. (1989). Linking the information technology structure with organizational competitive strategy: A survey. *MIS Quarterly, 13*(3), 309–317.

Tech America. (2013). CIO insights: Leading innovation in a time of change. *Tech America's Twenty-Third Annual Survey of Federal Chief Information Officers*, 1–28.

Ulrich, D. (1998). Intellectual capital = competence × commitment. *Sloan Management Review, 39*(2), 15–26.

Wang, E. T. G., & Tai, J. C. F. (2003). Factors affecting information systems planning effectiveness: Organizational contexts and planning system dimensions. *Information and Management, 40*, 287–303.

Ward, J., & Peppard, J. (2002). *Strategic planning for information systems* (3rd ed.). Chichester: John Wiley & Sons.

Wijnhoven, F., Spil, T., Stegwee, R., & Fa, R. T. A. (2006). Post-merger IT integration strategies: An IT alignment perspective. *Journal of Strategic Information Systems, 15*, 5–28.

Yang, K., & Melitski, J. (2007). Competing and complementary values in information technology strategic planning. *Public Performance & Management Review, 30*, 426–452.

II
TOPICAL ISSUES IN PUBLIC ADMINISTRATION AND INFORMATION TECHNOLOGY INNOVATION

Chapter 8

An Analysis of Failure in a Government-to-Government E-Government Context via the Updated Delone and McLean Model

Lies Van Cauter, Monique Snoeck, and Joep Crompvoets

Contents

Abstract

Implementation of e-government systems lags behind policy ambitions, and rhetoric of reform. Failure is omnipresent. In the context of information systems (IS), failure can be defined as an implemented system not meeting the user expectations (Ewusi-Mensah, 2003). In order to ensure future e-government system success, it might be useful for IS managers to assess failure, and base action on these assessments. One way to do so is by using the Updated Delone and McLean (D&M) model. This model has been widely investigated in the private sector, and a few studies have used it to assess government-to-government (G2G) e-government systems. This chapter aims to contribute to this research gap.

The transferability and the completeness of the Updated D&M model are tested by using a multiple case study of two failed intergovernmental IS. The case results show that the model is transferable to the G2G e-government domain. Nevertheless, the model is incomplete in detecting the causes of failure of the studied IS. It did not detect a lack of trust, path dependency on IS infrastructure, insufficient organizational competence, unfulfilled user expectations, and a lack of local management support. Future research might further explore and test the completeness of factors of the Updated D&M model, and its transferability to a G2G e-government context.

Keywords: e-government, Delone McLean, G2G IS, failure

8.1 Introduction

E-government induces the promise to lead to a better government. It could do so by, for example, stimulating efficiency and effectiveness, improving interorganizational collaboration, or enabling time, money, and personnel savings (Gil-Garcia, Chengalur-Smith, & Duchessi, 2007; Homburg & Bekkers, 2002).

Evaluations show that the implementation of e-government systems lags behind policy ambitions and the rhetoric of reform. E-government systems are difficult to monitor and prevent from failure. Managers try to streamline processes and data and search for consensus and aim to progress information streams in or between governments. They nevertheless face multiple challenges on the way toward successful systems (Rana, Dwivedi, & Williams, 2013). Technology evolves quickly, whereas implementation takes time. Goals, workflows, and data definitions of back offices do not necessarily overlap. Implementing e-government systems might shift power balances and may therefore be sabotaged (Homburg & Bekkers, 2002). What distinguishes information systems (IS) failure from other failures in the public sector is its overwhelming ubiquity (Goldfinch, 2007).

When trying to research public sector IS failure, several problems pop up.

First, a common definition of *failure* is lacking. Assessing the success/failure of e-government initiatives is a challenge, as the definition of outcomes is ambiguous. Success or failure is a question of judgment, representing different viewpoints of groups in (an) organization(s) (Elbanna & Linderoth, 2013). These judgments can change over time (Goldfinch, 2007). In the context of IS, failure can be defined as an implemented system not meeting user expectations (Dwivedi et al., 2014). In particular, we consider success and failure as the level to which system acceptance, usage, and experienced benefits meet or do not meet the expectations that motivated the development or acquisition of the software.

Second, it is difficult to define what exactly *e-government* is; there is no standard definition (Yildiz, 2007). E-government can be seen as an application of Internet in the government, as a goal in itself, as a tool for achieving public sector reform goals, or as an application of information and communication technology in the public sector to make the government work better (Heeks, in Yildiz, 2007). We share the latter view. This refers not only to the information and communication technology that an organization uses but also to the way people interact with it in support of business processes (Beynon-Davies, 2009). E-government research may be divided into four main domains: government to citizen (G2C), government to employee, government to business (G2B), and government to government (G2G) (Floropoulos, Spathis, Halvatzis, & Tsipouridou, 2010). Three additional domains could be added: government to civil societal organizations, citizen to government, and citizen to citizen (Yildiz, 2007).

Third, digital information sharing between governmental organizations remains elusive: Little is known about when and how efforts are likely to be successful (Chen, 2008). So far, the complex nature of challenges is not well understood, either in practice or in theory (Goldfinch, 2007; Scholl & Klishewski, 2007). Researchers into IS failure should devote more attention to the public sector (Dwivedi et al., 2014). To ensure success, more research is required to develop methods to assess and act upon failure. In e-government research, no standard methods to do so exist (Makedon, Sudborough, Pantziou, & Conalis-Kontos, 2003).

In the last three decades, IS success models have been widely investigated in the private sector. Although surveys have shown that many managerial-level challenges are similar for public and private IS, few studies have assessed e-government systems (Floropoulos et al., 2010; Scott, DeLone, & Golden, 2009; Wang & Liao, 2008). This chapter aims to contribute to this research gap.

Because many success models have been proposed and the e-government domain itself takes different forms, we narrow down the scope of the research to one IS success model and one domain of e-government studies: G2G. We will use the Updated DeLone and McLean (D&M) model (Figure 8.1), as it is the most commonly used and cited IS success model (Abdelsalam, Reddick, & El Kadi, 2012; Gable, Sedara, & Chan, 2008). This is further justified in Section 8.2.

The Updated D&M model has not been used much in e-government studies yet and especially not in G2G studies. Some e-government researchers use the model to focus on citizen adoption of governmental IS (G2C) (e.g., Chai, Herath, Park, & Rao, 2006; Gotoh, 2009; Rana et al., 2013; Scott et al., 2009; Sørum, Medaglia, Andersen, Scott, & DeLone, 2012; Thompson, Srivastava, & Jiang, 2008; Wang & Liao, 2008). Other studies take an employee perspective (government to employees) to gain insight on IS system success/failure and usage via the Updated D&M model (e.g., Abdelsalam et al., 2012; Floropoulos et al., 2010; Hussein, Karim, & Selamat, 2007; Hsu & Chen, 2007; Prybutok, Zhang, & Ryan, 2008). An example of a G2B study that uses the D&M model is that of Sambasivan, Wemyss, and Rose (2010). Gil-Garcia et al. (2007) focused on interorganizational collaboration with both business and government partners (G2B and G2G) but did not use the Updated D&M model as research framework. To narrow down the scope, we limit our focus to digital collaboration between governments (G2G) because this seems the area that has not been addressed before.

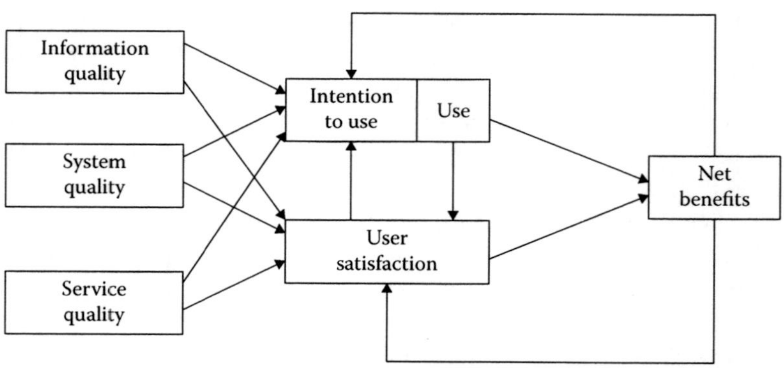

Figure 8.1 Updated information systems model. (From DeLone, W. H., & McLean, E. R. (2003), *Journal of Management of Information Systems*, 19(4), 9–30.)

We use the Updated D&M model in a basic explorative study to assess two e-government systems in Flanders. The efficient implementation of governmental IS is an explicit policy goal of the Flemish government (Belgium). The two IS are considered as failed, according to policy notes that they did not meet user expectations. As we use an IS success model to assess failure, it is necessary to clarify the "opposite" effect of the factors of the Updated D&M model. This means if the presence of a factor encourages success, the lack of it encourages failure (Gichoya, 2005). In this line of reasoning, it makes sense to assess causes of IS failure along the dimensions of a success model. The more we learn about these causes, the more we can aid managers in their quest for IS success (Petter, Delone, & McLean, 2013). As such, the main research questions are as follows:

- RQ1. Can the Updated D&M model be used to assess the causes of failure of two cases in a G2G e-government context?
- RQ2. If so, how complete is this model for assessing the causes of failure of two G2G e-government cases?

The rest of the chapter is structured as follows: Section 8.2 motivates our choice for the Updated D&M model. Section 8.3 describes the methodology. Section 8.4 investigates whether the Updated D&M model can assess the causes of failure of two e-government cases and, if so, how complete this assessment is. Results are discussed in Section 8.5. We conclude in Section 8.6.

8.2 The Updated D&M Model

Researchers have developed numerous models to look at IS success/failure (Dörr, Walther, & Eymann, 2013; Rana et al., 2013). Some widely discussed models are the Technology Acceptance Model (Davis, 1989), the Unified Theory of Acceptance and Use of Technology (Venkatesh, Morris, Davis, & Davis, 2003), the (Updated) IS success model (Delone & McLean, 2003), and the variance model of IS success (Seddon, 1997).

The Technology Acceptance Model and Unified Theory of Acceptance and Use of Technology focus mainly on acceptance of IS at a personal level, (non)acceptance is not the equivalent to success/failure, but a precondition. Delone and McLean's IS success model and Seddon's variance model take a broader perspective; they aim to explain IS success/failure. Ten years after Delone and McLean launched their model, they made an update. Today, the majority of IS researchers has switched to the updated version of the D&M model. The (Updated) D&M model is cited in thousands of publications and validated in more than 300. The Seddon model is much less investigated; we will use the D&M model in this chapter (Petter, DeLone, & McLean, 2008; Sørum et al., 2012).

The Updated D&M model contains of six main factors:

1. *System quality* measures the quality of information processing within the system. It looks at desirable characteristics of an IS: ease of use, integration and learning, system flexibility and reliability.
2. *Information quality* focuses on IS output and desirable characteristics of system outputs such as relevance of information, understandability, added value, data security, and completeness.
3. *Service quality* is the quality of system support that users get from the information technology (IT) department, such as responsiveness, empathy, communication, accuracy, or technical competence of staff.
4. *Intention to use* and *use* measure the user attitude. Use is seen as a behavior, the manner in which staff and customers use the capabilities of an IS, for example, amount and frequency of use, extent, and purpose of use.
5. *User satisfaction* describes the user's level of satisfaction with the IS.
6. *Net benefits* is the extent to which an IS contributes to the success of the individuals who use the system, for example, improved decision making, productivity, and efficiency (Dörr et al., 2013; Gable et al., 2008; Moges, Dejaeger, Lemahieu, & Baesens, 2013; Prybutok et al., 2008).

The Updated D&M model sees success/failure as brought about by causally linked factors. Underlying is the assumption that IS success/failure can be identified by the presence or absence of certain factors (Kautz & Cecez-Kecmanovic, 2013). The presence or absence of system, information, and service quality influences the intention to use an IS, the actual use, and the user satisfaction. These factors, in turn, influence the experienced net benefits. The presence/absence of net benefits for users impacts the further intention to use and user satisfaction. Finally, user satisfaction influences the intention to use, whereas use influences user satisfaction. In light of this model, the definition of IS success as given in the Introduction amounts to contrasting the actual usage and net benefits against the expected usage and benefits formulated by the client of the IS.

By using the Updated D&M model to reflect on success and failure, the appreciation of causes of failure will largely reflect the users' perceptions, given the central position of intention to use, user satisfaction, and net benefits in the model.

8.3 Methodology

8.3.1 Research Methodology

To investigate intergovernmental IS failure, two possible approaches can be considered. On the one hand is a bottom-up, inductive approach to identify the root causes of intergovernmental IS failure in different cases; see, for example, Lehtinen, Mäntylä, Vanhanen, Itkonen, and Lassenius (2014). These can be generalized to a theory for intergovernmental IS failure. Starting from scratch holds the disadvantage that many cases need to be studied to obtain a valid and reliable theory. On the other hand, we can follow a top-down, deductive approach to examine whether (proven) IS success models resulting mainly from research in private sector IS can be extended to examine G2G IS failure (Al Khatib, 2013; Scott et al., 2009).

We take a deductive approach to answer the first research question. We will test whether the Updated D&M model can be transferred to assess the perceived causes of failure of two G2G IS. An inductive approach will be conducted to answer the second research question. We will do a check of completeness of the Updated D&M model for assessing the perceived causes of failure of two G2G IS. The collected data are coded in NVivo to identify possible causes of failure. NVivo is a digital software tool that helps a researcher to connect, organize, and analyze qualitative data in depth. We will then eliminate the causes covered by the Updated D&M model and look whether additional causes are identified. An exploratory case study approach is an appropriate way to study both research questions. We study two cases to develop a more in-depth understanding of the phenomena than a single case could provide (Mills, Durepos, & Wiebe, 2010). This approach is well suited to IS research because it is helpful in identifying the causal chain that led to the success/failure of an IS (Benbasat, Goldstein, & Mead, 1987). The approach is suitable to study an area in which few previous studies have been carried out and to generate theories from practice. Inevitably, an element of subjectivity is involved in nonrandom sampling (Yin, 1994).

8.3.2 Selected Cases

In 2012, we created an inventory of Flemish intergovernmental IS because an overview was lacking (Van Cauter et al., 2013). From this inventory of 100 intergovernmental IS in Flanders, the cases of the Road Sign Database (RSD) and the Environmental Complaints Registration and Monitoring System (ECREMOS) were selected because of the following:

- They are operational for at least 5 years, and the added value of an IS reveals itself only after a number of years.
- The Flemish government asked municipalities to deliver data. How municipalities could be convinced to do so is far from evident and is dependent highly on the user's perspective of (the utility of) a system.
- Scientifically little is known about the voluntary use of systems (Elbanna & Linderoth, 2013).
- The cases showed interesting similarities: First, in both cases, project leaders of a troubled unpopular IS wanted to turn a failure into a success by renewing the IS. Second, both systems are for free. Third, the cases have the same kind of actors involved: A Flemish department created an IS and asked municipalities and another Flemish agency of the same policy domain to keep this up-to-date. Fourth, in both cases, the police might take up part of the municipal task.

8.3.2.1 The Road Sign Database

The RSD contains all road signs, their characteristics, and road positions. The opportunity to launch this IS politically was a huge traffic obstruction in Bruges. A bridge was hit by a truck as there was no road sign about the bridge's height. The Flemish department for Mobility and Public Works created the database and inventoried the road signs. It then asked its 308 municipalities (for the municipal roads) and the Flemish Agency for Roads and Traffic (for the regional roads) to keep the database up-to-date. However, they do not (Vlaams Parlement, 2008).

8.3.2.2 The Environmental Complaints Registration and Monitoring System

The Flemish environmental department created ECREMOS (MKROS in Dutch) because a central digital overview of environmental complaints was lacking. This IS contains complaints of Flemish municipalities that signed an agreement with the Flemish government. The police and the Flemish environmental inspection were asked to use ECREMOS. Environmental complaints cover odor nuisance, light/noise/soil, and air pollution. Registration in this IS was one of a series of environmental policy criteria: If municipalities met these criteria, they were rewarded financially. The agreement ended in December 2013. In 2014, municipalities were asked to keep ECREMOS up-to-date without financial rewards. The Flemish government assumed that ECREMOS was sufficiently accepted. However, the use of this IS dropped severely.

8.3.3 Data Collection

Researchers' bias can be counteracted by using multiple sources (Benbasat et al., 1987). We employed a multiple data collection method during our case studies: The sequence of events of the cases is reconstructed via legislation, policy documents, and interviews. The implementation of an IS typically takes place over a longer period of time, and it involves multiple actors and may be influenced by unexpected events. A case study methodology is therefore well suited to identify key events and actors and for linking them in a causal chain.

For both cases, the same 130 municipalities were questioned by the same researcher. They were asked about their system usage, their satisfaction with it, and if they wanted to change anything. We first interviewed 20 municipalities face-to-face. We noticed that respondents did not use the system or seemed not very fond of it. Was this also the case for other municipalities? We additionally questioned 110 Municipal Civil Servants by telephone with standardized open questions. We interviewed the actual users of the IS: mobility civil servants for the RSD and environmental civil servants for ECREMOS. Overall, we interviewed more than one third of the Flemish municipalities. Although the results could be slightly different for the whole population, we ensured that the chosen municipalities are representative in size and geographical distribution. At Flemish level, the project leaders were questioned face-to-face. Data collection was finished by February 2014. All interviews were transcribed and analyzed in NVivo.

To prevent being influenced by a theoretical lens, we explored the cases with open neutral questions; for example, municipalities were asked how frequently the IS was used. If they did not (often) use it, we asked why and if they employed any alternatives. Nonusers were asked if they desired to get (back) on board. Municipalities who used the system were questioned for what purpose and if they kept it up-to-date. Future intention to use was registered too. Table 8.1 summarizes the (intention to) use. Use of the RSD was totally voluntary, whereas use of ECREMOS became voluntary only in 2014.

In 2013, 38 of the 130 questioned municipalities used the RSD. An improved version of the database was launched in March 2013. The foresight of this improvement encouraged 23 other municipalities to use the database. At the moment our interviews were finished, 32 municipalities were still undetermined whether they would start to use a renewed RSD. Their future use is coupled to long-term conditions, for example, "if they find time" or "once they retire their successor probably will use the RSD."

In 2013, 114 of the 130 questioned municipalities used ECREMOS. An improved version of the IS will be launched around September 2014. Of the questioned municipalities, 44 plan to use ECREMOS in 2014. If the Environmental Department can convince the police and Flemish environmental inspection to use ECREMOS in 2014, 65 municipalities intend to use ECREMOS. At the moment our interviews were finished, around 24 municipalities were still undetermined. Their future use is

Table 8.1 (Intention to) Use of RSD and ECREMOS

	RSD	*ECREMOS*	
Municipal usage, 2013	38/130	114/130	
Degree of voluntariness	Voluntary	Mandatory until December 2013 via a cooperation agreement	
Intention to use, 2014		Without police and environmental inspection	With police and environmental inspection
Use	63/130	44/130	65/130
Nonuse	35/130	67/130	41/130
Undetermined	32/130	19/130	24/130

Note: ECREMOS = Environmental Complaints Registration and Monitoring System; RSD = Road Sign Database.

coupled to conditions, for example, "they will use it if they do not develop an own system," "if the municipal council agrees," or "if they are happy with the changes."

In IS research, use is seen as a precursor of success (Petter et al., 2008). In 2014, a maximum half of the questioned municipalities really intend to use the IS under study. What went wrong?

8.4 Data Analysis

8.4.1 The Updated D&M Model and the RSD

We analyze the collected data by using the Updated D&M model as theoretical lens. To appreciate the level to which a factor contributes to success/failure, we relied on how often a factor was spontaneously mentioned and the proportion of these responses that indicate an aspect as problematic or not. When the large majority of respondents indicated that they were dissatisfied about a certain factor, we scored it as low. Because the six main factors are quite general, labeling them as causes of failure would not be very informative. For each factor that seems to score low, we therefore identified the main underlying causes for that factor. Table 8.2 provides an overview. Each factor is concluded with the identification of the main underlying causes of failure (abbreviated C1.1 to C1.6).

Analysis of the factors suggests a negative influence on net benefits. The majority of municipalities does not experience the IS as a tool to make their work more efficient. They struggle with technical issues (C1.1) and experience a lack of end-user support (C1.3). Time investments (C1.2) appear to outweigh net benefits. A lack of net benefits appears to be caused mainly by a lack of purpose (C1.4), the option for cheap and quick alternatives (C1.5), and a bad reputation (C1.6). By analyzing the data along the factors of D&M, six main causes of failure could be detected. The Updated D&M model not only gives insight in six causes, it also seems to point out the consequences of this failure: poor information quality and low actual use.

8.4.2 The Updated D&M Model and the ECREMOS

We follow the same procedure to analyze the causes of failure of the ECREMOS. Table 8.3 identifies the causes (abbreviated C2.1 to C2.6) per factor.

By analyzing data along the factors of D&M, six main causes of failure could be detected. The causal interaction of these factors points to a negative influence on net benefits; ECREMOS scored low on system and information quality (C2.1–C2.5). This negatively influences (intention to) use and user satisfaction. A low intention to use caused by a lack of purpose and a low user satisfaction because of the absence of feedback (C2.6) made ECREMOS unpopular. However, it was used in 2013 by 114 of the 130 questioned municipalities. Because many municipalities

Table 8.2 The Updated D&M Model and the RSD

System quality	Reliability of the system appears to be low; it crashes often. Users have to log in four times to enter the database. The more users that enter, the slower the response time. Those who own a separate IS cannot upload their data. Respondents who do not use the system often have to find out how it works every time. The overall system quality is low.
	→ Technical issues (C1.1) and slow time performance (C1.2) seem to cause low system quality.
Information quality	92 of the questioned municipalities do not use the database (any more); 90 of these report their data to be outdated. Remarkably, two municipalities who do not use it for their own decision making keep it up-to-date once a year. On the other hand, one could expect that the 38 municipalities who use the database keep it regularly up-to-date. Only 12 of 38 users do. The remaining users update the data once a year. There are no alternatives; if a municipality does not keep track of its signs, no other party will. The overall information quality is low.
	→ The main underlying cause of bad information quality seems to be lack of usage.
Service quality	The Flemish government has a competent staff that supports the database. Municipalities find this staff difficult to reach when experiencing problems. A few nonusers remarked that they asked to join a training course or to receive a log-in code but never received an answer. Service quality appears to be low.
	→ A lack of technical end-user support (C1.3) seems to cause low service quality.
Intention to use	Intention to use is determined by the three previous factors and by user satisfaction and net benefits (see further). Some respondents intended to use the database until they heard how badly it was functioning. The Flemish Department for Mobility and Public Works does not have incentives to encourage updating. The police seldom used the system. The Agency for Roads and Traffic created their own database. Many politicians did not see it as a priority. Easy alternatives for gathering road sign data decreased the benefits of the database. With the foresight of a renewed database, more municipalities intend to use it. Some will use it only because they think it will be legally obliged to do so soon. We notice a low intention of voluntary use.

(Continued)

Table 8.2 (Continued) The Updated D&M Model and the RSD

	→ A lack of purpose (C1.4), cheap and quick alternatives (C1.5), and a bad reputation (C1.6) seem to cause low intention to use.
Use	Usage is determined by the three previous factors and by net benefits. Less than one third of the questioned municipalities uses the system; 7/130 do so on a monthly basis. Frequency of use is dependent partly on the number of new road signs. The overall use is low.
User satisfaction	User satisfaction is determined by the three quality factors, by use, and by net benefits (see further). Respondents who still use the database find it supportive for their mobility plans, to localize road signs, or to advice the municipal council. However, the interviews also indicate that low system quality (slow time performance) caused many users to abandon the ship. User satisfaction is low.
	→ Most municipalities who once used the system believe user satisfaction to be low because of a slow time performance (C1.2).

Note: D&M = Delone and McLean; IS = information systems; RSD = Road Sign Database.

did not experience any benefits apart from the financial reward, it is not stunning that about half of them will stop their use now that financial rewarding has ceased.

8.4.3 The Check for Completeness

To investigate the completeness of the analysis via the Updated D&M model, we conducted an inductive approach. On the basis of detailed interview observations, we could systematically develop a higher level of propositions. Underlying similarities and constant associations between interviewee statements are regrouped to more general causes. The causes detected by the model were compared with the causes discovered by coding interview data in NVivo. We additionally found three and four causes of failure for the two cases.

Three causes (abbreviated C1.7–C1.9) seemed to further determine the failing of the RSD:

- C1.7: Many municipalities remarked that the IS did not function as was pictured by the Flemish government. This disillusioned users and made two thirds turn their back on the system.

Table 8.3 The Updated D&M Model and ECREMOS

System quality	ECREMOS outgrew its technical teething problems, but most respondents describe it as a slow system with a bad search engine or too many pop-ups. The majority make updates once a year. They have to study the instructions to do so. Five municipalities upload data via xml; each time ECREMOS is updated, they reinvest in interoperability. The police refuses to invest for fear of duplication: It too receives environmental complaints but registers these in their own database.
	The system times out after 10 minutes of nonuse, and data often are lost as civil servants have to deliver front office services while updating. The system is not as complete nor flexible as hoped. Complaints cannot be pictured on geographic maps, and no attachments are allowed and the topic list does not cover all complaints. System quality appears to be low.
	→ Low system quality seems to be caused by technical issues (C2.1) and a user alienated design (C2.2).
Information quality	Registration cannot be done on the fly as complaints are reported. By the time the IS opens, complainants have left. Registration and monitoring in ECREMOS are therefore done afterward and only to receive the financial reward. Local politicians wanted the money.
	Some municipalities report all their complaints, where only others are important ones. This raises questions on reliability. Big reporters fear that benchmarks will damage their reputation.
	Municipalities do not know what happens with their data, e.g., by the Flemish environmental inspection. They also fear that politicians might use data for electoral purposes. Information quality in terms of data security is low. This seems to cause bad information quality in terms of completeness of data.
	→ A lack of purpose (C2.3), fear for a bad reputation (C2.4), and fear for data security breaches (C2.5) seem to cause a decrease in information quality.

(Continued)

Table 8.3 (Continued) The Updated D&M Model and ECREMOS

Service quality	Newsletters inform users on updates and common problems. When they experience problems, the project leader of ECREMOS responds quickly and provides technical support. ECREMOS has had only one project leader, and municipalities feel some empathy "as they know the man for years." Each time a municipality asked to expand the IS, this request was granted. Over the years, it became a ponderous system. Service quality is high.
	→ Users appreciated the system support and praised the service quality.
Intention to use	ECREMOS is a ponderous system. Mainly small municipalities paid a joint authority to fill in the data. Costs were not covered by the Flemish rewards. Small municipalities know their problem zones by heart and are satisfied when problems are solved. Cities have their own citywide complaint system, and the Flemish tool came as an afterthought. Some local governments will continue to use it because they believe it will be legally obligated soon. The police has an own database. At Flemish level, ECREMOS is used only for making figures by the environmental department. The environmental inspection does not use the IS and asks municipalities for other data for its functioning. The intention to use of some municipalities can get a boost if the police and environmental inspection would use ECREMOS. The intention to use appears to be low.
	→ A lowered intention to use seems caused by a lack of purpose (C2.3).
Use	In 2013, 114 of the 130 questioned municipalities used ECREMOS. In 2014, about half of them might stop doing so. This is influenced by the cancelled rewards. Having their own better IS or a lack of personnel capacity also plays a role. Those who use ECREMOS do so because they have put efforts into it, because they lack an alternative, or because they believe a Flemish overview is important. In 2013, use was high.

Table 8.3 (Continued) The Updated D&M Model and ECREMOS

User satisfaction	According to their own wordings, some respondents rate their satisfaction literally "below zero, because of a poor system quality." Others find the IS a useful tool to create an overview or reports for the municipal council. Some claim usage "has become a habit." The Flemish government controls municipalities: Those who do not register enough data could lose their financial reward. Those who perform well are never contacted. They, however, would like to know what the results of their efforts are. User satisfaction has dropped.
	→ Absence of feedback (C2.6) appears to decrease user satisfaction.

Note: D&M = Delone and McLean; ECREMOS = Environmental Complaints Registration and Monitoring System; IS = information systems.

- C1.8: Mainly small municipalities do not have enough personnel capacity in terms of time and number to keep the IS up-to-date. Internal quarrels can hamper use. Several municipalities who did not use the RSD claimed that the effort to catch up had become too big.
- C1.9: When large investments had been made in the past, it was difficult to take another path. Municipalities are mostly path dependent on a certain service supplier and his willingness to build applications.

Four additional causes (abbreviated to C2.7–C2.10) seem to further contribute to the failure of ECREMOS:

- C2.7: It became clear during the interviews that the Flemish government asks to use the IS but does not use it itself. Municipalities deliver data to the Flemish government. It is very hard to get data in return from the environmental inspection. Lack of transparency decreases trust of local governments.
- C2.8: Organizational capacity differs; mainly, small municipalities do not have enough personnel capacity to keep the data up-to-date. The police does not always want to use its capacity to communicate to municipalities the environmental complaints it received.
- C2.9: Municipalities are path dependent from investments made in the past. Some did not want to pay (several times) for developing an xml connection.
- C2.10: Local politicians do not see digitized environmental complaints as important; at the moment, sustainability is the hot topic. Some politicians demanded that the data were filled in to get the money.

8.5 Discussion

The Updated D&M model is a causal reasoning model for IS success/failure. The degree to which a factor is perceived positively by many users will lead to success, whereas the degree to which a factor is perceived negatively will lead to failure:

- If we evaluate these factors for RSD, we notice that it scores low on service and system quality. These negatively influence user satisfaction and (intention to) use. Combine this negative stream with a bad reputation and slow time performance and users drop out; they look for alternatives, which in turn creates a lack of purpose of the IS. The absence of net benefits will affect user satisfaction and intention to use (Delone & McLean, 2003).
- By following the causal logic of the Updated D&M model, we come to a diagnosis of failure for the ECREMOS. It scores low on information and system quality and high on service quality, which makes the balance for the three quality measures negative. This influences (intention to) use and user satisfaction. Combined with a lack of purpose, net benefits drop.

The Updated D&M model allowed for identification of factors of the causal chain that led to the failure of the studied IS. The inductive approach allowed one to find additional causes of failure that were not covered by the model. Petter et al. (2013) recognize that current research has possibly omitted important IS success/failure determinants. Recently, Petter et al. remarked that the factors of the Updated D&M model are influenced by independent determinants that cause success/failure. Relevant determinants that affect IS success/failure need to be described. After a literature review of 140 IS success studies, they detected 43 variables that are determinants of IS success. The most important influencers of IS success were pictured in "a road map for determinants of IS success." These determinants are divided into three groups of characteristics. The first group is "task characteristics" and addresses task compatibility and difficulty. The second group is called "project and organizational characteristics." Determinants are user involvement, relationships with developers, management support, extrinsic motivation, management processes, organizational competence, and IT infrastructure. The third and last group is "user and social characteristics"; it covers attitudes toward technology, enjoyment, trust, and user expectations. It is still unclear whether this study will require a further update of the D&M model; it is in its infancy.

If we compare the additional causes of failure of the two cases with the determinants of the roadmap from Petter et al., we see some matches (Table 8.4).

One must be cautious to generalize inductive data beyond the case setting. The inductively found additional causes of failure are nevertheless in line with previous research findings:

Table 8.4 Match between Additional Causes of Failure and the Road Map for Determinants of IS Success

Characteristics	*RSD*	*ECREMOS*
Task characteristics	—	—
Project and organizational characteristics	C1.8: Organizational competence	C2.8: Organizational competence
	C1.9: Path dependency on IT infrastructure	C2.9: Path dependency on IT infrastructure
		C2.10: A lack of local management support
User and social characteristics	C1.7: Unfulfilled user expectations	C2.7: A lack of trust

Note: ECREMOS = Environmental Complaints Registration and Monitoring System; IS = information systems; IT = information technology; RSD = Road Sign Database.

- Authors such as Scholl and Klishewski (2007) and Gil-Garcia & Pardo (2005) distinguish "organizational competence" as well.
- "Path dependency on IS infrastructure" is remarked by, for example, Rigg and O'Mahony (2013) and Heeks and Bhatnagar (1999).
- That "unfulfilled user expectations" can contribute to IS failure is recognized by Rigg and O'Mahony (2013), Chen (2008), and Dugdale, Daly, Papandrea, and Maley (2005).
- Gil Garcia and Pardo (2005), and Heeks and Bhatnagar (1999), among others, warn of "a lack of local management support."
- Finally, "a lack of trust" can be a recipe for failure, according to Thompson et al. (2008) and Chen (2008).

Based on the Updated D&M model and the road map for determinants of IS success, 14 causes of failure of RSD and ECREMOS were detected (Table 8.5). The analyzed cases have causes of failure in common: poor technical issues, a lack of purpose, reputation issues, insufficient organizational competence, and path dependency on IT infrastructure. The other causes are case specific.

Different stakeholders may validly come to different conclusions about the same IS's degree of success (Seddon, 1997). Information systems managers should be aware of different stakeholders' desires and perspectives. For both cases, the central government IS managers did not pay enough attention to the situation of the user, that is, local governments, before the IS were launched. Quick alternatives existed, and many invested in their own IT infrastructure and organizational competence as well

Table 8.5 Overview of the Causes of Failure (Delone, McLean)

	RSD	*ECREMOS*
System quality	Low (technical issues/slow time performance)	Low (technical issues/user alienated design)
Information quality	Low	Low (lack of purpose, fear for bad reputation, data security breaches)
Service quality	Low (lack of technical end user support)	High
Intention to use	Low (lack of purpose, cheap quick alternatives, bad reputation)	Low (lack of purpose)
Use	Dropped	Dropped
User satisfaction	Low (slow time performance)	Low (absence of feedback)
Net benefits	Negatively influenced	Negatively influenced
Project and organizational characteristics	The independent variables further caused failure (insufficient organizational competence/path dependency on IT infrastructure)	The independent variables further caused failure (insufficient organizational competence/path dependency on IT infrastructure/a lack of local management support)
User and social characteristics	The independent variables further caused failure (unfulfilled user expectations)	The independent variables further caused failure (lack of trust)

Note: ECREMOS = Environmental Complaints Registration and Monitoring System; IT = information technology; RSD = Road Sign Database.

as local management support for the planned IS diverged greatly. During the IS's implementation, it seemed that the central and local governments did not communicate enough: Local governments complain of a user alienated design, no eye for user expectations, and absence of feedback and support. Because local wishes were not taken into account, a lack of purpose for local governments arose. A lack of purpose, in combination with a lack of transparency and doubts about data security, leads to a low local trust in the central government IS and paved the path for IS failure.

Future managers of public sector IS could base action on these assessments to prevent future failure. Like in other impaired IS projects, failure is due to different interrelated factors (Dwivedi, Henriksen, & Rahul De', 2013). Managers should not only take into account the factors of D&M but should also be sensitive to other determinants of IS success/failure. Some factors or determinants are within the control of IS managers. Others may be outside their control, but managers may still be able to influence these (Petter et al., 2013).

The additional causes of failure point out that IS managers should try to get local management support, create trust, match user expectations, and create willingness to invest in interoperable IT infrastructures along the organizational competence. The Updated D&M model takes a rational project management view. One could wonder if this management technique is sufficient to influence factors of IS success/failure. Only focusing on step-by-step developments and on achieving goals, managerial or organizational structures within budget and time constraints, might not be sufficient (Fincham, 2002). Managing interorganizational IS requires attention to processes of "consensus building" and "cooperative behavior." A process management approach recognizes that organizations may be driven by self-interest and accepts that decision-making takes place in various not predefined arenas. Process managers seem more able to deal with the dynamics and nature of interorganizational relationships (Homburg & Bekkers, 2002). Petter et al. (2013) recognize that the impact of the project management technique to warrant IS success needs further research.

The research presented in this chapter has some limitations. Because the outcomes were based on a sample of two IS, this research can be considered only a basic exploratory case study. Furthermore, our analysis is a snapshot in time. Another limitation is that our scope was delimited to the causal lens of the Updated D&M model on IS success/failure. Other lenses such as a social process lens exist and might provide additional insights on G2G IS failure (Fincham, 2002).

8.6 Conclusion and Future Research

IS managers must be aware that G2G e-government systems face multiple challenges. Intergovernmental IS failure is multifaceted and omnipresent. There is a need to assess failure to base actions on these assessments.

Concerning RQ1, the Updated D&M model appeared to be transferable for assessing the causes of failure of the studied G2G IS. In answer to RQ2, although transferable, this model appeared to be incomplete. It did not detect a lack of trust, path dependency on IT infrastructure, insufficient organizational competence, unfulfilled user expectations, and a lack of local management support as causes of failure. These causes are in line with findings of other researchers. Additionally, several causes of failure that match determinants from "the road map for determinants of IS success" were found.

For future IS managers in the public sector, having insight on the potential causes of intergovernmental IS failure is a useful skill. The Updated D&M model as well as the road map of IS success determinants might be useful managerial tools. Being aware of different stakeholders' views and communication with the stakeholders are basic steps for preventing IS failure. Besides a project management technique, a process management technique might be valuable.

The findings are restricted to two Belgian cases; additional research could include similar cases in other countries. Transferability of results refers to the degree to which the results of qualitative research can be generalized or transferred to other contexts or settings. Future research could further explore the transferability and completeness of the Updated D&M model for a G2G e-government context. Assessing IS failure in e-government can enlarge our insight on how public IS managers can take action to prevent future IS failure.

Author Biographies

Lies Van Cauter is a PhD student at the KU Leuven, Public Management Institute. Her research focuses on failure of intergovernmental information systems.

Monique Snoeck is a full professor at the KU Leuven, Research Center for Management Informatics (LIRIS), and a visiting professor at the UNamur. Her research focuses on enterprise information systems engineering, enterprise modeling, requirements engineering, model-driven engineering, and business process management. Main guiding research themes are the integration of different modeling approaches into a comprehensive approach, the quality of models through formal grounding, models to code transformations, and educational aspects of conceptual modeling. She has published over 80 peer-reviewed papers.

Dr. Joep Crompvoets is an associate professor at KU Leuven Public Governance Institute and a secretary-general of EuroSDR—a European Spatial Data Research network linking national mapping agencies with research institutes and universities for the purpose of applied research. He has been involved in several international projects related to the development of spatial data infrastructures (SDIs) around the world. He has written numerous publications dealing with SDIs, GIS, and e-governance.

References

Abdelsalam, H., Reddick, C. G., & El Kadi, H. A. (2012). Success and failure of local e-government projects: Lessons learned from Egypt. In S. K. Aikins (Eds.), *Managing e-government projects. Concepts, issues and best practices* (pp. 242–261). Hershey, PA: Information Science Reference.

Al Khatib, H. (2013). *E-government systems success and user acceptance in developing countries: The role of perceived support quality* (Doctoral dissertation). Brunel Business School: London, 1–405.

Benbasat, I., Goldstein, D. K., & Mead, M. (1987). The case research strategy in studies of IS. *Management of Information Systems Quarterly, 11*(3), 369–383.

Beynon-Davies, P. (2009). *Business information systems* (pp. 1–482). Houndmills, Basingstoke: Palgrave-Macmillan.

Chai, S., Herath, T. C., Park, I., & Rao, H. R. (2006). Repeated use of e-gov web sites: A satisfaction and confidentiality perspective. *International Journal of Electronic Government Research, 2*(3), 1–22.

Chen, B. (2008). Assessing inter-organizational networks for public service delivery a process-perceived effectiveness framework. *Public Performance & Management Review, 31*(3), 348–363.

Davis, F. D. (1989). Perceived usefulness, perceived ease of use, and user acceptance of information technology. *MIS Quarterly, 13*(3), 319–340.

DeLone, W. H., & McLean, E. R. (2003). The DeLone and McLean model of information systems success: A ten-year update. *Journal of Management of Information Systems, 19*(4), 9–30.

Dörr, S., Walther, S., & Eymann, T. (2013). Information systems success—A quantitative literature review and comparison. *11th International Conference on Wirtschaftsinformatik*, Leipzig, 1813–1827.

Dugdale, A., Daly, A., Papandrea, F., & Maley, M. (2005). Accessing e-government: Challenges for citizens and organizations. *International Review of Administrative Sciences, 71*(1), 109–118.

Dwivedi, Y., Henriksen, H. Z., & Rahul De', D. W. R. (2013). Preface. In Y. Dwivedi, H. Z. Henriksen & D. W. R. Rahul De' (Eds.), *Grand successes and failures in IT public and private sectors* (pp. V–IX). Bangalore: Springer.

Dwivedi, Y., Wastell, D., Laumer, S., Henriksen, H. Z., Myers, M., Bunker, D., Elbanna, A., Ravishankaent, M., & Srivastava, S. (2014). Research on IS failures and successes: Status update and future directions. *Springer Information Systems Frontiers.* doi: 10.1007/s10796-014-9500-y [epub ahead of print].

Elbanna, A., & Linderoth, H. C. J. (2013). Tracing success in the voluntary use of open technology in organisational setting. In Y. Dwivedi, H. Z. Henriksen & D. W. R. Rahul De' (Eds.), *Grand successes and failures in IT public and private sectors* (pp. 89–104). Bangalore: Springer.

Ewusi-Mensah, K. (2003). Software Development Failures. MIT Press, USA.

Fincham, R. (2002). Narratives of success and failure in systems development. *British Journal of Management, 13*(1), 1–14.

Floropoulos, J., Spathis, Ch., Halvatzis, D., & Tsipouridou, M. (2010). Measuring the success of the Greek taxation information system. *International Journal of Information Management, 30*, 47–56.

Gable, G. G., Sedera, D., & Chan, T. (2008). Re-conceptualizing information system success: The IS-impact measurement model. *Journal of the Association for Information Systems, 9*(7), 377–408.

Gichoya, D. (2005). Factors affecting the successful implementation of ICT projects in government. *The Electronic Journal of e-Government, 3*(4), 175–184.

Gil-Garcia, J. R., & Pardo, T. A. (2005). E-government success factors: Mapping practical tools to theoretical foundations. *Government Information Quarterly, 22*, 187–216.

Gil-Garcia, J. R., Chengalur-Smith, I., & Duchessi, P. D. (2007). Collaborative e-government: Impediments and benefits of information-sharing projects in the public sector. *European Journal of Information Systems, 16*, 121–133.

Goldfinch, S. (2007). Pessimism, computer failure, and information systems development in the public sector. Perspectives on performance and accountability in public administration. *Public Administration Review, 67*(5), 917–929.

Gotoh, R. (2009). Critical factors increasing user satisfaction with e-government services. *Electronic Government an International Journal, 6*(3), 252–264.

Heeks, R., & Bhatnagar, S. (1999). Understanding success and failure in information age reform. In R. Heeks (Ed.), *Reinventing government in the information age: International practice in IT-enabled public sector reform* (pp. 49–74). London: Routledge.

Homburg, V., & Bekkers, V. J. J. M. (2002). The back-office of e-government. *Proceedings of the 35th Hawaii International Conference on System Sciences*, 1666–1674.

Hsu, F.-M., & Chen, T.-Y. (2007). Understanding IS usage behavior in e-government: The role of context and perceived value. *PACIS Proceedings*, 477–490.

Hussein, R., Karim, N. S. A., & Selamat, M. H. (2007). The impact of technological factors on IS success in the electronic-government context. *Business Process Management Journal, 13*(5), 613–627.

Kautz, K., & Cecez-Kecmanovic, D. (2013). Sociomateriality and information systems success and failure. In Y. Dwivedi, H. Z. Henriksen & D. W. R. Rahul De' (Eds.), *Grand successes and failures in IT public and private sectors* (pp. 1–20). Bangalore: Springer.

Lehtinen, T. A. O., Mäntylä, M. V., Vanhanen, J., Itkonen, J., & Lassenius, C. (2014). Perceived causes of software project failures—An analysis of their relationships. *Information and Software Technology, 56*(6), 623–643.

Makedon, F., Sudborough, C., Baiter, B., Pantziou, G., & Conalis-Kontos, M. (2003). A safe information sharing framework for e-government communication. *IT White Paper*, 1–11.

Mills, A., Durepos, G., & Wiebe, E. (2010). *Encyclopedia of case study research*. Thousand Oaks, CA: Sage Publications.

Moges, H.-T., Dejaeger, K., Lemahieu, W., & Baesens, B. (2013). A multidimensional analysis of data quality for credit risk management: New insights and challenges. *Journal of Information & Management, 50*(1), 43–58.

Petter, S., DeLone, W., & McLean, E. R. (2008). Measuring IS success: Models, dimensions, measures and interrelationships. *European Journal of Information Systems, 17*, 236–263.

Petter, S., DeLone, W., & McLean, E. R. (2013). Information systems success: The quest for the independent variable. *Journal of Management Information Systems, 29*(4), 7–61.

Prybutok, V. R., Zhang, X., & Ryan, S. D. (2008). Evaluating leadership, IT quality, and net benefits in an e-government environment. *Information and Management, 45*, 143–152.

Rana, N. P., Dwivedi, Y. K., & Williams, M. D. (2013). Examining the factors affecting intention to use of, and user satisfaction with OPGRS in India. In Y. Dwivedi, H. Z. Henriksen & D. W. R. Rahul De' (Eds.), *Grand successes and failures in IT public and private sectors* (pp. 240–260). Bangalore: Springer.

Rigg, C., & O'Mahony, N. (2013). Frustrations in collaborative working: Insights from institutional theory. *Public Management Review, 15*(1), 83–108.

Sambasivan, M., Wemyss, G. P., & Rose, R. C. (2010). User acceptance of a G2B system: A case of electronic procurement system in Malaysia. *Internet Research, 20*(2), 169–187.

Scholl, H. J., & Klischewski, R. (2007). E-government integration and interoperability: Framing the research agenda. *International Journal of Public Administration, 30*, 899–920.

Scott, M., DeLone, W., & Golden, W. (2009). Understanding net benefits: A citizen based perspective on e-government. *30th ICIS Conference*, Phoenix, Arizona. Red Hook, NY: Curran Associates.

Seddon, P. B. (1997). A respecification and extension of the DeLone and McLean model of IS success. *Information Systems Research, 8*(3), 240–253.

Sørum, H., Medaglia, R., Andersen, K. N., Scott, M., & DeLone, W. H. (2012). Perceptions of IS success in the public sector: Webmasters at the steering wheel? *Transforming Government People, Process and Policy, 6*(3), 239–257.

Thompson, T., Srivastava, S., & Jiang, L. (2008). Trust and electronic government success: An empirical study. *Journal of Management Information Systems, 25*(3), 99–131.

Van Cauter, L., Snoeck, M., & Crompvoets, J. (2013). *Flemish intergovernmental data collections: An inventory*. Technical Report, University of Leuven, Leuven, 1–127.

Venkatesh, V., Morris, M. G., Davis, G. B., & Davis, F. D. (2003). User acceptance of information technology: Towards a unified view. *MIS Quarterly, 27*(3), 425–478.

Vlaams Parlement. (2008). Vraag om uitleg aan Crevits, Vlaams minister van Mobiliteit en Openbare Werken, Energie, Leefmilieu en Natuur, over MKROS –551.

Wang, Y.-S., & Liao, Y.-W. (2008). Assessing e-government systems success: A validation of the DeLone and McLean model of IS success. *Government Information Quarterly, 25*, 717–733.

Yildiz, M. (2007). E-government research: Reviewing the literature, limitations, and ways forward. *Government Information Quarterly, 24*(3), 646–665.

Yin, R. K. (1994). *Case study research: Design and methods* (2nd ed.). Newbury Park, CA: Sage Publications.

Chapter 9

The Discretionary Space of Geo-Information and Communication Technology Professionals in Public Sector Cooperation Programs

Walter Timo de Vries

Contents

Abstract

Innovations in interorganizational e-government projects in the public sector (G2Gs) tend to be shaped by the discretionary decisions of information and communication technology (ICT) professionals. Public sector managers who are insufficiently knowledgeable about the implications of such discretions can therefore not fully control the innovation. As a result, they risk being confronted with unintended results. To fully understand this quandary, there is a need to investigate the type of, variation in, and reasons for such professional discretions. This chapter conceptualizes discretions as adverse reactions to coordination requirements. The context in which the staff works further determines when, why, and how discretions emerge. Combining coordination requirements and context leads to a taxonomy of six discretions types. These are empirically evaluated in four G2G case studies in the Netherlands that rely on a specific ICT, namely, geoICT. Such "geoG2G" projects are considered particularly prone to discretions because the professional field of geoICT is so specific that the potential for uncontrollable professional discretions is even larger than average. Empirically, these discretions are visible through a set of indicators for each discretion type. The observed variation in the cases suggests that discretions in this field emerge as a result of a perceived complex environment, which exhibits both the need for staff members to simplify one's own tasks and the need to closely adhere to the needs of external clients. These findings justify the need for a better understanding of discretions to decrease the dependency on these in ICT projects in the future, so that managing these projects effectively improves.

Keywords: discretions, discretionary space, discretionary behavior, geoICT, geoICT coordination

9.1 Introduction

Innovations in interorganizational e-government projects in the public sector (usually referred to as "G2G" projects, or simply G2Gs) exhibit practices where information and communication technology (ICT)-related agreements, coordination, and monitoring activities are handled largely by ICT professionals only (Bekkers, 2009). However, a fair amount of cases have demonstrated that it is difficult for public sector managers to control the development and decisions in these projects (Heeks, 2002; Kaniadakis, 2012; Tapia, Maldonado, Tchouakeu, & Maitland, 2012). In particular, the dependence on ICT decisions by a relatively small and technically oriented group of professionals raises the question of whether public sector cooperation is fundamentally changing and whether there is any significant influence of a set of (less controllable) professional decisions on these changes. This is a relevant question in the larger scheme of public sector management because it would suggest that a specific technology, and its related set of professional practitioners handling this technology, could influence if, when, how, and why public sector agencies cooperate and integrate. This harms the integrity of the public sector. However, this potential risk and how it relates to ICT-related decisions have received limited attention in public administration so far.

One way to evaluate the role of professional ICT decisions is through the concepts of "discretionary space" and "discretions." Discretionary space refers to the degree of freedom that individual actors claim to reach to decisions, which may be adverse to either institutional or (inter)organizational requirements. Discretionary space is utilized through discretions, which are decisions made by individual actors based on personal judgments rather than organizational procedures and fully rational assessments. Davis refers to discretions as the "freedom to make a choice among possible courses of action and inaction within the effective limits of someone's power" (1976, p. 4). Hupe and Hill (2007) argue that the presence of discretions is inevitable once such effective limits exist. Lipsky, one of the first authors to point to discretions of "street-level bureaucrats," argues that the sum of all discretions in a public sector system influences the overall governance of that system, and as a result, the sum of all discretions is also part of how the government functions: "The actions of most workers actually constitute the services 'delivered' by government. Moreover, when taken together the individual decisions of these workers become, or add up to, agency policy" (1980, p. 3).

In public administration discourses, discretions are usually associated with uncontrollable power of street-level bureaucrats (Hudson, 1989; Lipsky, 1983). Because certain processing procedures allow public sector professionals to make autonomous decisions on resources, they become a source and agency of uncontrollable or even invisible power. Lipsky (1983) argues that this is problematic because a large number of discretions or a wide range of discretions will hamper the possibility of verifying the accountability of an organization. Because the original discussions on discretionary space were confined to "good-and-bad" or "present-or-absent" dichotomies, the critique has been that "Discretions should be regarded as a series of gradations of freedom to make decisions and, therefore, the degree of freedom professionals have at specific conjunctures should be evaluated on a situation-by-situation basis" (Evans & Harris, 2004, pp. 871–872). Thus, public administration science needs more empirical evidence on the degree of discretion and the extent of the discretionary space to assess the changes in power relations and accountability of public sector organizations.

Bovens and Zouridis (2002) argue that given the rapid intrusion of ICT in public sector activities, evaluating discretionary space should no longer be focused only on street-level bureaucrats and their discretions, but more on system-level bureaucrats, who are key actors in the design and implementation of information systems. Zuurmond

(1994) notes that the informatization process in the public sector reflects diminishing uncertainty and tighter control of public sector staff discretions. For the delivery of social services, Zuurmond argues that "Checking information, previously an exclusive task of professional staff, is increasingly replaced by IT" and "Information systems not only ensure control over the actions of employees, but they also put limits on their thoughts. Through this electronic control, the 'masters' of an organization can determine what is seen as fact (as data) and what is not seen at all" (translated from Dutch). In line with these observations, van de Donk (1997) states that street-level bureaucrats find increasingly regulatory constraints in their discretions arising from ICT. Informatization in this case does not generate operational uncertainty but reduces this operational uncertainty and limits the discretions of individual servicing staff. Similarly, Jorna and Wagenaar (2007) argue that the relationship with the role of ICT tends to delete discretions at operational levels. Despite this apparent decrease, Jorna (2009) remarks that the informatization process in a number of Dutch cases did not exactly destroy operational discretion but instead obscured operational discretion. As operational staff may often not be entirely aware of the norms underlying the ICT systems, they have to define their own norms. These shape their personal discretions. It is therefore necessary to research the role of discretions at different levels.

This article zooms in to discretions in a specific subset of ICT projects, namely, those involving "geoICT." GeoICT refers to the "collection of ICT based systems allowing the study of natural and man-made phenomena with an explicit bearing in space" (de By, 2004, p. 15). Specific to geoICT is that it creates and manipulates "geospatial data." Similarly to other types of ICT, "Geo-ICT has the potential to influence the public organisation beyond the operational core of public administration" (Zouridis & Thaens, 2005). GeoICT has a relatively confined set of professionals; hence, the potential that professional discretions occur, which may not be remarkable beyond the professional field, is larger than in projects with widely used ICT applications. Furthermore, it makes public sector managers, in particular, vulnerable to such uncontrollable discretions and makes them even more dependent on such discretions than any other ICT application. Most geoICT-related projects also exhibit relatively high uncertainties, which affect potential discretions (de Vries, Crompvoets, Stoter, & VandenBerghe, 2011). An effect of having to work with various types of uncertainties is that individual actors create a certain degree of freedom to handle these uncertainties or to deal with the internal restrictions needed to combat uncertainty. As a result, discretions are assumed to emerge.

The specific research problem thus can be summarized as follows: How can one recognize discretionary actions and behavior in interorganizational geoICT projects (shortened by the term *geoG2Gs*) and how can emerging discretions be explained?

This problem is dealt with through an explorative research aimed at synthesizing theory and deriving a research strategy. The next section relates discretions to (geoICT) coordination. Second, because individual staff members operate in a context of both internal working relation and external networks of peers, possible contextual reasons are summarized for how and why discretions may arise. Third, as different coordination types have different aims, and given that discretions relate to coordination, one can also make a distinction between different types of discretions. This will complete the typology of discretions. The typology is then converted into a set of empirical indicators to evaluate whether any type of discretion is present and to evaluate the contextual reasons for each occurrence of discretions. It shows furthermore how to analyze the discretion types empirically, which is followed by an evaluation of discretions in four cases. The interpretative sections compare the results of the evaluation and explain the findings with theory wherever fitting and appropriate.

9.2 Discretions and Public Sector ICT Coordination

The study of discretionary behavior has been given a new impetus by the question to which extent technical innovations in the public sector also affect public sector behavior. One perspective is to relate discretionary behavior to coordination actions in the public sector. Chisholm (1992) defines coordination as "mechanisms through which communications take place and solutions are sought and implemented" (p. 65). In this view, coordination is not only an interactive and operational activity of communication and implementation but also a normative activity of solution seeking and choosing of priorities among possible solutions. These address public sector problems of accountability at large or organizational efficiency and service provision (Webb, 1991). When treated as an organizational problem, coordination is an activity that connects organizational structure (formal or informal) to organizational tools (including technology).

When intersecting coordination with ICT, the coordination solutions are represented as a form of alignment between different options. Alignment can be the activity of connecting strategy to information technology (IT), but may also be the activity of finding the middle ground between the views and values from different actors with respect to the use of a technology. In either case, the activity intervenes in current organizational routines and may involve a restructuring of interpersonal or interorganizational relations. Emerging technology may thus be the start of coordination (activities conducted to adapt the organizational routines to the emergent changes in technology) or may be the effect of the coordination (the emerging technology from the activities between different people or organizations). In both cases, coordination is a goal-setting and goal-implementing activity, whereby goals may be shaped in the form of either rules or norms (such as standards), processes (chains or product lines), products/services (specific output requirements), or impacts (such as envisaged externalities).

Combining coordination with ICT in the public sector or public administration brings about the generic characteristics of ICT coordination: a normative activity associated with (inter)organizational tools geared at certain outcomes that are relevant in a particular (inter)organizational context. Specific for the public sector is its context and relations. Specific for ICT is the set of tools that the set of activities applies to. This makes ICT coordination relevant for ICT norms, ICT activities, ICT outcomes, and ICT organizational contexts. In sum, ICT coordination can be considered "goal-setting and goal-implementation action which aims at aligning ICT activities and choices in at least two organisations, and which usually results in a change intervention in how public sector agencies cooperate with ICT" (de Vries, 2013).

If ICT coordination actions involve prescribing and/or regulating behavior, then professional discretions can be regarded as any behavior that deviates from the prescribed coordination objectives. Strictly speaking, discretions can be conceptualized as decisions and associated actions of individual actors based on personal judgments rather than organizational procedures and fully rational assessments. Davis refers to discretions as the "freedom to make a choice among possible courses of action and inaction within the effective limits of someone's power" (1976, p. 4). Thus, if ICT coordination prescribes certain courses of action, then discretions can be regarded as the actions and inactions of people when dealing with those prescriptions.

Because coordination is an activity that connects organizational structure (formal or informal) to organizational tools (including technology), discretions may also arise in relation to framing the objectives for both aspects as well as on the preferred implementation strategy. Hence, discretions may relate to (a) the preferred structure, usually reflected as choices on preferred responsibilities and hierarchies (strategic, alignment, and operational); (b) the technology, visible through the preferred types of ICT and choices on relevant ICT components; and finally, (c) the choices made when executing the alignment, that is, the implementation choices when aligning either rules or norms, processes, products, or intended impacts.

Summarizing, ICT-related discretions can be conceptualized as a reaction to three coordination aspects:

1. Preferred structure
2. Preferred technology
3. Preferred implementation strategy of coordination

9.3 Conceptualizing Discretions and Discretionary Space

So far, only few publications make discretions in the public sector the core subject of a conceptual study or even formulate a theory of discretions. Migué, Bélanger, and Niskanen (1974) aim at such a "grand theory"; however, they focus primarily on managerial discretions. Similarly, the classic work of Lipsky (1980) focuses primarily on street-level discretions only. Thus, the larger context of any kind of discretions and the conceptualization of interrelated or interdependent discretions in an organizational context have remained underdeveloped. Evans and Harris (2004) critique these studies with the argument that, in practice, there are often gradations in discretions that have to be evaluated on a case-by-case basis. However, to evaluate such case-by-case discretions, some form of conceptual basis that acknowledges the existence and the variety of discretions within an organization is required.

Although discretions refer to personal judgments, they differ at the respective levels at which staff work. Such personal judgments are not entirely personal but can relate both to the environment (through the links that individual actors may have to hierarchical networks and representative networks) and to resource limitations and opportunities. Thus, discretions are actions of individual staff members that rely on a personal appraisal of what is appropriate, given the socio-organizational circumstances and preferences of that particular staff member. Discretions, as a result, are a kind of regulatory agency, which result in a change in organizational behavior and/or structure and possible a change in the stability of cooperative arrangements with geoICT.

Several authors identify the specific reasons why individual staff discretions exist or emerge. Moe (1984) argues that discretions in bureaucratic organizations arise because of the different interests in the interaction between those who manage and those who execute. As individual staff members operate in a context of both internal working relation and external networks of peers, the role of context on discretions is considered important for how discretions may arise. Finkelstein and Hambrick (1990) expand on three sets of factors that determine the variation in discretions of individual actors:

1. Organizational environment
2. Available resources
3. Ability to envision alternatives

The first factor concerns the degree to which the environment in which actors operate allows the actors variety and change in their daily work. In most cases, many regulations and resource limitations exist, which prevent actors from deriving any personal discretion. However, most actors are also aware of the space "granted" by the environment. Hay and Wincott (1998) refer to the existence of cognitive filters within actors. Cognitive filters are "the perceptions [of actors] about what is feasible, legitimate, possible and desirable in the institutional environment in which they find themselves and existing policy paradigms and worldviews" (Hay & Wincott, 1998). It is through such cognitive filters that actors exercise certain discretions. The discretions are in line with the degree of freedom that actors perceive to have within their environment.

The second factor concerns the degree to which the organization is amenable to an array of possible actions and empowers the executive to formulate and execute those actions. Andersson, Lindgren, and Henfridsson (2008) claim that it is not the technological solutions themselves or the enforcement of technological solutions (such as standards) that leads to the confinement of discretionary space, but the limitation in actors' resources to implement the technological solutions that lead to this. Even if staff members were willing to seek personal discretions for their work, the constraints in resources (and capacity) prevent them from doing so. Unlimited resource opportunities might lead to personal discretions.

The third factor concerns the degree to which the staff would be able to personally envision or create multiple courses of action. A possible reason why operational staff may not have discretions is when they would prefer to work under routinized procedures rather than to have freedom to interpret procedures for themselves. Quane, Su, and Joshi (2009) find that operational staff at the front end of service delivery were "more comfortable with a rigid approach to service delivery when it relieved them of having to make difficult judgment calls." Indeed, Lens (2006) found that even if operational staff have the opportunity for more discretionary space, they would prefer to rely on routine processes.

In sum, discretions arising out of contextual networks are dependent on the degree to which staff members:

1. Pertain a cognitive filter of the environment
2. Have personal access to resources
3. Carry the ability to personally envision courses of action

9.4 Types of Public Sector ICT-Related Discretions

In addition to the previous types of discretions arising directly in response to coordination and contextual constraints and opportunities, professional staff members may also have personal reasons to reach to discretions. In this case, personal satisfaction executing work plays a dominant role. This may be indirectly connected to coordination requirements, yet it seems less dependent on the specific ICT coordination details or on the regular contextual networks in which the staff members operate.

Van de Donk (1997) notes that many public sector staff members find themselves increasingly constrained in executing the tasks that they perceived as their personal domains and contributions. The increase in ICT seems to reduce their discretions when executing their daily tasks. Similarly, Jorna and Wagenaar (2007) argue that the relationship with the role of ICT tends to delete discretions at operational levels. Jorna (2009) remarks that, despite this apparent decrease, the uptake and adoption of ICT in a number of Dutch cases did not destroy operational discretion but instead obscured operational discretion. Additionally, because operational staff is often not aware of the norms underlying the ICT systems, they have to define their own norms. These are reflected in their personal discretions.

Maynard-Moody and Musheno (2000) differentiate two contrasting perspectives on how frontline officers may have discretions and how they would use their discretions. They may employ individual discretions as a way to simplify their own tasks and/or to handle ambiguous policies. Although almost every aspect of operational work relies on multiple rules and instructions, the extent to which all these rules are applied depends on the ability of supervisors of operational staff to enforce those rules (Lipsky, 1980). However, in practice, even this direct link from operational staff to supervisors has little constraining influence over operational staff decisions and actions (Brehm, Gates, & Gomez, 2003).

Conversely, operational staff could seek discretionary space when acting as engaged advocates for their deserving clients. At some occasions, certain staff members engage in an ad hoc relation with customers and clients, outside the hierarchy and administrative procedures of the organization. The relation with the external environment is thus based more on personal acquaintance and may even be based on personal sympathies:

> A defining characteristic of street-level work and what distinguishes it from other work at the bottom of bureaucracies is the street-level worker's direct contact with citizens. Unlike elected and other top government officials they do not see citizens as abstractions but as individuals: as clients, students, criminals, suspects, victims and so on. Their relationships with these various citizen clients are personal and emotional, rarely cold and rational. (Maynard-Moody & Musheno, 2000)

As a result of this personal relation, they may favor the specific client's interest. The emergence of such a relation and associated discretionary favoring may be especially reenforced when staff members feel curtailed by far-reaching automization of their tasks. As Snellen and Van de Donk (2002) noted, "The technostructure confines their decision space. Their jobs tend to be downgraded." The effect may be contrary to the first mentioned type of discretion. Rather than reaching to discretions that simplify the execution of a multitude of tasks, discretions arise in this case to enrich the tasks with more content and meaning.

In sum, in the execution of discretion, there are two major types to satisfy personal task motivation:

1. Discretions that simplify tasks
2. Discretions that enrich the tasks

Combining this dichotomy of autonomous/self-interest versus joint/external interests with the types of discretions at different levels leads to six possible discretions, as shown in Table 9.1. Discretions aimed at self-interest can be labeled "autonomous" discretions, and discretions aimed at supporting joint or external interests can be labeled "joint" discretions.

Table 9.1 Types of Discretions at Various Organizational Levels

	To Facilitate Own/Autonomous (Self-) Interest	*To Support Joint or External Interests*
Strategic discretions	Autonomous strategic discretions	Joint strategic discretions
Alignment discretions	Autonomous alignment discretions	Joint alignment discretions
Operational discretions	Autonomous operational discretions	Joint operational discretions

9.5 How Discretions Are Visible

The conceptual section identified six types of discretions: (autonomous/joint) strategic discretions, (autonomous/joint) alignment discretions, and (autonomous/joint) operational discretions. The next step is how to find and explain these discretions empirically. A case study methodology is appropriate to do so (Benbasat, Goldstein, & Mead, 1987; Yin, 1994). "Multiple-case designs are desirable when the intent of the research is description, theory building, or theory testing… Multiple-case designs allow for cross-case analysis and the extension of theory. Of course, multiple cases yield more general research results" (Benbasat et al., 1987). The first step in selecting cases is to choose cases that are likely to exhibit different kinds of discretions yet in a similar field of ICT with similar ICT professional values. Hence, the domain of geoICT was chosen. Typical of the geoICT domain is furthermore that many cases of shared geoICT applications are on strongly coordinated by laws, rules, and resources, yet a variety of professional discretions occur in relation to the different types of coordination activities. In addition, it was assumed that cases that involve multiple partners or complex cooperative arrangements would exhibit different kinds of discretions. Given that the origin of discretion types and significance may lie in the difference in how the projects are coordinated, four comparative cases were chosen to reflect these coordination differences. Hence, cases were chosen that were different in the number of partners involved, had different coordination agreements, and had different types of geoICT to manage. All of these cases were in the Netherlands to ensure a similar institutional background in which the geoICT professionals were operating.

1. Cadastre-municipalities (Cadastral geoG2G): Two types of organizations are involved: the national agency Kadaster and 443 municipalities (in 2010). Both types of organizations have to adhere to a national geoICT coordination strategy guiding the implementation of authentic registers of addresses

and buildings. The implementation is extensively documented in professional magazines, regularly maintained websites, and newsletters of the Kadaster agency and the Netherlands Ministry of environment and infrastructure. Further evidence could be collected through staff members working on either side of the Cadastral geoG2G.

2. Actual Heights Netherlands (AHN): This geoG2G is constructed by the national agency of public works (RWS) and 12 regional water management agencies. Both types of organizations follow a mutually agreed geoICT coordination strategy on the production and distribution of accurate height data. Both agencies maintained regular newsletters and publications on the developments and progress in AHN, which provided the documentary evidence. In addition, staff members from both RWS and several regional water agencies were interviewed on AHN.
3. Dataland: This geoG2G involves 300 municipalities who agreed to cooperate on the specifications and distribution of local house and building data. All municipalities adhere to a geoICT coordination strategy relying on national "other government" objectives. Information on Dataland activities and decisions was available through the regular Dataland newsletters and congresses. Access to this information was possible through registration to both and through personal contacts with the Dataland actors.
4. Sabimos: This geoG2G constitutes of 15 municipalities and a regional government, who agreed to collect, monitor, and distribute data in relation to performance indicators and coordination of public transport. Both types of organizations were collaborating on a geoICT coordination strategy on regional mobility. The majority of data on this case could be collected through interviews and observations of the main responsible actors in this geoG2G.

Based on the differentiation of discretions, empirical indicators needed to be formulated for discretionary aspects. In total, there are five aspects that enable the description of discretion characteristics: personal task simplification, adherence to client interests, cognitive filter to the environment, personal access to resources, and ability to envision alternative courses of action. Table 9.2 provides the details of these aspects. The indicators of these aspects were evaluated through qualitative analyses of interview transcripts, grey literature, progress reports, and other types of documentation. A value of "high" was given if, in the different data sources (interviews, documents, references), there was a relatively frequent and/or consistent mentioning or reference to the particular indicator. Similarly, if this was lacking or people systematically and consistently denied the presence of this aspect, then the value was labeled as "low." This process derived values that are, by nature, qualitative and subjective and not quantitative. Hence, the discrete values low and high describe the perceived degree to which the aspects are present given the available evidence to qualify the variation in aspects.

Table 9.2 Aspects and Empirical Indicators of Discretions

Characteristics of Discretions	*Aspects*	*Indicators*
Autonomous/ joint discretions	Personal task simplification	The perceived degree to which individual staff members modify their own tasks to facilitate their own schedules and activities
	Adherence to client interests	The perceived extent to which staff members refer to other organizations or alternative coordination mechanisms as a justification for their actions
Strategic/ alignment/ operational discretions	Cognitive filter environment	The perceived degree to which individual staff members—at a certain level within the geoG2G—can formulate alternatives for geoICT coordination requirements
	Personal access resources	The perceived degree to which individual staff members—at a certain level within the geoG2G—(can) start up activities to acquire additional funds, equipment, information, and people to execute or to bypass geoICT coordination requirements
	Ability to envision courses of action	The perceived degree to which individual staff members—at a certain level within the geoG2G—show the ability to formulate alternative solutions for given problems

9.6 Comparison of Professional Discretions in geoICT Programs in the Netherlands

Table 9.3 summarizes the assembled scores for each of the five aspects in each case. The qualifications high and low refer to the perceived degree of occurrences of discretions. The method of deriving the scores can be explained by a detailed description for one case. In the Cadastral case, a change in behavior of project staff members that emerged as a reaction to having to implement a national policy of base registers on addresses and buildings (BAG) could be qualified as discretions because it

Table 9.3 Empirical Findings of Discretions Aspects

Characteristics	*Aspects*	*Cadastral*	*Actual Heights Netherlands*	*Dataland*	*Sabimos*
Strategic/alignment/ operational discretions	Cognitive filter to environment	High (for strategic staff)	Low, but increasing (for strategic staff)	High	High
Strategic/alignment/ operational discretions	Personal access to alternative resources	Low	Low	Low	Low
Strategic/alignment/ operational discretions	Ability to envision courses of action	High (for strategic staff); low (alignment and operational)	High (for alignment staff)	High (for alignment staff)	High (for alignment staff)
Autonomous/joint discretions	Degree of personal task simplification	Low	Low	High	High
Autonomous/joint discretions	Degree of adherence to client interests	High	Low	Low	Low

fundamentally deferred from the original intended or required behavior by the BAG policy. The BAG law fundamentally challenged the monopolist role of the Kadaster as a key geo data provider. Much more than before, the Kadaster's role in the BAG became a servicing role toward the Ministry and the municipalities. The expectation was that the uncertainty arising from these fundamental changes would have an effect on the extent and type of discretions. During interviews, in particular, strategic Cadastral staff felt the fundamental changes resulting from the BAG. They were, on their own account, searching for new avenues of cooperation to sustain their role in the future. They indicated seeing themselves acting in larger cooperative arrangements. Being aware that the BAG law would fundamentally change their organization vis-à-vis the environment and changes the autonomy of the Kadaster, the strategic staff members actively investigated possible scenarios. Such behavior can be interpreted as a form of discretion. It shows that, in particular, strategic staff members are cognitive of the changes in the environment. The aspect of "cognitive filter to the environment" was therefore qualified as high for strategic staff members. As for other types of staff members, such behavior was absent, and the score for this aspect was qualified as low.

The aspect of "personal task simplification" and "adherence to client interests" scored low for the Cadastral case. The fact that, toward the end of 2010, municipal geoICT staff became increasingly satisfied with the standardization process resulting from BAG, even though it decreased their ability for discretionary decisions, implies in retrospect that their earlier alignment and operational discretions within municipalities had been to facilitate and simplify their internal work processes. Thus, there was a shift in purpose of discretions. Before BAG, the regular activities gave rise to alignment and staff discretions for the sake of personal task simplification. With BAG standard procedures making the staff's working life easier, they became more willing to accept the new BAG regulations, thereby implicitly accepting less room for discretionary actions. For strategic staff members, the BAG rules also decreased their room to maneuver, yet this did not necessarily lead to more complex or more simplified tasks. All in all, the degree to which in 2010 discretions still existed with the intention to simplify their personal tasks can be considered as low.

In sum, the Cadastral case exhibits high scores for "cognitive filter to the environment" and "ability to envision courses of alternative actions" for strategic staff members, implying the presence of discretions among strategic staff members. For alignment and operational staff members, discretions were present in the past (before the extensive influence of BAG), but given the limited degree of alignment and operational influence, it is fair to say that most of these discretions gradually disappeared and that most alignment and operational staff—in particular those working within municipalities—concurred with the clarity and certainty of operational processes provided by BAG coordination. Thus, BAG resulted in a personal task simplification for those staff members. Finally, the client/external interests are a motivation to reach to discretions for all staff members.

For all three other cases, the process of deriving values for the discretions indicators was similar. By systematically assembling observations on behavior in relation to coordination requirements, and by deriving perceptions and reasons for the discretions in a comparable way as for the Cadastral case, it was possible to derive different scores for each indicator in each case. These values are summarized in Table 9.3. Overall, the values in Table 9.3 exhibit the following characteristics.

The aspect of "cognitive filter to the environment" is relatively high in three cases but relatively low for the AHN case. In the AHN case, it is increasing only for strategic staff members. The AHN case is distinctive from other cases because the collaboration in AHN does not involve municipalities but involves only national and regional agencies dealing with water management. This would suggest that AHN partners are relatively confined in their views of alternative geoICT requirements (namely, only those relevant for water) and as a result do not aim to expand their geoICT application beyond the limits of this water management (visible in the low score on the aspect of "cognitive filter to the environment"). Hence, in cases where the geoICT is deliberately applied by geoICT professionals in a single sector only, it seems unlikely that discretions occur on the basis of alternative ICT requirements but occur only as a reaction to the specific geoICT requirements in that field.

The aspect of "personal access to resources" is consistently low in all cases. Staff members in geoICT-related projects have a common perception of having no influence on resource allocations of the organizations they work for. They tend to feel uninvolved in strategic organizational, financial, and political deliberations or decisions. Reversely, those who set requirements on geoICT apparently do not provide enough freedom to operational staff members to alter their environment according to their own needs and wishes. The consistently low scores in "personal access to resources" relate to the findings that, in all cases, there is a rather tight and heavily regulated budget available for the geoICT activities. The geoICT staff members claim to have limited financial responsibilities for their individual tasks. As a result, they feel little incentive or obligation to look for alternative funds.

The degree that individual staff members—at any level within the geoG2G—show the ability to formulate alternative solutions for given problems reflects another aspect of discretions. The score for this aspect is high for all cases but is different for staff members at different organizational levels. In the Cadastral case, it was particularly high for strategic staff members, whereas in all other cases, there are mainly discretions of alignment staff members. This difference is likely to relate to the size of the organizational unit. The Kadaster is by far the largest organizational partner engaging in a partnership when comparing with the partners within AHN, Dataland, and Sabimos. Although in some cases municipal organizations of larger cities may have more staff than the Kadaster organization, they are seldom a single partner but part of an umbrella partnership. Therefore, the one-to-many organizational relation of the Kadaster is unique and might therefore also explain the exception when it comes to the emergence strategic discretions. Strategic staff needs to be more aware of possible strategic bottlenecks. That is why in the Cadastral case, they choose to be actively involved in the new interorganizational service developments.

The high score for discretions of alignment staff members can be related to unawareness of the technical geoICT possibilities of strategic staff members in these three partnerships. In contrast, in all these three cases, the alignment staff members are in active contact with the technical experts in this field (such as private IT companies or universities). This active contact provides them the opportunity to maneuver around or even outside higher-level coordination requirements. One theory that could explain this behavior is social network theory, for example, which emphasizes that the networks in which individual staff members are active are influential and have the capability to make organizations opt for certain solutions.

The scores for the aspect of "personal tasks simplification" reveal that in two cases, Dataland and Sabimos, it was the dominant justification to bypass the coordination requirements. In both these cases, the task simplification actions consisted of avoiding constructing geoICT interoperability (with other existing geoICT systems). This finding has be put in context with the fact that the Dataland and Sabimos cases are different from the AHN and the Cadastral cases in both the amount of partners and resources available to construct interoperable database, workflow, and IT system models. With fewer resources, tackling interoperability constraints is possible through making tasks simpler. Instead, they opt to either maintain two parallel systems or utilize one of these systems at their convenience. In contrast, for the staff members in the AHN and the Cadastral cases, interoperability of systems is so crucial that simplification of tasks is not an issue.

Comparing the scores on "client interests" reveals that only in the Cadastral case did discretions emerge with client interests in mind. In all other cases, this aspect scores low, implying that practitioners in these cases are less willing to adapt the requirements of specific geoICT coordination in favor of requirements of particular clients. An obvious difference between the cases is that the client base for the Cadastral case is much more confined to a particular domain than the client base for all other cases. Clients from the Cadastral data are primarily actors in the land and property sector, whereas clients in all other cases are from a larger variety of sectors (AHN data are used by actors in both water and environment; Dataland data are used by small- and medium-sized enterprises; Sabimos data are used by government, citizens, and transport companies). Adhering to needs of a specific client is thus easier in the Cadastral case as compared to all other cases.

9.7 Interpretation of Discretion Findings with Theory

The comparison of scores reveals that discretions vary with the context of individual staff members and with the justification that staff members utilize to reach to discretions. The comparison of the discretion aspect scores across the cases can be further explained using a number of theoretical frameworks.

The similar values of "cognitive filter to the environment" reflect a practice of mimicking or isomorphism across the geoICT domain. The mimicking can not only be the resultant of coercing institutional rules on geoICT, as these differ in each case, but also may, in particular, be the resultant of frequent professional interaction. Through mimicking, individual staff and organizations "fit in" and become accepted in the professional network of geoICT. The result of mimicking is that many staff members have similar professional behavior and thus perceive their environment in a similar way. This behavior is in line with isomorphism theory (DiMaggio & Powell, 1983; Meyer & Rowan, 1977), which posits that mimicking is a survival strategy, on the one hand, and provides stability on the other. Apparently, the professional network of geoICT peers upholds a certain set of technical preferences that steer the network members toward certain solutions. This isomorphic behavior of geoICT practitioners tends to favor the technical solutions of their peers rather than possible alternative technical solutions. Hence, the emergence of alternative discretions, rooted outside the narrow cognitive filter of the environment of the geoICT practitioners, is unlikely.

The scores in "personal access to resources" suggest that cooperation processes reflect resource dependency behavior (Pfeffer & Salancik, 1978), whereby organizations seek to establish relationships with others to obtain the resources that they lack. For the geoICT cases, it seems that as long as actors remain dependent on few human geoICT resources, they are more likely to ensure the access to these resources by complying with the rules to obtain these resources. In the realization that dependency also implies the loss of control and freedom to pursue their daily business, organizations endeavor to minimize their dependence or to increase the dependence of other organizations on them. This pursuit drives organizations to adapt their structure and behavior to optimally acquire and maintain the needed resources. Acquiring the external resources comes by decreasing the organization's dependence on others and/or by increasing others' dependency on it, that is, modifying an organization's power with other organizations.

The emergence of discretions for reasons of personal task simplification reflects uncertainty avoidance. The uncertainty in interorganizational relations arises if the relation is insufficiently regulated, insufficiently enforced, or insufficiently complied with (Williamson, 1998). The result of this insufficient regulation is that each partner incurs a certain cost to enforce or comply with the relation. In the geoG2G cases, the building in and having to maintain the interoperability between geoICT systems can be regarded as an increase in the transaction cost of individual partners. By avoiding maintaining this interoperability continuously, they incur transaction costs only when needed. When this need is not frequent, the transaction costs remain relatively limited. They would become more inclined to invest in more interoperability if the sum of transaction costs of all incidental conversions between systems becomes higher than the transactions cost of converting the whole system at once.

Uncertainty avoidance also explains discretions for clients or other external users. With a wide variety of clients, adaptation would imply having to invest frequently in acquiring information about the client's needs. Instead, it would then be easier, and thus having to invest less transaction cost, to adhere to certain coordination requirements and not to opt for discretions to suit client needs. In the Cadastral case, the strategic staff members already have to invest in acquiring information about the needs and wishes of their partners on a regular basis, so there is no additional transaction cost. This makes discretions favoring such needs easier and less limited by any potential increase in transaction costs.

The combination of these theoretical explanations of discretionary behavior provides a first step toward deriving a more "grand theory" on discretions. This may help in understanding how and e-government projects evolve and how and where projects managers can control and steer projects in a desirable direction.

9.8 Conclusions

Being able to recognize discretionary actions and behavior in interorganizational geoICT projects is not evident, yet it is necessary to better understand good and bad practices in interorganizational (geo)ICT projects in the public sector. Discretions should be taken into account when designing and implementing ICT innovations. Informatization can generate uncertainty or reduce it by limiting discretions of staff. An important part of the recognition of discretionary practices is acknowledging that they tend to arise as reactions to coordination requirements. The extent to which professionals reach for discretions relates to the degree of freedom that the environment grants to actors, the access to resources, and the ability of individual actors to seek and benefit from this freedom. The first and most prominent expression of discretion is in the simplification of tasks. A second expression of discretions is modifying prescribed tasks to cater to the needs of external clients. The first promotes self-interest; the second promotes interest of others. An additional dimension is that discretions vary based on their organizational function: strategic, alignment, or operational.

Empirical evidence of discretions in geoG2Gs shows that strategic discretions originate in the ability to envision alternative courses of action. They are often justified by a perceived need to adhere to external users or clients. The ability to envision alternative courses of action also exists in other cases but is visible mainly among alignment staff rather than strategic staff. Specific to the findings in the AHN case is the relatively low score in the "cognitive filter to the environment" aspect and the relatively high score in the "operations orientation" aspect. This suggests that the functional domain of the geoICT partners can influence the potential of discretions.

The Dataland and Sabimos cases do not exhibit any distinctive scores in discretion aspects. In both AHN and Sabimos, the discretions are reached mainly by alignment staff members and are reached mainly to simplify personal tasks. These findings suggest that either the coordination requirements lack specifications, requiring staff to create their own set of instruments and results, or the context in which the coordination requirements need to be applied is too complex, requiring the staff to simplify their own tasks.

A comprehensive explanation for all discretions may require a combination of theories. This study shows that adoption of ICT may derive various types of discretions. Although the geoICT domain is small, it is prone to discretions because of a relatively limited group of professionals being aware of the geoICT technical choices and opportunities. As a result, project outcomes may differ significantly from original project coordination objectives, as public sector managers are not sufficiently capable of managing the origins or the consequences of each of the discretions. A first attempt in understanding the reasons for emerging discretions combines isomorphic theory, resource dependency theory, and transaction cost theory. Refining such a theory with more empirical evidence is recommended, linking discretions to internal organizational and external-domain specific functions. There is not a best practice of researching or evaluating discretions or of executing geoG2Gs without any discretions. However, acknowledging and understanding that discretions may still emerge regardless of how strictly projects are managed or regulated is a better practice than ignoring such behavior altogether.

Author Biography

Walter Timo de Vries (w.t.devries@utwente.nl) is an assistant professor and course coordinator for land administration at the faculty of geo-information science and earth observation, University of Twente, Enschede, Netherlands. He has a geodetic engineering (MSc) degree from Technical University Delft and a PhD degree in public administration from Erasmus University Rotterdam. His research addresses cooperation and coordination needed to align geoICT and information services within the public sector.

References

Andersson, M., Lindgren, R., & Henfridsson, O. (2008). Architectural knowledge in inter-organizational IT innovation. *The Journal of Strategic Information Systems*, *17*(1), 19–38.

Bekkers, V. (2009). Flexible information infrastructures in Dutch e-government collaboration arrangements: Experiences and policy implications. *Government Information Quarterly*, *26*(1), 60–68.

Benbasat, I., Goldstein, D. K., & Mead, M. (1987). The case research strategy in studies of information systems. *MIS Quarterly*, *11*(3), 369–386.

Bovens, M., & Zouridis, S. (2002). From street-level to system-level bureaucracies: How information and communication technology is transforming administrative discretion and constitutional control. *Public Administration Review*, *62*(2), 174–184.

Brehm, J., Gates, S., & Gomez, B. (2003). Donut shops, speeds traps, and paper work: Supervision and the allocation of time to bureaucratic tasks. In G. A. Krause & K. J. Meier (Eds.), *Politics, policy, and organizations. Frontiers in the scientific study of bureaucracy* (pp. 133–159). Ann Arbor, MI: University of Michigan Press.

Chisholm, D. (1992). *Coordination without hierarchy: Informal structures in multiorganizational systems*. Los Angeles, CA: University of California Press.

Davis, K. C. (1976). Discretionary justice in Europe and America. *The Annals of the American Academy of Political and Social Science, 428*(1), 134.

de By, R. A. (Ed.). (2004). *Principles of geographic information systems* (3rd ed.). Enschede: ITC.

de Vries, W. T. (2013). *GeoICT uniformity in flexibility. Analysis of the influence of geoICT coordination on the cooperation between public organisations with geoICT* (PhD dissertation). Erasmus University Rotterdam, Enschede.

de Vries, W. T., Crompvoets, J., Stoter, J., & VandenBerghe, I. (2011). Atlas of INSPIRE—Evaluating SDI development through an inventory of INSPIRE experiences of European national mapping agencies. *International Journal of Spatial Data Infrastructures Research, 6*(special issue—selected articles from INSPIRE 2010), 126–144.

DiMaggio, P. J., & Powell, W. W. (1983). The iron cage revisited: Institutional isomorphism and collective rationality in organizational fields. *American Sociological Review, 48*(2), 147–160.

Evans, T., & Harris, J. (2004). Street-level bureaucracy, social work and the (exaggerated) death of discretion. *British Journal of Social Work, 34*(6), 871–895.

Finkelstein, S., & Hambrick, D. C. (1990). Top-management-team tenure and organizational outcomes: The moderating role of managerial discretion. *Administrative Science Quarterly, 35*(3), 484–503.

Hay, C., & Wincott, D. (1998). Structure, agency and historical institutionalism. *Political Studies, XLVI*(46), 951–957.

Heeks, R. (2002). Information systems and developing countries: Failure, success, and local improvisations. *The Information Society: An International Journal, 18*(2), 101–112.

Hudson, B. (1989). Michael Lipsky and street-level bureaucracy: A neglected perspective. In L. Barton (Ed.), *Disability and dependency*. East Sussex, UK: Psychology Press.

Hupe, P., & Hill, M. (2007). Street-level bureaucracy and public accountability. *Public Administration, 85*(2), 279–299.

Jorna, F. B. A. (2009). *De autobureaucratie. Informatisering en leren van uitvoering* (Doctor). Leiden. Retrieved from https://www.openaccess.leidenuniv.nl/handle/1887/13782?mode=more.

Jorna, F., & Wagenaar, P. (2007). The 'iron cage' strengthened? Discretion and digital discipline. *Public Administration, 85*(1), 189–214.

Kaniadakis, A. (2012). ERP implementation as a broad socio-economic phenomenon: The agora of techno-organisational change. *Information Technology & People, 25*(3), 259–280.

Lens, V. (2006). Examining the administration of work sanctions on the frontlines of the welfare system. *Social Science Quarterly, 87*(3), 573–590.

Lipsky, M. (1980). *Street-level bureaucracy: Dilemmas of the individual in public services*. New York: Russell Sage Foundation.

Lipsky, M. (1983). *Street-level bureaucracy: Dilemmas of the individual in public services*. New York: Russell Sage Foundation.

Maynard-Moody, S., & Musheno, M. (2000). State agent or citizen agent: Two narratives of discretion. *Journal of Public Administration Research and Theory, 10*(2), 329–358.

Meyer, J. W., & Rowan, B. (1977). Institutionalized organizations: Formal structure as myth and ceremony. *The American Journal of Sociology, 83*(2), 340–363.

Migué, J.-L., Bélanger, G., & Niskanen, W. (1974). Toward a general theory of managerial discretion. *Public Choice, 17*(1), 27–47.

Moe, T. M. (1984). The new economics of organization. *American Journal of Political Science, 28*(4), 739–777.

Pfeffer, J., & Salancik, G. R. (1978). *The external control of organizations: A resource dependency perspective.* New York: Harper and Row.

Quane, J. M., Su, J. H., & Joshi, P. (2009). "It's not business as usual any more": Workers' reflections on discretion, trust and power dynamics at the frontlines of welfare reform. Working paper 09-04. Retrieved from https://orchid.hosts.jhmi.edu/wfp/files/WP%2009-04%20It's%20Not%20Business%20As%20Usual%20Any%20More.pdf.

Snellen, I., & Van de Donk, W. (2002). Electronic governance: Implications for citizens, politicians and public servants. *International Review of Administrative Sciences, 68*(2), 183–198.

Tapia, A. H., Maldonado, E., Tchouakeu, L.-M. N., & Maitland, C. F. (2012). Coordinating humanitarian information: The problem of organizational and technical trajectories. *Information Technology & People, 25*(3), 240–258.

van de Donk, W. B. H. J. (1997). *De arena in schema: Een verkenning van de betekenis voor de verdeling van middelen onder verzorgingshuizen (The arena in a framework: An exploration of the meaning of the distribution of resources under welfare houses).* Lelystad: Koninklijke Vermande.

Webb, A. (1991). Coordination: A problem in public sector management. *Policy & Politics, 19*(4), 229–242.

Williamson, O. E. (1998). Transaction cost economics: How it works, where it is headed. *De Economist—Quarterly Review of the Royal Netherlands Economic Association, 146*(1), 23–58.

Yin, R. K. (1994). *Case study research: Design and methods* (2nd ed.). Thousand Oaks, CA: Sage.

Zouridis, S., & Thaens, M. (2005). Reflections on the anatomy of e-government. In V. H. V. Bekkers (Ed.), *The information ecology of e-government. E-government as institutional and technological innovation in public administration.* Amsterdam: IOS Press.

Zuurmond, A. (1994). *De infocratie. Een theoretische en empirische heroriëntatie op Weber's ideaaltype in het informatietijdperk (The infocracy. A theoretical and empirical reorientation on Weber's ideal type of the rational legal bureaucracy in the information age.).* Den Haag, Netherlands: Uitgeverij Phaedrus.

Chapter 10

The Evolution of Virtual Currencies: Analyzing the Case of Bitcoin

Cecilia G. Manrique and Gabriel Manrique

Contents

Abstract

Major newspapers have carried items on the growing use and acceptability of the digital currency known as bitcoin in electronic transactions that take place without government or central bank intervention. This paper attempts to explain the history and progress of virtual currencies, of which bitcoin is the most widely known. The authors used the case study approach to analyze both the promise and the difficulties associated with the increased use and acceptance of virtual currencies such as bitcoin. The widespread use of virtual currencies, if it does materialize, will have implications for public administration, laws and regulations, business, and economic policy. In public administration, allowing the proliferation of bitcoin as "currency" will challenge the existing premises that surround the sovereignty of the state with respect to what constitutes "legal tender." Bitcoin also has the potential to redefine not only business transactions but also transactions between citizens and government agencies. Likewise, it has the potential to change international transactions among countries and businesses and all types of cross-border developments. The authors will trace and explain the workings of bitcoin and the development of its national and international reach. Issues regarding its constitutionality and economic viability as well as its public policy implications will be addressed.

Keywords: virtual currencies, bitcoin, international exchanges, hashrate, mining

10.1 Introduction

Major newspapers such as *The Wall Street Journal* and magazines such as *Newsweek* and the *Economist* have carried numerous items on the growing use and acceptability of digital currencies, particularly one known as bitcoin, in electronic payment transactions. One reason for its increasing popularity is the freedom it has from government and central bank intervention and regulation. This raises the question of whether the existence of such virtual, unregulated "currencies" poses a challenge to the premises surrounding the sovereignty of the state in declaring what constitutes "legal tender." Should such unregulated virtual currencies be allowed to proliferate for domestic and international transactions, the methods by which citizens of any country conduct transactions with businesses and government agencies are bound to change significantly.

This chapter takes a look at the role of virtual currencies, of which bitcoin happens to be the most well-known example. The chapter will trace bitcoin's national and international reach and raise issues with regard to constitutionality, economic viability, and feasibility as well as its political relations/public policy implications. Both the possible positive and negative effects of the widespread use of bitcoins will be examined. It is hoped that information provided by the authors will allow for the critical evaluation of the role of bitcoin and other virtual currencies in both domestic and international matters.

Because the focus of this book is on government processes and the adoption and adaptation of information technology and modern communication, this chapter also discusses the possible implications on such matters as payment mechanisms by citizens and residents, accounting systems for multiple payment mechanisms, the valuation and taxation of transactions, and the accurate identification of transaction participants.

Thus, this chapter deals with the unregulated spread of virtual currencies, specifically bitcoin, and its impact on public transactions. A systematic literature review is used to place bitcoins within the context of the growing types of virtual currencies and the case study method is used to analyze where bitcoin can be placed in this unregulated spread and within the realm of public transactions.

10.2 Virtual Currencies: A Background

For the most part, societies create money for the purpose of facilitating the exchange of goods and services. Because members of society agree to the value of money based on the backing by the strength of the economy measured by the goods and services it produces, there is a tendency to take the existence of such for granted. In some societies, the creation of money is determined by government through a central bank. The designation of the currency can even be constitutionally determined.

In the United States, the controversy regarding the dollar, its value, and its existence can spark a never-ending debate. Suffice to say, there are those who believe that so long as people agree, anything can serve as money, even cowry shells as long as they maintain reasonable purchasing power over time. In this sense, an official government pronouncement is not necessary to create money. However, there are those who contend that money creation is constitutionally determined and look toward sections of the U.S. constitution to reinforce this view. Section 8 permits Congress to coin money and to regulate its value. Section 10 denies states the right to coin or to print their own money. Some see this as proof that the framers of the constitution clearly intended to create a national monetary system based on coinage and for the power to regulate that system to rest only with the federal government (Rozeff, 2010).

Regardless of perspective, we do know that aside from the dollar, various forms and names for a medium of exchange have sprouted up in the United States and all over the world. During the Great Depression of the 1930s, one result of the Federal Reserve's policy to initially allow the money supply to contract was a substantial shortage of currency. Because of this shortage, about 150 communities experimented with printing their own money to grease the wheels of commerce and get the economy going again. According to Brian Wesbury (2013) in *The Wall Street Journal* article "How Much Does That Burger Cost in Bitcoins?"

none of those Great Depression currencies worked, and all have disappeared. Today, because of worries that the Federal Reserve is printing too much money, which would result in the loss of purchasing power, a private online global currency called bitcoin has been created to protect society from inflationary monetary policy. The introduction of bitcoin or any other form of virtual currency raises the question of whether bitcoins can become an alternative to the money or monies that currently exist.

According to a *BusinessWeek* article by Kharif (2009), virtual currencies have been gaining in popularity since 2009. To date, there are more than 80 kinds of virtual currencies, with an array of odd names like feathercoin, bbqcoin, fireflycoin, and zeuscoin. These have supplanted the previous attempts to create virtual currencies in the 1990s. Table 10.1 lists some of today's virtual currencies, including a few that have been created within virtual worlds.

Understanding virtual currencies is important, according to Guo and Chow (2008), who examined WoW Gold, QBcoin, Linden Dollar, and Acebucks as virtual money systems as shown in Table 10.1. Their importance lies in their ability to help the world understand the economic relations among virtual community members. It allows for the build-up of economic and social order, which could bind the behavior of those in the virtual world into legal virtual activities. That should enable the creation of a bridge between the virtual world and the real world for healthy e-commerce transactions (Guo & Chow, 2008).

The existence of four other virtual currencies mentioned above, the Linden Dollar, Project Entropia Dollar, dotori, and simolean, has raised, for Bray and Konsynski (2006), the following vital questions: Will the formation of virtual worlds undertaking virtual transactions "see the rise of endogenously formed virtual governments?" If such does take place, how will real-world governments respond to the formation of virtual businesses, governments, and realities? Will real and virtual world laws and law enforcement intersect? Will they collide? (Bray & Konsynski, 2006).

It is quite obvious that the real world and the virtual world can intersect and can conflict. People join virtual worlds because they are generally lawless at present and represent an extreme form of liberalism. Such freedoms lead to the "profit-driven abuses of virtual currencies," explained by Huang (2014). Such abuses can include theft, money laundering, exploitation, and market disruption (Huang, 2014).

Given the existence of virtual currencies and the notoriety that bitcoin exhibited during the time period at the end of the 2013 and early 2014, it made the topic ripe for a case study of how a virtual currency that can be used for real-world transactions can exist within the two worlds. Given the melding of the two boundaries with the ability to freely transact, the question is raised with regard to which laws, if any, prevail. What is the impact of the existence of bitcoin on markets, people, and laws especially when it is breaking existing boundaries? This is the focus of the next section of this chapter.

Table 10.1 Some of Today's Virtual Currencies

Name	*Characteristics*
Altcoins	This is a generic name for alternative virtual currencies, most of which use the same software but change a number of economic parameters for it to be generated.
Litecoin	Litecoin has the biggest market capitalization next to bitcoin. It is created every 2.5 minutes, on average, compared with every 10 minutes for bitcoins. It is cheaper in value and easier to generate.
Peercoin	Peercoin is mined, distributed, and validated differently from bitcoins.
Freicoin	Freicoin encourages users to continuously spend this type of money. The fixed supply of bitcoin encourages hoarding, which freicoin does not by keeping it circulating.
Ripple	Ripple has a distinct protocol that allows it to bill itself as a payment system as well as a currency and distributed currency exchange.
Linden Dollar	The Linden dollar was first used in 2003 for the three-dimensional virtual world called Second Life for buying goods and services.
PED or Project Entropia Dollar	This virtual currency of the virtual world Entropia Universe, popular in Europe, does not float but is fixed to an exchange value with the dollar.
Dotori	Dotori, literally meaning "acorns," is the currency in another virtual world called Cyworld.
Simolean	This is the online currency in The Sims Online, a massive multiplayer online real-life game.
QBCoins	This is a virtual currency distributed by the Chinese Internet conglomerate Tencent Holdings, Ltd.
WoW Gold	This is the currency used in the WorldOfWarcraft .com gaming community.

(*Continued*)

Table 10.1 (Continued) Some of Today's Virtual Currencies

Name	*Characteristics*
AceBucks	This currency is used in the Facebook social networking community.
Other names	Greekyogurtcoins, weezercoins, alphacoin, fastcoin, namecoin, hobonickels, gircoin, fireflycoin, zeuscoin, bbqcoin, anoncoin, zerocoin, worldcoin.

Source: Vaishampayan, S. (2013, November 18). Five virtual currencies other than bitcoin. *MarketWatch*. Retrieved from http://www.marketwatch.com/story/five-virtual-currencies-other-than-bitcoin-2013-11-18; Bray, D. A., & Konsynski, B. R. (2006, November 1). *Virtual worlds, virtual economies, virtual institutions*. Retrieved from http://papers.ssrn.com/sol3/papers.cfm?abstract_id=962501; Guo, J., & Chow, A. (2008). *Virtual money systems: A phenomenal analysis*. Retrieved from http://citeseerx.ist.psu.edu/viewdoc/summary?doi=10.1.1.155.1443.

10.3 The Case of Bitcoins: Legal Tender in the Making

Over the period 2009 to 2014, from among the virtual currencies in existence, bitcoins have captured the attention of academics and popular media alike based on the number of articles that has been written about them. Evans-Pughe (2012) tried to analyze bitcoins as a viable alternative to cash when bitcoin was 3 years old. However, even prior to that, Wallace and Jacobs wrote articles that were skeptical of the rise of bitcoin as tantamount to a pyramid scheme (Wallace, 2011) and the challenges to regulatory authorities when it comes to enforcing anything on bitcoin (Jacobs, 2011). There are others who are more optimistic. Grinberg (2011) discussed bitcoins as an innovative alternative digital currency, and Bollen (2003) looked at them as being the future of online currencies.

So what are bitcoins exactly and how do they work? Bitcoins are billed as the first decentralized peer-to-peer payment network or digital, virtual, or "crypto" currency of its kind. It is the first implementation of a concept called "crypto-currency," which was first described in 1998 by Wei Dai on the cypherpunks mailing list. At that time, it was suggested that a new form of money can be made that uses cryptography (the study or the science of the techniques of transforming information to make it secure from unintended recipients or users) to control its creation and transactions via the Internet and solely online (*Merriam-Webster Dictionary*, 2014).

In fact, bitcoins are so "crypto" that nobody knows who developed it. Various sources have indicated that the first specification and proof of the bitcoin concept were published in 2009 in a cryptography mailing list by an unknown person or group that goes by the name Satoshi Nakamoto. A person named Dorian Satoshi

Nakamoto has since been alleged to be the originator of bitcoin, but he has denied being connected with bitcoin. He has sued *Newsweek* for incorrectly linking him to bitcoin (Lattman, 2014; *Winona Daily News*, 2014).

Rather than having a central authority or bank to regulate them, bitcoins are powered by their users, with no authority, organization, government, or middlemen to control them. Because there is no mint either for bitcoins, users "mine" bitcoins by using computer algorithms to solve complex mathematical problems. Bitcoins are mined by running a computer algorithm that creates a digitized code that can then be used to complete online transactions. The algorithm makes it progressively harder to mine bitcoins as the total number increases. There is supposed to be a preset limit to the number of bitcoins that can be mined. That limit is 21 million, and as of the first quarter of 2014, about 12 million bitcoins were in circulation. The tremendous power and speed at which computers attempt to solve the problems and accumulate more bitcoin were initially measured in terms of "hashrate," the speed at which blocks of code are solved to mine virtual coins. With the exponential increase in computer power and speed, the measure has been raised to "petahash," with *peta* standing for a quadrillion times the power to mine a virtual coin (Sidel, 2013a). Because of the anticipation of a wider acceptance of bitcoins, the price of a bitcoin soared to more than $1,200 at the beginning of 2014. However, its price fell significantly due to a number of factors, including the declaration of Baidu.com, a Chinese web services company, that it would not accept bitcoin for payments; the hacking that had taken place on the server of one of the major exchanges; and the bankruptcy of Mt. Gox, one of the major exchanges for bitcoin (Galston, 2014). It is important to note that the volatility in the price of bitcoin is one factor that prevents its greater acceptance.

Bitcoins are transferred from person to person via the Internet, without going through a bank or clearing house. This lowers the transaction fees when using bitcoin compared with using traditional money. Bitcoins can be used in every country as long as private parties engaged in the transactions agree to its use. Bitcoin accounts cannot be frozen by government agencies, and there are no arbitrary limits on the size and frequency of transactions.

Bitcoin's underlying software code, known in developer circles as "the protocol," is believed to keep track of every transaction using a special marker that can be traced via an online ledger. The bitcoin network is sharing a public ledger, called the "block chain." This ledger contains every transaction ever processed, allowing a user's computer to verify the validity of each transaction. The authenticity of each transaction is protected by digital signatures corresponding to the sending addresses, allowing all users to have full control over sending bitcoins from their own unique bitcoin addresses and wallets.

Users trade the coins among themselves, usually through exchanges like the now defunct Mt. Gox, or spend them on various items like tickets to Sacramento Kings basketball games and at online retailer Overstock.com.

When compared with money, virtual currencies can perform the usual functions of money. They are a medium of exchange, a unit of account, a store of value, and a standard used for deferred payment. From the perspective of many institutionalists though, it is lacking the one characteristic required for a "currency" like bitcoin to truly become money: It is not yet generally accepted by economic entities in the performance of the above-mentioned functions. It is widely accepted by societies that believe that what ultimately gives money its value is that it is accepted by others in exchange for something else believed to be of equal value. To become a true alternative currency, bitcoins need to be accepted by a wide enough segment of society so that normal business transactions can be undertaken. Short of such widespread acceptance, bitcoins will always trade at a discount of its potential value. Currently, total currency in circulation and deposits in the U.S. banking system (the money supply, M2) is roughly $11 trillion. Assuming 21 million bitcoins are mined and they become an accepted currency, each one could be worth as much as $524,000 (Wesbury, 2013).

It is estimated that only a miniscule one-hundredth of 1% of companies accept bitcoin as payment for all transactions. If we were to apply the same proportion to all spending in the economy (the gross domestic product), then bitcoin would be worth only $52.40, compared with its potential of $524,000. This is consistent with the notion that anything to be used as money will only get its value from the goods, services, and assets that it can purchase (Wesbury, 2013).

Among the factors prohibiting bitcoin from becoming generally accepted as money are the computer-related costs it requires. Bitcoins require computer storage space, the electrical power to run the computer, a sense of security from hacking, and the computational power that allows for serious encryption on both sides of a transaction to make users believe in its staying power and ability. There are firms that act as middlemen in bitcoin transactions and firms that make a market in bitcoins, but because they are new and have no reliable financial track record, many potential users are wary of using them especially after the fall of Mt. Gox. Another factor working against bitcoin is the perception that it facilitates illegal commerce particularly because it has been linked with the website Silk Road, where it is alleged that online, anonymous, nontraceable transactions for illegal goods along with legal goods take place. Thus, it can be said that the bitcoin world is not friction-free or clean (Wesbury, 2013).

To become a real alternative currency, bitcoins must be recognized by a vast majority of users including businesses and consumers. They must be deemed as safe as, if not safer, than the currency issued by a central bank. And they must be easily transportable. Currently, because bitcoins do not meet any of these requirements, it is trading for much less than its actual convertible U.S. dollar value (Wesbury, 2013).

Supporters of this virtual currency, sometimes called bitcoinistas, view bitcoin as more than just a currency used to store value. For them, bitcoin is an open financial platform that could house scores of types of data in a secure and universal ledger. Mundane tasks from payments for cars and houses to online purchases of tickets can be transacted in bitcoin if it can evolve into an independent, secure, and reliable host of financial and personal information. And such is the hope of those who now transact in bitcoins.

However, even bitcoin evangelists concede that for bitcoin to replace the dollar, three conditions need to exist, some of which undermine the underlying reason why users gravitated to it in the first place. For one, the existing infrastructure, which is largely anonymous, anchored by unregulated overseas exchanges, and vulnerable to manipulation by criminals, must be overhauled through the creation of U.S. exchanges overseen by financial watchdogs, the same watchdogs that undermine the freedom from regulation that bitcoinistas boasted about. Second, institutional money like pension funds will have to be invested in bitcoin to curb its wild price volatility. Third, banks will have to view bitcoin as a legitimate mode of transaction and enable customers to exchange it for dollars and cents. None of these are a reality for today's bitcoin, which makes it a long way from serving as a replacement for the dollar. There is a real possibility though that bitcoin will play more than a bit part in the financial industry of the future (Guerrera, 2014).

In fact, when David Woo of the Bank of America, Head of Global Rates and Currencies Research, came out with an assessment of bitcoin in November 2013, it became tantamount to an endorsement of bitcoin as a major means of payment for e-commerce transactions. Woo even identified the three things that needed to happen for bitcoin to be worth what it currently is. He said that to justify the current value of bitcoin, it will need to (1) account for at least 10% of all global e-commerce business to consumer transactions, (2) become one of the top three players in the money transfer industry, and (3) acquire a store of value reputation close to silver (Bank of America, 2013).

10.3.1 Bitcoin Uses—Positive and Negative Impacts

Bitcoin has had its impact in the economic and political realm during the 5 years that it has been in existence. It can be used in buying video games, gifts, books, and others. Table 10.2 lists some of the items that can be paid for using bitcoins. For smaller businesses and freelancers, the willingness to accept bitcoin is an effective way to get noticed.

In its website, the Bitcoin Foundation provides users with various reasons for the positive impact of bitcoins in society in much the same way that it elaborates on the negative impact of bitcoins.

Positive Effects

Payment freedom: It is possible to send and receive any amount of money instantly anywhere in the world at any time.

Very low fees: Bitcoin payments are currently processed with either no fees or extremely small fees.

Fewer risks for merchants: Bitcoin transactions are secure and irreversible and do not contain customers' sensitive or personal information.

Table 10.2 Some Companies That Accept Bitcoins for Transactions

Company	*Items*
Sacramento Kings basketball team	Tickets, jerseys, hot dogs, and beer
Car company	Luxury buying such as Teslas and Lamborghinis
University in Cyprus	Tuition
Bitcoin Foundation	Pays employees in bitcoin
Zynga	Farmville 2
Overstock.com	Items it sells such as shirts, appliances, etc.
OKCupid	Dating site
EVR	Swanky bar in Manhattan
Bitcoin Shop	Cupcakes and sushi; sells everything from soap to cellphones

Source: Information culled from various *The Wall Street Journal* articles such as Casey, M.J., & Sidel, R. (2014, February 14). *The Wall Street Journal.* p. C2; Fowler, G. A. (2014, February 19). *The Wall Street Journal.* pp. D1, D3; Light, J. (2014, February 7). *The Wall Street Journal.* p. C4; Wesbury, B. (2013, December 16). *The Wall Street Journal.* p. A13.

Security and control: Bitcoin users are in full control of their transactions; it is impossible for merchants to force unwanted or unnoticed charges as can happen with other payment methods. Bitcoin payments can be made without personal information tied to the transaction.

Transparent and neutral: All information concerning the bitcoin money supply itself is readily available on the block chain for anybody to verify and use in real-time.

Negative Effects

Degree of acceptance: Many people are still unaware of bitcoin.

Volatility: The total value of bitcoins in circulation and the number of businesses using bitcoin are still very small compared with what they could be. Therefore, relatively small events, trades, or business activities can significantly affect the price.

Ongoing development: Bitcoin software is still in beta with many incomplete features in active development (Bitcoin Foundation, 2009).

Despite the negative aspects of bitcoin outlined by the Bitcoin Foundation above, an interesting article in *The Wall Street Journal* written by Vigna (2013) talks about a husband and wife team, the Craigs, for whom bitcoin represented a chance at adventure and an independent movie career. They embarked on a road trip in the United States and abroad, during which they not only introduced a number of people to the fledging currency but eventually even persuaded them to use it. After scouring Provo, UT, they found one grocery store, LoLo's Fresh Food Warehouse, that would accept the currency. They also found one auto insurance company that would take bitcoin. The single hardest thing to get regularly was someone who would accept bitcoins for gasoline.

Mr. Craig's employer agreed to pay him in bitcoin. The couple found travel agencies in Germany that accepted bitcoin and, through them, arranged their hotels and flights for the world tour. They rented a car with bitcoin from Five Star Auto Direct in Orem, UT, which also helped sponsor the film. At every stop, they found at least one bitcoiner willing to help so that the Craigs managed to live on bitcoin for over 3 months engaging in their bitcoin-affirming adventure (Vigna, 2013).

There has been sufficient interest in the currency that it is now being discussed in investing conferences. There are even international conferences that focus on bitcoin and virtual currencies. Table 10.3 provides a partial list of such conferences for the year 2014 alone.

The fame or notoriety of bitcoin is usually measured by the news about Tokyo-based Mt. Gox, the most publicized of bitcoin failures. In *The Wall Street Journal*, Galston (2014) describes the failure as a virtual-wallet bankruptcy. Users store their bitcoins in virtual wallets that can reside in exchanges like Mt. Gox. If such exchanges are to be successful, they must be able to avoid the following: fund seizures by governments and regulatory agencies, the malfunctioning of poor payment processors, and weather the storm of technical accounting incongruities. Mt. Gox's failure was not unique, and it will not be the last one. Other bitcoin exchanges may be likely to fail unless operational problems such as the prevalence of gray-market processing, possible government seizures of funds because of the unclear regulatory environment, and the lack of ring-fencing around customer deposits are eliminated. This was evident in the hacking of Mt. Gox, which occurred before the bankruptcy. The hacking caused it to lose a substantial amount of bitcoin owned by its customers (Galston, 2014).

Bitcoin's value, when expressed in U.S. dollars, has been subjected to wild swings. This prevents bitcoin from being widely accepted as money even though using virtual electronic currency can be easier than carrying cash and more secure than using a debit or credit card. Because most people still cannot grasp the derivation of bitcoin from complex algorithms mined by computers, most ordinary people do not partake in its derivation and use. Without a large base of users, virtual currency values will continue to be subjected to unpredictable swings in value. Therefore, it cannot be reliably counted on as a store of value, a means of deferred payment, a unit of account for the valuation of goods and services, and a medium of exchange—the four functions of money referred to previously.

The volatility of bitcoin's value will continue as long as its acceptance remains very limited because relatively small events can trigger a big change in its value. Over time, as the usefulness of virtual currency becomes more widely accepted,

Table 10.3 Bitcoin Events and Conferences in 2014

Date	*Conference Title*	*Place*
April 3	Swedish bitcoin conference 2014	Stockholm, Sweden
April 7	Inside bitcoins NYC	New York, NY, United States
April 11	Bitcoin expo 2014	Toronto, ON, Canada
April 23	Bitcoin conference Russia	Moscow, Russia
May 10	Global bitcoin summit 2014	Beijing, China
May 15	Bitcoin foundation bitcoin 2014	Amsterdam, the Netherlands
May 31	Central Europe bitcoin expo	Vienna, Austria
June 20	Bitcoin in the beltway	Washington, DC, United States
June 24	Inside bitcoins conference & expo	Hong Kong, China
July 9	Inside bitcoins conference & expo	Melbourne, Australia
August 25	Camp bitcoin	Black Rock City, NV, United States
September 1	World bitcoin forum 2014	Bonn, Germany
September 15	Inside bitcoins London	London, England
November 2	Bitcoin world at money 2020	Las Vegas, NV, United States

Source: The Bitcoin Foundation. *Bitcoins*. Retrieved from https://bitcoin.org/en/.

such volatility will decline particularly if more institutional investors include virtual currency in their portfolios.

Furthermore, even as bitcoin affords users with additional means to practice economic and political freedoms, the same freedoms are used by some to engage in illegal activities, as shown by some of the traffic that went on in Silk Road. The use of bitcoin in transactions involving money laundering, illicit drugs, narcotics, and child pornography (to name just a few) also leads to resistance from legitimate businesses.

However, the fact that all bitcoin transactions are publicly available (and therefore can be tracked, in theory, by law enforcement agencies) and that every bitcoin is defined by its unique transaction history (making it difficult for criminals to cover their tracks) may limit the growth of its use in the black market/underworld. In addition to its role as a means of payment for online commerce, bitcoin can be used for transfers of money (e.g., immigrant worker in the United States sending remittances back home). This can be done very cheaply and fast (online settlement in under 10 minutes if the sender is trustworthy, like a family member, or 50 minutes when the settlement is with strangers). Its potential role in the fast-growing realm of global remittances is significant.

10.3.2 International Exchanges

The bitcoin industry has spawned a whole new set of vocabulary and acronyms for institutions and organizations engaged in various activities. Table 10.4 provides

Table 10.4 Country and Bitcoin Organization

Country	*Organization Name (Some in Native Language)*
Argentina	Fundación Bitcoin Argentina
Australia	Bitcoin Australia
Austria	Bitcoin Austria
Belgium	Belgian Bitcoin Association
Canada	Bitcoin Embassy, Bitcoin Alliance of Canada
Denmark	Dansk Bitcoinforening
Germany	Bundesverband Bitcoin e.V.
India	Bitcoin Alliance of India
Ireland	Irish Bitcoin Foundation
Israel	איגוד הביטקוין הישראלי
Italy	Bitcoin Foundation Italia
Netherlands	Stichting Bitcoin Nederland
Sweden	Svenska Bitcoinföreningen
Switzerland	Bitcoin Association Switzerland

Source: The Bitcoin Foundation. *Bitcoins*. Retrieved from https://bitcoin.org/en/.

information about the international reach of bitcoin by showing the reader what countries it operates in and what the name of the organization in that country is (sometimes even in the native language).

In addition, shown in Table 10.5 are some of the bitcoin exchanges in existence worldwide that earn income from trading commissions and other service fees.

The demise of Mt. Gox and the extreme fluctuations that were experienced by investors in bitcoin who are now looking for ways to sue and get their money back have brought bitcoin to the attention of venture capitalists, computer programmers, merchants, investors, entrepreneurs, bitcoin enthusiasts, technology experts, and fledgling firms as well as governments.

10.3.3 Regulation

What types of regulation need to be developed, and who will need to be involved? Because of the span of the impact of transacting in bitcoins, various countries have been trying to come up with their perspectives on how to deal with such a phenomenon. Table 10.6 provides a snapshot of a few international views on the regulation of bitcoins.

As one can tell from Table 10.6, the perspectives on the regulation of virtual currencies, including bitcoins, can be quite controversial. Those who advance the freedom to transact in virtual currencies are afraid that regulation will stifle or stall innovation and send jobs overseas.

The events in early 2014 with the hacking of computers and the bankruptcy of Mt. Gox have been forcing U.S. regulators to confront the many questions regarding their authority and responsibility toward overseeing the fledgling virtual currency (Table 10.7). Very often, it becomes a question of determining whether their oversight extends to bitcoins. Some, like the Federal Reserve, have indicated that they cannot oversee bitcoin without legislative action, that the central bank does not have the authority to regulate bitcoin as long as the currency remains separate from the banking system that the Federal Reserve oversees. Other regulators, including the Federal Trade Commission, appear more inclined to act.

Some government officials think that virtual currency companies should be subject to more regulation, including rules governing what risks companies must disclose to customers. A study of 40 bitcoin exchanges set up over the past 3 years found 18 closed, with customer accounts often wiped out. Because bitcoins can be moved quickly between anonymous accounts and transactions are not reversible, there is little recourse in recovering that money.

Owners of bitcoin can send and receive bitcoin using software and swap it on exchanges for dollars or other fiat currencies. A small but rising number of merchants accept it. Unlike with banks, there is no backing by a federal regulatory agency like the Federal Deposit Insurance Corporation. Moreover, because no

Table 10.5 Bitcoin Exchanges, National and International

Exchange Name	*Country*	*Description If Available*
BitInstant	New York	Popular website allows users to buy and sell bitcoin; one of the first exchanges to register with the Treasury Department.
BTC	China	Has introduced vouchers allowing users to buy and sell bitcoin without having to pay the exchange through a third party.
OKCOIN	China	Allows users to make payments directly to bank accounts.
BitStamp	Slovenia	N/A
BTC-e	Bulgaria, Cyprus	N/A
Bitomat	Poland	N/A
Mt. Gox	Tokyo, Japan	Tokyo-based platform that started as an exchange for Magic trading cards but quickly became the world's dominant platform for exchanging bitcoins. Has since gone bankrupt.
CoinDesk	United States	Leading provider of bitcoin-related data.
BitPay Inc.	Atlanta	Processes bitcoin transactions on behalf of merchants.
Coinbase	New York/ San Francisco	Digital wallet helps users buy and accept bitcoins. Startup that seeks to make it easier for merchants to take bitcoin as payment.
BitCoin foundation	Washington, DC	A trade organization that promotes virtual currency with regulators, law enforcement, and other groups.
Coinsetter	New York	Bitcoin-trading platform.
Genesis block		New York research and data firm that tracks the bitcoin industry.

(*Continued*)

Table 10.5 (Continued) Bitcoin Exchanges, National and International

Exchange Name	*Country*	*Description If Available*
SecondMarket Holdings Inc.		Creates a market for trading of private holdings of bitcoin.
The Winklevoss Bitcoin Trust		Proposed exchange traded fund that would hold bitcoins; registered with the Securities and Exchange Commission.

Source: Material for this table was culled from various *The Wall Street Journal* such as Deng C. (2014, January 9). *The Wall Street Journal*. p. C2; Sidel, R. (2013b, November 29). *The Wall Street Journal*. p. C4; Sidel, R. (2014, February 12). *The Wall Street Journal*. pp. C1–C2; *Economist*; and *Newsweek* issues.

Table 10.6 International Views Regarding Bitcoin

Country	*View*
Finland	Central bank views bitcoin as a digital commodity.
Germany	Bitcoin is treated as private money.
China	Clamped down and restricted banks' use of virtual currencies. People's Bank of China stated that financial institutions should not take deposits from bitcoin-related businesses or provide them insurance or custody services. Chinese regulators will require bitcoin exchanges to file trading records and take measures against money-laundering risks associated with the virtual currency.
United States (New York)	Regulation is slow in coming[a] (BitLicence—set up rules for virtual currencies).
Russia	Regulations are being tightened.
Canada	Regulating bitcoin usage is being considered.
European Union	European Commissioner in charge of financial regulation said that government penalties should apply if unregulated currencies such as bitcoin are being used to commit fraud or other criminal activities.

Source: The Economist. (2014, February 1). Bitten. p. 62.

[a] Discussed further in the text.

Table 10.7 Government Agencies Involved

Agency	*Unit*	*Task*
Treasury	Financial Crimes Enforcement Network	The agency has registered as a money-transmission business since issuance of guidelines in March that encouraged them to do so.
	Secret Service	U.S. Secret Service, which investigates counterfeit currencies, needs to ensure that companies dealing in bitcoin comply with existing regulations on money laundering and other matters.
Federal Trade Commission	Consumer Financial Protection Bureau	The agency has held meetings to study the issue.
Securities and Exchange Commission		The agency is still evaluating whether bitcoin falls under their jurisdiction.
Federal Reserve		The Fed has no authority to regulate the virtual currency as long as it "doesn't touch" banks that the Fed oversees.

Source: Various *The Wall Street Journal* such as Tracy, R. (2013, November 19). *The Wall Street Journal.* pp. C1–C2; Tracy, R., & Armour, S. (2014, March 3). *The Wall Street Journal.* pp. C1–C2; Warnock, E., & Mochizuki, T. (2014, February 25). *The Wall Street Journal.* p. C3; Yadron, D., & Barrett, D. (2013, October 23). *The Wall Street Journal.* p. A3; *Economist,* and *Newsweek* articles.

central bank or government backs the currency, there is no assurance that depositors who use bitcoin exchanges or electronic wallets will get their funds back should something happen to the currency (Tracy & Armour, 2014).

The Consumer Financial Protection Bureau and the Federal Trade Commission have held meetings to study the issue. Market regulators, including the Securities and Exchange Commission, are still evaluating whether bitcoin falls under their jurisdiction. Agencies that oversee banks and payment systems are monitoring bitcoin, but Federal Reserve Chairwoman Janet Yellen has already indicated that the Fed has no authority to regulate the virtual currency as long as it "doesn't touch" the banks that the Fed oversees (Sidel, Warnock, & Mochizuki, 2014).

10.4 Conclusions

If the use of bitcoin, or any other virtual currency, becomes widespread, government agencies will also need to rethink the way they conduct business with citizens, with companies, and with other agencies, both national and international. The changes that it might bring about will create challenges that are not too early to address. Some of the questions it raises for government agencies are addressed in the following discussion. In addition to the new laws and regulations that will have to be written, there will have to be major work by professionals in information technology to respond to the challenges.

The first issue we raise concerns taxes. Should sales and excise taxes apply to bitcoin transactions the way they do to usual currency transactions? If so, how should taxes be calculated? What reporting mechanism will be utilized to record bitcoin transactions? With regard to income, how should the income tax on transactions that involve bitcoin be calculated, assessed, and paid? For example, if a company agrees to pay its employees their salary in bitcoin, how will the withholding taxes be taken and sent to the Internal Revenue Service? Not reporting transactions in bitcoins to the appropriate government agencies will be tantamount to sanctioning more of the underground economy.

The second issue concerns what methods of payment will be accepted by government agencies. For example, if a municipality charges its resident for water use, will it be possible to pay in bitcoin? If the answer is yes, this will have ramifications in keeping records, as accounting entries will have to be done in bitcoin as well as in the local currency. As a result, there will also be a need to develop procedures for converting bitcoins to local currency and vice versa.

The anonymity of the parties involved in bitcoin transactions poses major challenges to government agencies as well. Regulations and processes will have to be developed to allow the relevant information to be collected. This will necessitate even more government gathering of cyber data even as many around the world have deplored the massive collection of telephone and email data by the National Security Agency of the United States.

Cross-border transactions will also pose challenges if virtual currencies become widely accepted for international transactions. Virtual currency use by criminals for criminal activity can be expected to rise. However, even for legitimate international transactions using virtual currencies, processes will need to be developed to track the flow of bitcoin for data collection purposes, for taxation, and for the development of appropriate public policies that address cross-border transactions.

Bitcoin's most ardent friends and foes have predicted that it might one day replace the dollar and other traditional currencies. Bitcoin's strength is predicated on three supposed qualities: It is anonymous, or at least pseudonymous (transactions are recorded but the identities of the parties are encrypted); it is difficult to hack; and it cuts out financial middlemen like banks.

That is the crucial question. To thrive, bitcoin has to be more useful than current payment systems. Two possible applications could embed virtual currencies into the financial infrastructure in a way that is complementary to existing forms of payments. The first is as a conduit for small international transactions like remittances from foreign workers. Bitcoin could reduce both the cost and the time required for such payments (Guerrera, 2014).

Because money is nothing more than a common numeraire that reduces the search costs associated with conducting beneficial trades, and also because money is a psychological abstraction, literally anything can serve in this capacity as long as people are willing to accept it as a medium of exchange, so long as it maintains its purchasing power over time, and as long as it can serve as a convenient unit of measure. A government edict is technically not necessary to create money.

At this particular stage, one of the attractions of bitcoins and virtual currencies is its freedom from government regulation, as shown by its not being under the purview of the Federal Reserve and the reluctance of any government agency to take on its regulation. As such, bitcoins have taken on the form of an economic disruption likened to a stock market investment. Thus, its value rises and falls as a result of the decisions of those in the market. In addition, at this stage, aside from the questioning of the government's role in protecting those who dabble in bitcoins, its existence has not produced a political disruption because the trade in bitcoins is but a minute part of financial transactions nationwide or worldwide to cause an economic upheaval that leads to a political one.

The field of virtual currencies and bitcoins as an area of research is a fertile one because material on it is unfolding right before our very eyes. There are therefore numerous areas for further research, beginning with knowledge about the various international virtual currencies in existence and their usefulness. The "cryptocurrency" craze has spawned more than 80 entrants to the field wanting to be recognized as legitimate media for exchange. Where will such lead? Will it lead to the proliferation of currencies? Or will it lead to the acceptance of only a few? Will it lead to national regulation? Will it lead to international regulation? How it will change the way nations deal with each other in terms of foreign transactions?

Bitcoins have brought to the forefront of the agenda several crucial questions that are ironically begging for responses from the very government that bitcoins was supposed to be immune from.

Acknowledgments

We acknowledge the research help provided by Chelsea Fischer and Tiana Goglio, undergraduate students at the University of Wisconsin–La Crosse.

Author Biographies

Dr. Cecilia G. Manrique is a professor and former chair of the Political Science/Public Administration Department, University of Wisconsin–La Crosse. She received her doctorate in political science from the University of Notre Dame, her MA degree in international affairs from Ohio University, and her BA degree in economics from Maryknoll College in the Philippines. She has published in the fields of immigration (*The Multicultural or Immigrant Faculty in American Society*) and incorporating technology in the political science classroom (*The Houghton Mifflin Guide to the Internet for Political Science*).

She has been a member of the American Political Science Association (APSA) since 1988, where she was the president of the Computers and Multimedia Section (now Information Technology and Politics) and has been the Treasurer since 1995. She is also the Treasurer of the Wisconsin Political Science Association. In 2012, she completed a second three-year term as the President of the International Leadership Council of the Golden Key International Honour Society.

She is married to Gabriel and they have two children, Patrick and Michelle.

Gabriel Manrique is a professor and chairman of the Department of Economics at Winona State University. He has also served as the dean of the College of Business and the director of Outreach and Continuing Education. His industry experience includes working as the globalization specialist of Fastenal Company, with the responsibility of setting up its foreign subsidiaries. He received a PhD degree in economics from the University of Notre Dame, a MA degree in international affairs from Ohio University, and a BA degree in psychology from Ateneo de Manila University in the Philippines. He has published in the areas of international trade, immigration, and economic development.

References

Bank of America. (2013). *Bitcoin: A first assessment.* Retrieved from http://cryptome.org/2013/12/boa-bitcoin.pdf.

The Bitcoin Foundation. (2009). *Bitcoins.* Retrieved from https://bitcoin.org/en/.

Bollen, R. (2003). *The legal status of online currencies: Are bitcoins the future?* Retrieved from http://papers.ssrn.com/sol3/papers.cfm?abstract_id=2285247.

Bray, D. A., & Konsynski, B. R. (2006, November 1). *Virtual worlds, virtual economies, virtual institutions.* Retrieved from http://papers.ssrn.com/sol3/papers.cfm?abstract_id=962501.

Casey, M. J., & Sidel, R. (2014, February 14). Firms bank on bitcoin bounce back. *The Wall Street Journal*, p. C2.

Deng, C. (2014, January 9). China bitcoin exchanges face a new reality. *The Wall Street Journal*, p. C2.

The Economist. (2014, February 1). Bitten, p. 62.

Evans-Pughe, C. (2012). *From megabytes to megabucks.* Retrieved from http://www.timelex.eu/frontend/files/userfiles/files/publications/2012/Bitcoin_article.pdf.

Fowler, G. A. (2014, February 19). Want to buy everyday stuff with bitcoin: A user's guide. *The Wall Street Journal*, pp. D1, D3.

Galston, E. (2014, February 28). The bitcoin paradox that undid Mt. Gox. *The Wall Street Journal*, p. A13.
Grinberg, R. (2011). *Bitcoin: An innovative alternative digital currency*. Retrieved from http://hstlj.org/wp-content/uploads/2011/12/8-Grinberg-159-208.pdf.
Guerrera, F. (2014, February 18). Turning point for virtual money. *The Wall Street Journal*, pp. C1, C3.
Guo, J., & Chow, A. (2008). *Virtual money systems: A phenomenal analysis*. Retrieved from http://citeseerx.ist.psu.edu/viewdoc/summary?doi=10.1.1.155.1443.
Huang, D. Y. (2014). *Profit-driven abuses of virtual currencies*. Retrieved from http://www.scribd.com/doc/191989114/Profit-Driven-Abuses-of-Virtual-Currencies.
Jacobs, E. (2011). Bitcoin: A bit too far? *Journal of Internet Banking and Commerce, 16*(2), pp. 1–4. Retrieved from http://goo.gl/mLYvRV.
Kharif, O. (2009, May 6). Virtual currencies gain in popularity. *BusinessWeek*. Retrieved from http://www.docstoc.com/docs/75596198/Virtual-Currencies-Gain-in-Popularity.
Lattman, P. (2014, March 18). Man denies bitcoin role described by *Newsweek*. *International New York Times*, p. 17.
Light, J. (2014, February 7). New perch in markets for bitcoins. *The Wall Street Journal*, p. C4.
Merriam-Webster Dictionary. (2014). s.v. "cryptography." Retrieved from http://www.merriam-webster.com/dictionary/cryptography?show=0&t=1419294141.
Rozeff, M. (2010). *The US constitution and money*. Retrieved from http://mises.org/books/rozeff_us_constitution_and_money.pdf.
Sidel, R. (2013a, November 6). Race to mine bitcoin gathers steam. *The Wall Street Journal*, pp. C1–C2.
Sidel, R. (2013b, November 29). Bitcoin puzzle: Prices vary widely on exchanges. *The Wall Street Journal*, p. C4.
Sidel, R. (2014, February 12). Hackers launch attacks on bitcoin exchanges. *The Wall Street Journal*, pp. C1–C2.
Sidel, R., Warnock, E., & Mochizuki, T. (2014, March 2). Almost half a billion worth of bitcoins vanish. *The Wall Street Journal*, pp. A1, A6.
Tracy, R. (2013, November 19). Authorities see worth of bitcoin. *The Wall Street Journal*, pp. C1–C2.
Tracy, R., & Armour, S. (2014, March 3). Losses mobilize the bitcoin police. *The Wall Street Journal*, pp. C1–C2.
Vaishampayan, S. (2013, November 18). Five virtual currencies other than bitcoin. *MarketWatch*. Retrieved from http://www.marketwatch.com/story/five-virtual-currencies-other-than-bitcoin-2013-11-18.
Vigna, P. (2013, November 14). Bitcoin couple leave wallet behind. *The Wall Street Journal*, pp. C1–C2.
Wallace, B. (2011, November 23). The rise and fall of bitcoin. *Wired Magazine*, pp. 1–4.
Warnock, E., & Mochizuki, T. (2014, February 25). Wanted: A bitcoin regulator. *The Wall Street Journal*, p. C3.
Wesbury, B. (2013, December 16). How much does that burger cost in bitcoins? *The Wall Street Journal*, p. A13.
Winona Daily News. (2014, March 18). Nakamoto hires lawyer in *Newsweek* bitcoin fight, p. A2.
Yadron, D., & Barrett, D. (2013, October 23). Bitcoin poses test to law enforcement. *The Wall Street Journal*, p. A3.

Suggested Readings

Casey, M. J. (2013, January 17). Coin of kings: Bitcoin scores. *The Wall Street Journal*, p. C1.
Casey, M. J. (2014, February 16). Bitcoin fix sees delays. *The Wall Street Journal*, p. B2.
Deng, C. (2013, December 5). Bitcoin miners dig a path to China. *The Wall Street Journal*, pp. C1–C2.
Light, J. (2013, November 21). Bbqcoin? Virtual money is smoking. *The Wall Street Journal*, pp. C1–C2.
Mochizuki, T., & Warnock, E. (2014, February 17). Mt. Gox shows bitcoin's growing pains. *The Wall Street Journal*, p. C3.
Money and the Constitution, © 1994–2012 GMW—University of Groningen—Humanities Computing. Retrieved from http://www.let.rug.nl/usa/essays/general/a-brief-history-of-central-banking/money-and-the-constitution.php.
Morganteen, J. (2013). *BofA: Bitcoin can become 'major means of payment' for e-commerce.* Retrieved from http://www.cnbc.com/id/101250070.
Queenan, J. (2014, January 5). I'm up on bitcoins, even if they're down. *The Wall Street Journal*, p. C11.
Sidel, R. (2013, November 14). Hire boosts bitcoin credibility. *The Wall Street Journal*, p. C2.
Sidel, R. (2014a, January 29). Bitcoin advocates speak up. *The Wall Street Journal*, p. C3.
Sidel, R. (2014b, February 6). Hard times for a bitcoin evangelist. *The Wall Street Journal*, pp. C1–C2.
Sidel, R. (2014c, February 9). Another hiccup for bitcoin. *The Wall Street Journal*, p. B2.
Sidel, R., & Casey, M. J. (2014, February 11). Fragility of bitcoin uncovered by glitch. *The Wall Street Journal*, pp. C1–C2.
Sidel, R., Casey, M. J., & Warnock, E. (2014, February 26). Shutdown rattles bitcoin market. *The Wall Street Journal*, pp. A1, A8.
Sidel, R., Deng, C., & Horobin, W. (2013, December 6). Bitcoin warnings multiply. *The Wall Street Journal*, p. C2.
Weisenthal, J. (2013, December 5). *This is, quite simply, the biggest endorsement that bitcoin has ever received.* Retrieved from http://www.businessinsider.com/david-woo-on-bitcoin-2013-12#ixzz2xf8J1Ioy.
Yadron, D. (2014, January 2). Tech maverick sets sights on bitcoin. *The Wall Street Journal*, p. A4.

Chapter 11

Gov 2.0, Mobility and Inclusion: A Critical Examination of Social Assistance Reform in Ontario, Canada

Jeffrey Roy

Contents

Abstract

The purpose of this chapter is to examine the relationship between mobility, Gov 2.0, and digital inclusion. By digital inclusion, we refer to not only Internet access (via traditional desktop or mobile channels) but also an expansion of developmental opportunities for a widening proportion of the populace within a given jurisdiction (thereby reducing what has been commonly referred to as the digital divide). As some jurisdictions such as Great Britain and Denmark explicitly target digital inclusiveness, this chapter critically examines the Canadian context in this comparative light. We draw specifically from the Province of Ontario, which has been home to several interrelated and relevant initiatives in this regard, including (a) a recent Provincial Auditor General examination of ServiceOntario, the lead entity responsible for online and integrated service delivery for the Province; (b) an external review panel of provincial social assistance policies and governance arrangements; and (c) a major external research study in the academic sector (supported by, but separate from, the Province) focusing on the impacts of mobile technologies on all segments of Ontario (including government, industry, and society at large). Taken together, these three elements provide a useful case study to better understand if and how digital inclusion is likely to be addressed within Ontario's evolving parameters of mobility, electronic service delivery, and wider Gov 2.0-stylized reform efforts.

Keywords: mobility, digital, inclusion, government, social, assistance, engagement, empowerment, participation, technology

11.1 Introduction

Gov 2.0 denotes a more participative and networked era driven by the Internet's "web 2.0" evolution featuring user-driven content (including text, imagery, and video), social networking, big data, and cloud computing. The advent of the mobile Internet further accentuates this paradigm with, for example, the Organization for Economic Co-operation and Development (OECD, 2011) highlighting the transformational potential of mobile government, provided that technology be viewed and understood as a means toward greater ends: "Focus should be indeed on the needs of the public sector and of the end-users, be these citizens or businesses, to ensure that technology is exploited to reorganize the way civil servants work and to meet the needs of citizens through improved service delivery" (p. 12).

However, such change does not come easy, and almost all organizations are struggling with mobility: One 2013 survey across sectors, for instance, stated that that nearly 80% of business and technology executives viewed their mobile efforts as low to medium in maturity levels, whereas 40% define their organization's mobility strategy as weak (Roy, 2014b). In a public sector grounded in traditional work cultures and legacy infrastructures and underinvesting worldwide in mobile relative to most other industries, such findings undoubtedly find resonance (Roy, 2014b). The government is deeply conflicted between traditional democratic and administrative structures and more Gov 2.0-stylized governance associated with mobile (Roy, 2013, 2014a).

Further societal complexities are also apparent with ramifications for government service delivery. Although the trends presented above lend credence to the importance of mobility for government performance and legitimacy, they likewise underscore stark cleavages and corresponding challenges. Nearly one half of the population, for example, does not presently own a mobile device, and almost all public sector service providers in Canada continue to struggle with a fragmented citizenry in terms of comfort levels and preferences with online channels (both traditional, i.e., PC based, and mobile). Accordingly, with the exception of a small number of countries where digital ubiquity is at the very least viewed as an attainable objective (such as Scandinavian nations that regularly top global digital government surveys and now likewise lead in mobile penetration rates), various studies and audits in Canada and elsewhere reveal a highly uneven trajectory in terms of uptake of online service delivery and a persisting set of fault lines with respect to Internet and technological usage (Auditor General of Canada, 2013; Auditor General of Ontario, 2013). Online and now mobile banking trends in Canada are similar in many respects.*

Within such a context, the aim of this chapter is to examine the relationship between mobility, Gov 2.0, and digital inclusion. By digital inclusion, we refer not only to Internet access (via traditional desktop or mobile channels) but also an expansion of developmental opportunities for a widening proportion of the populace within a given jurisdiction (thereby reducing what has been commonly referred to as the digital divide). As some jurisdictions such as Great Britain and Denmark explicitly target digital inclusiveness, this chapter critically examines the Canadian context in this comparative light. Nonetheless, we specifically examine the Province of Ontario, which has been home to several interrelated and relevant initiatives in this regard, including: (a) a recent Provincial Auditor General examination of ServiceOntario, the lead entity responsible for online and integrated service delivery for the Province; (b) an external review panel of provincial social assistance policies and governance arrangements; and (c) a major external research study in the academic sector (supported by, but separate from, the Province), focusing on the impacts of mobile technologies on all segments of Ontario (including government, industry, and society at large). Taken together, these three elements provide a useful case study to better understand if and how digital inclusion is likely to be addressed

* In 2013, 70% of Canadians owning a smart phone reportedly used financial apps. For a wider review of the survey findings please see: https://newsroom.bmo.com/press-releases/bmo-mobile-banking-survey-70-per-cent-of-canadian-tsx-bmo-201309170898511001.

within Ontario's evolving parameters of mobility, electronic service delivery, and wider public sector reforms centers on human and developmental assistance.

After this introduction, the chapter proceeds as follows. Section 11.2 provides a further conceptualization of Gov 2.0 and mobility, and Section 11.3 examines some explicit strategies for digital inclusion undertaken elsewhere, notably in Great Britain and Denmark. Building on these conceptual and empirical reviews, Sections 11.4 and 11.5 present a critical case study—the Province of Ontario, Canada; Section 11.4 examines three recent initiatives with ramifications for digital inclusion, and Section 11.5 seeks to apply a specific Gov 2.0 template (drawn from the Australian State of Victoria) as a basis for critiquing Ontario's recent social assistance review exercise. Section 11.6 concludes this chapter with a summary of key lessons learned.

11.2 Gov 2.0 and Mobility

There is a growing imperative for governments at all levels to strategically and proactively address the advent of mobility. Although the most tangible signs of the mobile era are the explosion of smart devices and user-devised apps, as the above OECD passage implies, there are much larger consequences and ramifications for the public sector as new patterns of information exchange and communications take hold, new expectations arise, and new governance models are formed. Although smart devices and apps are undoubtedly the most tangible and recognizable symbols of mobility, this movement arguably encompasses four interrelated dimensions that reflect important shifts in technology, demography, and governance: (a) the explosive growth of broadband capacity; (b) the proliferation of personal, handheld devices of all kinds (by 2015, there is estimated to be nearly as many mobile smart devices as living people on Earth); (c) the rapid expansion of both front and back end virtualization (most notably social media and cloud computing respectively); and (d) the emergence of a more digitally ubiquitous society and workplace where traditional boundaries between personal and professional spaces are increasingly blurred (Roy, 2013, 2014a).

New participatory mechanisms, systemic openness, and virtualization are underpinning an emerging governance ethos that, for the public sector, is often termed as the emergence of Gov 2.0. At the heart of Gov 2.0 are drivers of collective intelligence and more collaborative forms of governance that are typically associated with a widening online universe and less hierarchical and control-minded forms of governance (Maier-Rabler & Huber, 2011; Shirky, 2008; Wyld, 2010). From both external vantage points on new societal formations (such as Wikipedia and a myriad of social media-driven movements) as well as internal to the public sector (what Lips characterizes as "public administration 2.0"), governments are increasingly challenged to move beyond a typology of hierarchies and markets and embrace usage of networks typically more open and collaborative in formation and execution (Kostakis, 2011; Lips, 2012; Roy, 2013, 2014a; Stoker, 2005; World Economic Forum, 2011).

Governments are embracing such changes, which present significant structural and cultural shifts. The State of Victoria in Australia, for example, has articulated a Gov 2.0 Action Plan (further adopted as a basis for examining the Ontario case study below) premised on the following principles: leadership, participation, transparency, and performance. Aligned with the spirit of such principles, the spreading of web 2.0 experimentation within government is specifically meant to foster collaboration and democratize the creation and exchange of ideas:

> The role of citizens in an open government environment—enriched by open government data—can be one of democratic innovators. In an ongoing open innovation process, citizens can draw on open data, and propose both policy-areas to tackle and technical approaches to take. (Maier-Rabler & Huber, 2011, p. 186)

This new front-end of service delivery and more active forms of citizen engagement is enjoined by evolving back-end cloud computing infrastructures reflecting the dimensions of virtualization and mobility. As Young (2012) points out, the cloud and mobility are intimately related through four trends similar to what is described above:

> First, electronic sensors got smaller and better. Second, people started carrying powerful computing devices, typically disguised as mobile phones. Third, social media made it normal to share everything. And fourth, we begin to get an inkling of the rise of a global super-intelligence known as the cloud (p. 11).

In a likeminded, albeit more forward-looking and prospective examination of mobility and government, Accenture (2013) offers four claims as to the potential of mobility to drive value and benefit for government and society:

1. Mobile technology can bridge the digital divide.
2. Mobility can change relationships between citizens and government.
3. Mobility can help government become more insight-driven.
4. Mobility can increase workforce productivity—and satisfaction.

Within this wide spectrum of prospective beneficial impacts, our primary focus in this chapter lies in exploring the evolution of the digital state within a 2.0 and mobile context and whether opportunities for enhancing individual development opportunities and in a corresponding manner, lessening the digital divide. The notion of digital inclusion stems from all three of these dimensions and has been increasingly recognized by some jurisdictions as a major component of digital governance projects for the public sector and for society as a whole, with mobility as a prominent feature in this regard.

11.3 Mobility and Digital Inclusion

The notion of digital inclusion stems from the interrelated concepts of the digital divide on the one hand and digital citizenship on the other hand (Mossberger, Tolbert, & McNeal, 2008). The essence of digital citizenship is that online access is necessary but insufficient, as it must also be accompanied by the skills and capacities to develop both individually and collectively within socioeconomic, cultural, and civic and political processes that increasingly comprise a virtual dimension. By contrast, digital divides denote an absence of inclusion with online cleavages across both access and usage tied to factors such as education, income, culture, and geography (Jaeger, 2012).

For the public sector, moreover, digital and mobile reforms and performance and legitimacy are increasingly intertwined. The World Economic Forum (2011), for instance, in their 2011 report on "The Future of Government," links a technology-enabled public sector to the creation of public value as defined in the following four manners: (a) quality public services are delivered; (b) socially desirable outcomes are achieved; (c) citizens are satisfied with the services and outcomes; and (d) trust in government is created and/or increased. Accordingly, fostering digital inclusion is one important source of public value creation because it is not only socially desirable but also an important determinant of public trust in government.

With respect to the nexus between the first two points—the delivery of quality services and socially desirable outcomes—mobility is viewed by many governments as an important inflection point. The U.K. Digital by Default strategy, for instance, explicitly targets mobile as a significant source of unrealized savings to be exploited over the next several years (Kernaghan, 2013; Roy, 2014b). The strategy goes on to underscore that in addition to cost savings for governments (savings that can facilitate a redeployment of resources and/or in place of reduced overall expenditures), a move toward digital self-service can also represent an important basis for individual development and a lessened risk of outright social and economic exclusion. The British report also acknowledges that the "Assisted Digital" approach cannot be a government endeavor alone but instead denotes a complex policy and service agenda entailing partnerships with both industry and civil society.

Beyond service delivery, current efforts to foster more open government face similar questions and debates. Open government and open data strategies are based upon the notion that public sector information is a resource, the release of which will maximize its social and economic value as citizens use it individually and collectively. In this regard, for example, the OECD rightly underscores open data and open government as a driver of greater self-empowerment social participation (Ubaldi, 2013). Such efforts, however, have also sparked fresh concerns about the digital divide and accentuating new forms of "data divides" (Halonen, 2012). As societies become more mobile in manners that reflect both preexisting digital and socioeconomic divides and more contemporary forms of unevenness in terms of abilities and interests, governments will, at the very least, need to orchestrate more coordinated efforts to better align strategies for service delivery, infrastructure development, and open and participatory government.

Within this shifting context for digital government and social development, the Danish Government provides one prospective window on what it defines as the "digital path toward future welfare"—an integrative and holistic strategy explicitly linking digital and mobile government to wider social and economic developmental efforts and objectives (cataloguing such measures from 2011 to 2015*). Denmark's unique degree of digital inclusion stems, in part, from an aggressive rollout of digitized administration and mandatory online interactions with citizens (mandatory in electronic form by 2015) and companies (mandatory since the beginning of 2013). The compete suppression of paper-based transacting creates transitional challenges akin to Great Britain's Assisted Digital approach that the Danish Government has sought to recognize and address:

> This major step towards eGovernment will require considerable changes to the way public authorities work, and a certain degree of acclimatization from citizens. However, the transition will take place gradually, as user-friendly eGovernment solutions are introduced in more and more areas. Help will be available for citizens who find it hard to use the new solutions. (Halonen, 2012)

Even in the relative digital ubiquity of Denmark, such help requires recognizing and addressing a number of barriers that stand in the way of achieving full digital inclusion (Jaeger, 2012). A better understanding of the skill deficiencies and varying motives of those presently displaced from online processes can thus enable an appropriate mix of incentives, both positive and negative, with the critical twofold finding from Denmark being that digital inclusion necessitates political attention and specific policy actions linking together traditional e-government strategies with education and social and developmental assistance and, accordingly, that the digital divide is unlikely to eradicate itself naturally through technological and mobile evolution (Jaeger, 2012).

The advent of mobility and Gov 2.0—and the specific efforts being developed by jurisdictions such as Singapore, Great Britain, and Denmark—would suggest that digital inclusion is increasingly recognized as a challenge meriting recognition and action. A closely related point is that the maintenance of a multichannel service apparatus may not only be increasingly costly but also detrimental to social cohesion and wider socioeconomic developmental capacities going forward. Finally, one can postulate that any proposed reforms to the existing apparatus of social development and assistance can be understood and examined within an integrative prism of traditional government supports and programs and the digital and participatory contours of Gov 2.0 and mobility. These broad lessons provide the basis for examining the evolution of mobility and digital inclusion in Canada's largest province.

---

* For details of the Danish strategy, please see http://www.digst.dk/Servicemenu/English/News/-/media/Files/Digitaliseringsstrategi/Engelsk_strategi_tilgaengelig.ash.

11.4 Case Study: Ontario, Canada

The Canadian landscape, broadly speaking, provides evidence for both optimism and concern in terms of digital inclusiveness. More than one half of Canadians in 2012 went online via a wireless handheld device, whereas nearly two thirds of Canadians are regular users of social media (with a growing proposition of such usage happening via mobile devices) (Statistics Canada, 2013). In addition, and as noted in the Introduction, the uptake of mobile banking in Canada has been stronger than initial levels of online banking nearly two decades ago, with a healthy proportion of mobile device users making use of mobile channels and applications.

On the other hand, there remains a notable divide between the roughly one half of the population owning smart phones or tablets and those without such devices. In one 2013 survey, nearly one quarter of Canadians indicated that they never shop online (Infantry, 2013) a figure rising to 31% in a more comprehensive 2012 Internet usage survey published by Statistics Canada (Statistics Canada, 2013). Statistics Canada also reports that 17% of Canadians did not go online for any reason in 2012, with persisting digital divides in Internet usage correlated to income, education, and urban–rural dwellings (Roy, 2014b).

As with the examples reviewed above, notably the British and Danish efforts, a critical issue for all governments in these settings is the degree to which mobility can serve as an accelerated platform for those previously offline to undertake more of their informational and transitional service offerings online—via mobile devices. A related issue is the degree to which mobile devices and interrelated platforms and strategies for Gov 2.0 can serve as the basis for a more holistic and integrated paradigm of citizen-centric services encompassing not only information and transactions but also a wider array of support and developmental services.

Within such a context, as noted in the Introduction, Ontario has seen three important initiatives in recent years that can provide insight as to whether digital inclusion is effectively recognized and addressed today and how it might be pursued going forward. As previously noted, these three initiatives include the following: first, a 2013 audit of ServiceOntario by the Provincial Auditor General; second, a 2012 external review of social assistance delivery in Ontario (the Commission for the Review of Social Assistance in Ontario, henceforth referred to as the Commission or the Review Commission); and third, a 2012 research study conducted independently in the academic sector with some financial support by the Province (Taking Ontario Mobile).

The latter initiative provides the most relevant starting point because it most directly addresses the themes of mobility and inclusion. The final report of this research initiative defines inclusion as "the removal of barriers to mobile access experienced by residents of Ontario due to socioeconomic factors, lack of mobile infrastructure and unmet functional needs (e.g., a screen magnifier for users with low vision)" (Diamond & Roberts, 2012, p. XXI). The report further notes that "barriers due to markets, costs and functions mean that the groups most likely to be excluded from the rapidly evolving mobile marketplace are the same groups that would disproportionately benefit from the increased access to public services, economic inclusion, jobs and productivity that these new technologies will offer" (Diamond & Roberts, 2012, p. XXI).

An essential point that bears noting here is that unlike many of the initiatives reviewed in the previous section, this Ontario study is separate from the provincial government (albeit one undertaken with some public sector financial support). Consequently, there are no direct ties between the directions and reforms espoused by the researchers and governmental action, and there has been no explicit response by the Government of Ontario in terms of endorsing a specific mobility action plan with a dimension that might be reasonably associated with digital inclusion and the main challenges sketched out above. With respect to economic and technological development, the Province has undertaken key initiatives to bolster research and industry capacities, often underscoring employment and innovation benefits in doing so, but the obvious risk in doing so is an accentuation of the digital divide that already exists.

The shifting landscape of online service delivery is also telling in this regard, as a 2013 audit of the Public Auditor of Ontario revealed that the Province has fallen well short of its targets in fostering the usage of online channels (and corresponding reduction in usage of traditional delivery channels, namely, telephony and in-person facilities). The following findings are indicative of this unmet potential in terms of potential efficiency savings:

> In the 2012/2013 fiscal year, only 30% of ServiceOntario transactions were done online, well short of ServiceOntario's forecast of 55% to 60%. Further savings could be achieved if ServiceOntario had an effective strategy to encourage people to switch to doing business online instead of in person. For instance, we estimated that ServiceOntario's operating costs would decrease by $2.9 million annually if 50% more license plate sticker renewals were done online. (Auditor General of Ontario, 2013, p. 249)

There is a close relationship here between the Ontario situation and the British analysis of their own digital gap—and what the British Government perceives as an opportunity to leverage mobility as a means of better reaching proportions of the populace that regularly use mobile devices but have yet to access government services online (particularly when many low-income and otherwise marginalized groups are rapidly adopting mobile devices even while demonstrating limited interest in and usage of traditional PC-based online channels and resources). The key point here with respect to the twin British efforts, Digital by Default and Assisted Digital, is recognition of the need for explicit mechanisms and strategies to target such groups (thereby also recognizing the rising mobile adoption rates are likely to be insufficient in this regard).

Because this audit of the Ontario Government's flagship service delivery entity (ServiceOntario) was focused more on assessing what has been done than what might be done in the future, mobility is a theme completely unaddressed by the auditor in terms of findings and recommendations. There is every reason to believe, however, that going forward, if Ontario is to better progress in prioritizing electronic service delivery to the citizenry as a whole (as well as to businesses both large and small), mobile channels must play an important role in doing so.

The third initiative to consider in terms of Ontario's efforts and prospects in the realm of mobility and digital inclusion examined inclusiveness through a social and human lens rather than through a technological one. From 2011 to 2012, an external review process of the entire social assistance apparatus of the Provincial Government (including municipal governments that are constitutionally subservient to the provinces and often intertwined with many aspects of social assistance and human development) was undertaken by a formal Review Commission with the aim of generating comprehensive and holistic reforms for a set of policies and supports widely regarded as well-intended but disjointed, dated, inefficient—and thus ineffective.

Interestingly, although perhaps not surprisingly, the findings from this exercise are remarkably consistent with many service reform strategies associated with e-government reforms in recent years: notably an emphasis on citizen-centric outcomes and a shift away from process toward integrative mindsets and delivery mechanisms. The difference with respect to the social assistance reform is the specific addressing of those constituencies in need of special assistance and supports, especially those facing various degrees of socioeconomic exclusion due to unemployment, underemployment, and cognitive and physical disabilities.

Moreover, the central point is that despite the analogous visions, the near total absence of a digital or mobile dimension to the social assistance review exercise suggests, to put it mildly, a missed opportunity. Such a missed opportunity has not gone entirely unnoticed. In a brief yet insightful commentary on the Ontario social assistance review, Toby Fyfe, editor of Canadian Government Executive, asks:

> Could the government turn Ontario's benefits system into a fully integrated model with a single point of access for benefits and a coordinated interface for clients? This would make programs more effective, reduce costs in the long run and improve service. Mobile, social media, the cloud and big data/analytics technologies could be part of an integrated, enterprise model for a common platform that would manage programs across all key departments and stakeholders. (Fyfe, 2013)

It is undoubtedly easier to pose such a question than to devise detailed strategies to foster the sort of integrative, forward-looking model that Fyfe espouses. However, it does bear noting that many other jurisdictions are seemingly moving in this type of direction, explicitly aiming to orchestrate some alignment between the realms of e-government and online service delivery on the one hand and social assistance and the welfare state on the other hand. At a minimum, we can conclude that the recent initiative aimed at social assistance review has failed to make any such link, which constitutes an important handicap going forward in addressing the very sorts of challenges and opportunities that have been flagged by other stakeholders in Ontario (namely, the research community and the Auditor General of Ontario in terms of the examples discussed previously in this chapter).

The next section aims to provide some insights and directions as to how a province such as Ontario can address this handicap going forward, through a more integrated and participatory lens of digital inclusion.

11.5 Gov 2.0 Stylized Reforms: From Assistance to Engagement

The purpose of this section is to go beyond the general albeit important observation that the recent review of social assistance in Ontario seems to have all but ignored matters of digital government and mobility (a point further underscored in terms of consequences by the audit of ServiceOntario that suggests a highly fragmented citizenry in terms of those utilizing electronic channels and those not doing so). In doing so, we leverage Fyfe's question put forth above regarding a wider and more ambitious digital adaption of social and developmental assistance and what this might entail. Although other jurisdictions, notably the examples reviewed in the earlier sections of this chapter, offer some insights in this regard, our primary source is the social assistance review conducted in Ontario and how some of its central proposals can be understood and pursued through a Gov 2.0 lens.

As discussed earlier in the chapter, Gov 2.0 denotes the evolution and adaption of electronic government for an era shaped by web 2.0 capacities and mobile technologies and a resulting pursuit of what some researchers have termed as *ubiquitous engagement* (Lee & Kwak, 2011). The State Government of Victoria in Australia defines Gov 2.0 as encompassing four main dimensions: leadership, participation, transparency, and performance. We draw upon these four dimensions to better understand the challenges and opportunities facing Ontario.

11.5.1 Leadership

The centrality of leadership in the Victoria Gov 2.0 Action Plan is meant to encompass not only senior management and political support for undertaking a structural and cultural reorientation of how government functions but also fostering whole of government capacities for pursuing and implementing specific reforms. A similar sentiment is echoed by the report "Taking Ontario Mobile," a point that is further accentuated by leading jurisdictions such as Singapore and Estonia that have sought to place digital and mobile reforms at the center of a political and societal project.

With respect to the governance of social assistance in Ontario, despite the absence of any focus on digital reforms, the external task force makes a number of likeminded recommendations for holistic and high-level leadership to oversee the realization of a more integrated, user-centric, activity and insertionist-based system (in place of a system today characterized as disjointed, complex, inefficient, and thus ineffective). Such a finding is not surprising: Any significant reform exercise of this sort requires strong leadership to overcome the invariable inertia of large organizations and engrained interests.

With respect to Fyfe's question mentioned previously in this chapter, and whether such reforms can be pursued through a digital and Gov 2.0-inspired lens, there are reasons for both concern and cautious optimism. On the one hand, a recent study of senior managers in the Ontario Public Service suggests resistance to embracing 2.0-stylized work patterns, particularly the deployment of social media and participatory tools (Bermonte, 2011). The sobering findings of the audit of ServiceOntario and the failure of the Government of Ontario to respond specifically to the Taking Ontario Mobile recommendations further suggest an absence of Gov 2.0 leadership at the political and senior management ranks (an observation seemingly supported by a likeminded languishing of the Canadian government's performance in various international surveys in recent years*).

On the other hand, the audit of ServiceOntario has led to a number of public recommendations made to the government, as well as corresponding pledges by the government to address them, perhaps a source of a political and managerial impetus for deepening digital reforms, at least with respect to service delivery. At the same time, the Premier of the Province has created an Open Government Advisory Panel meant to instill greater openness and transparency in the operations of the provincial government (a point in keeping with the third principle from the State of Victoria, transparency, further addressed below).

However, although there may be some oxygen for the pursuit of Gov 2.0 stylized changes overall, with respect to social assistance delivery, there remains a stark void. There is no fundamental vision politically of the sort we see in Great Britain to embrace a path of digital by default, assisted digital, or the digitization of the welfare state—all suggesting a fundamental leadership shortcoming that requires redress.

11.5.2 Participation

In keeping with the invocation of the previously mentioned ubiquitous engagement, participation is arguably the central driver of the web 2.0 era as social networks and other online platforms and tools enable user-created content and networking to

* For example, a decade ago, Accenture ranked Canada at the top of their global e-government surveys for several consecutive years, whereas a 2014 study by this same organization of 10 leading digital jurisdictions does not even include Canada. A similar decline is evident in other global surveys, notably the UN Global E-Government Survey. Details of the 2014 Accenture survey may be seen at http://newsroom.accenture.com/news/accenture-10-country-study-finds-singapore-norway-and-uae-lead-in-digital-government.htm.

influence decision-making both within and among organizations. The Government of Victoria, for example, thus aims for "using Government 2.0 initiatives to put citizens at the centre and provide opportunities for co-design, co-production and co-delivery" (Roy, 2013, 2014a).

Again, despite the absence of any specific mention or examination of technological capacities for doing so, the Ontario Review Commission similarly embraces participatory principles of this sort indirectly in their call for more collaboration within the government and more direct involvement by social assistance recipients in charting their course toward self-development and employment. The report even goes as far as recommending that provincial or local government agencies hire select social assistance recipients to work as peers with other recipients, a horizontal relationship strongly in keeping with the spirit and logic of web 2.0 capacities.

Separate from the Ontario Commission's specific findings (which largely ignores the technological dimension to its work), this alignment of social and technological participation with respect to assistance and development is increasingly recognized as a source with significant positive potential in realizing the very sorts of objectives sought by the Commission, namely, to create social and relational capital to lessen exclusion and foster individual and collective development. One researcher argues that social media, in this regard, carries much potential:

> Social media may support the enhancement of citizens' social capital as it enables interactions between offline and online sociability and the enrichment of social relations by creating and maintaining links through the use of social networks. In addition, social media can also contribute to the development of cultural capital of disadvantaged people as it broadens the access to digital content and other opportunities which facilitate (informal) learning processes. (Verdegem, 2011, p. 34)

Nonetheless, this same source further acknowledges that social media carries offsetting risks for contributing to isolation and passive forms of activity less conducive to furthering developmental opportunities. With the spread of social media usage, and its growth tied to the advent of mobile devices that are expanding their reach into those proportions of the population generally not online for government-related service and activity, the main point here is that social media cannot be ignored. Going forward, any meaningful socializing dimension to human developmental assistance is likely to be at least partially online, creating new opportunities for peer-based networking and shared learning and also new linkages between public servants and their clients.

In terms of this latter point, within the public sector as well, participation is a central element of any reform trajectory. This point is heavily underscored by the Ontario Commission in its lamenting of disjointed policies and information requirements that greatly limit the ability of case workers to engage directly with recipients and to focus on bettering integrative outcomes rather than a myriad of reporting requirements. Greater collaboration both within and among organizational units is thus called for by the Commission, and a precursor to such collaboration is dialogue and information sharing. The Commission thus calls for the creation of a new working group of government bodies and the Information and Privacy Commissioner to examine and address barriers to information sharing, identified as an enabler to fostering specific community pilot initiatives for alternative delivery mechanisms (more in keeping with the language and principles of codesign and codelivery). This sort of dialogue is critically important because privacy is frequently invoked by public servants as a barrier to exchanging information and resources, also an analogous and wider point for tension for the evolution of e-government and integrated and online service delivery (Roy, 2013, 2014a).

Here is where social assistance and service delivery (of the sort led by ServiceOntario) are destined to be increasingly enjoined, with tremendous potential for mobile devices and electronic service channels to facilitate greater self-service options that can translate into platforms for the sort of participatory governance models sought by the Review Commission. Indeed, the Ontario Government is already undertaking similarly likeminded steps with small businesses, in trying to better engage the small business community in a dialogue on how to improve the regulatory environment for creating and growing commercial ventures. Nevertheless, such logic, despite the calls of the Review Commission, has yet to find resonance within apparatus of social assistance delivery.

Shifting to a more participatory and citizen-centric mindset within the social assistance governance apparatus is a laudable and well-justified objective put forth by the Review Commission, albeit one done so without recognition of the potential of Gov 2.0-stylized platforms and tools in enabling such a shift. Part of the challenge here is a disconnect between the scope of the Commission's efforts, as well as a separate and more encompassing framework for digital and mobile government reforms (akin to the previously mentioned British and Danish examples and the State of Victoria Action Plan for that government as a whole). Despite the participatory overtones to all three Ontario initiatives examined in this chapter, the Review Commission reflects distinct and engrained boundaries rather than synergistic thinking.

11.5.3 Transparency

Greater transparency and openness are central to the Gov 2.0 project, a sentiment embraced by the State of Victoria in their Action Plan: "making government more open and transparent through the release of public sector data and information." Ontario has followed many other governments at all levels in Canada in embracing open data strategies, and as noted previously in this chapter, the Premier has sought further reforms to enlarge the open government concept in a manner echoed by many jurisdictions across the country (Roy, 2013, 2014a).

With respect to social assistance, the Review Commission, while remaining largely within its confines of social and human development, notably embraces transparency, recommending, for example, the creation of new online reporting mechanisms compelling communities to provide a more open and regularized accounting of activities and outcomes. A similar appeal is made to broaden existing report card mechanisms aimed at poverty reduction (particularly for children) to include progress on the transformation of social assistance. Information is unquestionably a prerequisite to the realization of more participative governance and anything resembling a state of ubiquitous engagement (Lee & Kwak, 2011). Unfortunately, the Commission's recommendations fall well short of a Gov 2.0 participatory logic and instead emphasize the importance for the public sector to do a better job of communicating its efforts and accomplishments. Such linear reporting measures fall well short of meaningful collaboration and processes of coproduction and codesign. Complementing the Commission's own caution in this regard is the trepid response of the government in the 2013 Budget to its own recommendations, pledging to consult stakeholders on prospective reforms but providing no detail as to how such consultation shall take place.

The Commission's failure to address transparency more systemically and proactively is all the more troubling when coupled with the thrust of open data initiatives and likeminded app contests by many governments, which seemingly stand to engage and benefit the most technologically literate citizens already well versed in new technologies and mobile devices specifically. The caveat here is that, as discussed previously in this chapter, some governments such as those of Britain and Denmark feel that mobility can serve as a platform to better reach and engage those currently disenfranchised digitally, but the central point in this regard is that those governments are articulating specific strategies and targeted mechanisms in doing so. A very different reality presents itself in Ontario.

11.5.4 Performance

In linking its Gov 2.0 Action Plan to better and more integrated outcomes, the State of Victoria emphasizes the need to build strong government-wide capacities and capabilities both individually and organizationally. Closely enjoined with the notion of leadership (forging a form of virtuous cycle in this regard), these new capabilities encompassing all levels of government organizations are achieved through training and dialogue, pilot experimentations, and a greater acceptance of risk than traditionally permissible within a model of traditional public administration and Westminster democracy (Roy, 2013, 2014a). In doing so, it is notable the Victoria Action Plan assigns ultimate responsibility for the plan to a whole of Government Implementation Task Force, a network committee itself supported by the Department of the Premier and the Cabinet.

Such a recipe reflects the governance tension that all Westminster jurisdictions face, namely, systemic and often historically rooted pressures for centralized authority within central agencies (and dictated by premiers and prime ministers), and more contemporary calls for collaboration and empowerment across all levels of the workplace and polity. In a point with relevance for both the State of Victoria and the Province of Ontario, it stands to reason that in the absence of some measure of political innovation to encourage more participation and collaboration democratically, the organizational reforms put forth by the public service are likely to be constrained by traditional practices rooted in hierarchy and control (Roy, 2013, 2014a). The Victoria Gov 2.0 Action Plan presents a uniquely important test case in this regard, although any assessment of its impacts and progress is outside of the scope of this chapter.

In terms of the relevance for Ontario and its social assistance review, we find that the Renew Commission faced similar tensions, calling for new coordinating bodies both within and outside the public sector while also recommending the creation of a new Provincial Commissioner for Social Assistance, a sort of new organizational czar to orchestrate the complex set of changes and interrelationships at play. Moreover, consistent with this Commission's own reporting relationship, it has thus fallen to a single minister responsible for social assistance to act (or not) on its recommendations.

This fragmented performance and accountability regime from the point of view of the government as a whole is further underscored by the separation of agendas between social assistance and general service delivery (i.e., the Ministry housing ServiceOntario). One can go further and add to this parceled regime the equally separate ministry most directly responsible for economic and technological development for the Province, including support for high-profile and flagship investments in a wireless and mobile technology industry cluster rooted primarily in the Greater Toronto Area.*

In contrast, a genuinely holistic mobility strategy for the province as a whole (province here is not capitalized with reference to not only the provincial public sector but the economy and society more broadly) necessitates the provincial government becoming a model user of mobile technologies in its own operations. Such an effort, in turn, requires some form of collective leadership and accountability mechanism to drive more integrative performance outcomes—at the very least akin organizationally to the government-wide task force deployed by the State of Victoria. In the absence of such a mechanism enjoining relevant ministries and ministers (and thus their senior officials), Ontario instead faces separate performance agendas for social services reform, general service delivery reform, and technological innovation.

---

* In 2013, the Province of Ontario announced $220 million of assistance to Cisco in its plans to invest in Toronto as a major hub of wireless and mobile research over the next decade (including the creation of an estimated 1700 jobs in doing so). Such plans underscore a theme addressed in the Taking Ontario Mobile study and an ongoing struggle for the provincial government, namely, the disparity between economic development growth in southern and more urban parts of the province and the northern regions where, in some communities, even affordable and reliable broadband infrastructure remains elusive. This geographic element to digital inclusion merits recognition, although it is outside of the confines of this chapter.

11.6 Conclusion

The emergence of Gov 2.0 has created much enthusiasm for new forms of online service delivery and citizen engagement in democratic and administrative governance, with the present advent of mobile, the latest stage in the digitization of governance both within and outside the public sector. In parallel to this evolution, stemming partly from longstanding digital divides and partly from newer concerns over accentuating online cleavages, some jurisdictions such as Great Britain and Denmark have begun to put forth explicit strategies for digital inclusion.

Such strategies may be viewed partly as critical enablers of expanding the usage of online service channels offered by public sector entities, thereby generating efficiency savings through channel reallocation and the gradual virtualization of traditional paper-based processes and place-based facilities. At the same time, there is an equally powerful argument for social development and assistance and the growing risk of deeper forms of socioeconomic exclusion for those citizens unable or unwilling to embrace online resources and activities (or, similarly, for those citizens gravitating to the Internet for malicious or criminal purposes). For both sets of reasons, an increasingly digital and now mobile public sector must place principles of inclusiveness and social equity at the heart of any meaningful calculus for realizing public value (World Economic Forum, 2011).

Conversely, the Ontario case study presented in this chapter underscores the dangers of a piecemeal approach to segmenting digital and social development and pursuing them in largely separate manners, despite the relevance of many aspects of the participatory and engagement-laden logic of Gov 2.0 for social assistance reforms aimed at shifting from passive assistance to more action-oriented programs and initiatives. In addition, Ontario is likely to struggle to achieve its own online service delivery milestones due to persisting divides in terms of Internet access and usage. Jurisdictions valuing and pursuing digital innovation and inclusion in a holistic manner, moreover, benefit from strong political commitment to government-wide and societal reforms, thus far absent in the Ontario context.

In sum, governments at all levels must understand not only the basic contours of digital divides within their jurisdictions but also the policy and governance linkages across an increasingly digitized and mobile public sector and those most disenfranchised from online opportunities for self-empowerment and collective support. Any effective reformation of social and developmental assistance necessitates a digital and mobile lens conducive to a more connected and participatory era.

Author Biography

Jeffrey Roy is a professor in the School of Public Administration at Dalhousie University, Halifax, Nova Scotia, Canada.

References

Accenture. (2013). *Four ways that mobility can transform government solutions*. Retrieved from http://www.accenture.com/SiteCollectionDocuments/PDF/Accenture-Four-Ways-that-Mobility-Can-Transform-Government-Solutionsv2.pdf.

Auditor General of Canada. (2013). *Access to online services* (chapter two). Ottawa: Office of the Auditor General. Retrieved from http://www.oag-bvg.gc.ca/internet/English/parl_oag_201311_02_e_38796.html.

Auditor General of Ontario. (2013). *ServiceOntario* (chapter three). Toronto: Office of the Auditor General. Retrieved from http://www.auditor.on.ca/en/reports_en/en13/2013ar_en_web.pdf.

Bermonte, A. (2011). *Senior leaders' use of Web 2.0 and social media in the Ontario Public Service* (Theses and dissertations). Paper 680. Ryerson University, Toronto. Retrieved from http://digitalcommons.ryerson.ca/dissertations/680.

Diamond, S., & Roberts, V. (2012). *Taking Ontario mobile*. Oshawa: Ontario College of Art and Design University.

Fyfe, T. (2013). How do you ensure efficient social assistance at an affordable cost? Retrieved from http://www.canadiangovernmentexecutive.ca/category/item/1401-how-do-you-ensure-efficient-social-assistance-at-an-affordable-cost%3F.html.

Halonen, A. (2012). *Being open about data analysis of the UK open data policies and applicability of open data*. London: Finnish Institute. Retrieved from http://www.finnish-institute.org.uk/images/stories/pdf2012/being%20open%20about%20data.pdf.

Infantry, A. (2013). Not all Canadians love to shop online: Study. Retrieved from http://www.thestar.com/business/tech_news/2013/08/21/not_all_canadians_love_to_shop_online_study.html.

Jaeger, B. (2012). *New frontiers in the digital divide: Revisiting policy for digital inclusion*. Retrieved from https://www.scss.tcd.ie/disciplines/information_systems/egpa/docs/2012/Jaeger.pdf.

Kernaghan, K. (2013). Changing channels: Managing channel integration and migration in public organizations. *Canadian Public Administration, 56*(1), 121–141.

Kostakis, V. (2011). The advent of open source democracy and wikipolitics: Challenges, threats and opportunities for democratic discourse. *Human Technology: An Interdisciplinary Journal on Humans in ICT Environment, 7*(1), 9–29.

Lee, G., & Kwak, Y. (2011). *An open government implementation model: Moving to increased public engagement*. Washington, DC: IBM Center for The Business of Government.

Lips, M. (2012). E-government is dead: Long live Public Administration 2.0. *Information Polity, 17*, 239–250.

Maier-Rabler, U., & Huber, S. (2011). "Open": The changing relation between citizens, public administration, and political authority. *JeDEM: Journal of Democracy, 3*(2), 182–191.

Mossberger, K., Tolbert, C. J., & McNeal, S. (2008). *Digital citizenship: The internet, society, and participation*. Cambridge, MA: MIT Press.

OECD. (2011). *M-government: Mobile technologies for responsive governments and connected societies*. Paris: Organization for Cooperation and Economic Development.

Roy, J. (2013). *From machinery to mobility: Government and democracy in a participative age*. New York: Springer.

Roy, J. (2014a). Cloud computing and Gov 2.0: Traditionalism or transformation across the Canadian public sector? *International Journal of Public Administration in the Digital Age, 1*(1), 74–90.

Roy, J. (2014b). *Mobility and government: Opportunities and challenges for service delivery and information management*. Toronto: Institute for Citizen-Centred Service.

Shirky, C. (2008). *Here comes everybody: The power of organizing without organizations*. New York: Penguin Group.

Statistics Canada. (2013). Individual Internet use and e-commerce, 2012. Retrieved from http://www.statcan.gc.ca/daily-quotidien/131028/dq131028a-eng.htm.

Stoker, G. (2005). Public value management—A new narrative for networked governance? *American Review of Public Administration, 36*(1), 41–57.

Ubaldi, B. (2013). *Open government data: Towards empirical analysis of open government data initiatives* (OECD Working Papers on Public Governance, No. 22). OECD Publishing. Retrieved from http://dx.doi.org/10.1787/5k46bj4f03s7-en.

Verdegem, P. (2011). Social media for digital and social inclusion: Challenges for information society 2.0 research & policies. *tripleC, 9*(1): 28–38.

World Economic Forum. (2011). *The future of government: Lessons learned from around the world*. Geneva, Switzerland: World Economic Forum, Global Agenda Council on the Future of Government.

Wyld, D. C. (2010). *Moving to the Cloud: An introduction to Cloud Computing in government*. Washington, DC: IBM Center for The Business of Government.

Young, N. (2012). *The virtual self*. Toronto: McClelland & Stewart.

Chapter 12

SPCData: The Italian Public Administration Data Cloud

Giorgia Lodi, Antonio Maccioni, and Francesco Tortorelli

Contents

Abstract

Recently, governments have started sharing a large volume of public data sets on the Web. However, to enable full reuse of the data and meet semantic interoperability requirements, it is crucial to provide them in a standard and machine-readable form, with possible interlinks directly exposed. This is a feasible way to foster the valorization of public sector information.

In the context of the Italian Interoperability Framework, a linked data cloud, named SPCData (http://spcdata.digitpa.gov.it), is gathering the attention of European Union functionaries, public administrations, (linked) open data practitioners, enterprises, and start-ups. These stakeholders identify SPCData as an open laboratory where a set of best practices takes place, demonstration services are showed, and high-valuable data are exposed to end-users. SPCData comprises several data sets (e.g., the National Public Administration Registry, contract data, and classifications) and a relevant amount of external links to form a linked data cloud within the Italian public administration. This chapter aims at giving a broad overview of SPCData project and of the initiatives that derived from it, including the development of three private sector applications. SPCData opens future perspectives in the Italian context of data management within the public administration.

Keywords: web of data, Italian public administration, linked data, semantic interoperability

12.1 Introduction

12.1.1 Scenario and Motivations

In response to recent international and European directives and strategies on open data (e.g., Digital Agenda and Open Government Initiative), governments have started sharing public data on the web. This has resulted in a number of initiatives of so-called open government data (Data.gov, 2012; Governo Italiano, 2012; Opening up Government, 2012), where a variety of heterogeneous data sets, which were previously hidden in applications or in databases exclusively managed by public administrations (PAs), showed up online. As a consequence, citizens, enterprises, and also PAs can now easily access these data and have the opportunities to build crowdsourcing, advanced mash up, and services in general. This aspect is fostered if data are provided in a standard format and using well-known schemas (Archer et al., 2013; Lodi et al., 2014).

However, despite these evident advantages of open government data (OGD), in the majority of current initiatives, data sets are not properly clean and up-to-date and are still published in heterogeneous, typically proprietary and nonstandard formats. This causes interoperability and usability issues and requires significant efforts by end-users to use and integrate them in the construction of services. To facilitate the large-scale integration and sharing of distributed data government sources and thus foster a wider usage of those data, it is crucial that PAs start becoming aware that data must be provided in a *standard, human, and machine-readable* form, where possibly interlinks among data are directly exposed on the Web (Ding et al., 2011).

In the specific Italian public sector scenario, heterogeneous PA systems are able to interoperate with each other by means of the Sistema Pubblico di Connettività e Cooperazione–Public Connectivity and Cooperation System (SPC).* The SPC can be defined as the Italian e-government interoperability framework; it consists of the set of technologies and services used by all Italian PAs to cooperate and exchange data for their everyday administrative procedures. Within this system, a large volume of public data are generated; these data can be made available online and contribute to the creation of a "web of data" within SPC and for the Italian PA.

12.1.2 Contributions

This chapter presents our experience in creating the linked data portal SPCData, belonging to the Italian Interoperability Framework SPC. Specifically, we will show how we generated the first interlinked data set. This data set is built from the National Public Administration Registry, named IPA; we consider the IPA as the nucleus of SPC's open data because it includes all the contact information of Italian PAs (e.g., tax codes, e-mail and postal addresses, and e-payment references).

SPCData appeared online in February 2012 and is one of the first linked government data clouds in the world. It was useful to show to citizens and enterprises the benefits of 5-star data and to apply the best practices in data management, which, too often, remain on paper and are intentions only. The results obtained by SPCData (i.e., the creation of different mobile applications) make it suitable to be presented as a best practice in the public sector information.

Additionally, it is worth mentioning that SPCData was mentioned as a case study in the recent study of the European Commission on the business models on Linked

---

* http://www.agid.gov.it/cad/sistema-pubblico-connettivita-spc.

Open Government Data. For this study, it was selected among many cases together with other 13 cases. The portal has been designed to (a) publish the data set and its ontology, (b) provide a freely accessible environment for querying the data, and (c) offer a set of demonstrative public services that exploit the IPA data set and created interlinks. This chapter presents an extended version of the paper by Lodi, Maccioni, and Tortorelli (2012), and the main contributions can be summarized as follows:

- We describe the methodology that has been adopted to construct SPC's web of data. In our view, the core of that web is represented by the data set that univocally identifies PAs. In SPC, the IPA* can be the nucleus because it may be used as the data set from which other SPC data and relevant data sets owned and managed by PAs are connected (e.g., territorial data). The IPA contains reference information of PAs such as tax codes, e-mail, and postal addresses; official websites; electronic payment contacts; and other data, of Italian PAs, and it is currently used as base registry for other national databases (e.g., the databases owned by Ministry of Economics and Finance).
- We describe how to convert SPC raw data in a standard, machine-readable, and queryable format using Linked Open Data (LOD) technologies to enable the construction and reuse within SPC of a set of high-quality open data. Specifically, we discuss how to perform this conversion in the case of IPA, presenting the main interlinks that have been created within it toward other public well-known data (e.g., DBpedia; Auer et al., 2007).
- We demonstrate the feasibility, effectiveness, and powerfulness of the LOD approach in the e-government context by introducing the design of the SPC open data web portal where we (a) publish the IPA data set with a Creative Commons' unrestricted license, (b) provide a freely accessible environment for querying the data set and created interlinks, and (c) provide a set of public demonstrative services over the data set and interlinks to be directly used by end-users.

12.1.3 Related Work

To the best of our knowledge, just two relevant works can be found (CNR, 2012; Dati Camera dei Deputati, 2012) in Italy that use standard LOD technologies when opening data of PAs. In other cases, data are published mostly in proprietary and nonstandard formats (a catalogue of them is Governo Italiano, 2012). In contrast, foreign countries, especially the United Kingdom,† the United States,‡ and Ireland (Linked Data Research Centre, 2012), appear to be more active in this field as they tend to share structured, machine, and human-readable data sets produced

* IPA is the Italian acronym for Indice della Pubblica Amministrazione.

† http://data.gov.uk.

‡ https://www.data.gov.

by using Semantic Web technologies. They have published initiatives of data clouds at the time SPCData was created. On the contrary, they aimed at the creation of a comprehensive catalogue for national data instead of focusing on high-quality data only, that is, 5-star data. Thus, Halb, Raimond, and Hausenblas (2008), Ding et al. (2010), and Auer, Lehmann, and Hellmann (2009) are examples of works that describe real initiatives of LOD in the e-government context. These systems are devised to transform specific data sets (e.g., in Halb et al., 2008, the Eurostat data set containing the statistical data about the European Union is discussed) in LOD where interlinks between them are provided. In Goodwin, Dolbear, and Hart (2008) and Auer et al. (2009), an approach similar to ours was applied to other specific types of data, that is, geo data.

12.1.4 Structure of the Chapter

The rest of the chapter is structured as follows. Section 12.2 provides the background on the LOD scenario. Section 12.3 presents the Italian e-government interoperability Framework and IPA. Section 12.4 describes the work that has been carried out and the methodology adopted to produce the Linked Open IPA. Section 12.5 introduces the SPC open data portal and a set of public services built on the Linked Open IPA, and finally, Section 12.6 concludes the chapter and highlights future works.

12.2 Background

Tim Berners-Lee introduced a new vision concerning the information available on the web where a "web of links" among documents could be replaced with a proper web knowledge base. This new web architecture depicts a linked information space (web of data) where data are being enriched and inferred (web of meaning). These two concepts of web of data and web of meaning are often referred to as Semantic Web. In the Semantic Web, data are not considered merely as bytes, but rather they are seen as knowledge because they provide their own semantics.

In the following, we discuss the main tools used within this scenario and the assumptions that have affirmed the development of the so-called LOD initiative.

12.2.1 Resource Description Framework

The infrastructure to enable the semantic web as well as every other interoperable infrastructure has been defined through a definition process of standard protocols and rules. *Errore: sorgente del riferimento non trovata* shows those protocols that are part of the so-called Semantic Web stack.

Among all the protocols depicted in Figure 12.1 we focus here on the Resource Description Framework (RDF) data model (W3C Semantic Web, 2012). Introduced in 1999, the RDF framework has been the first step realized in the context of

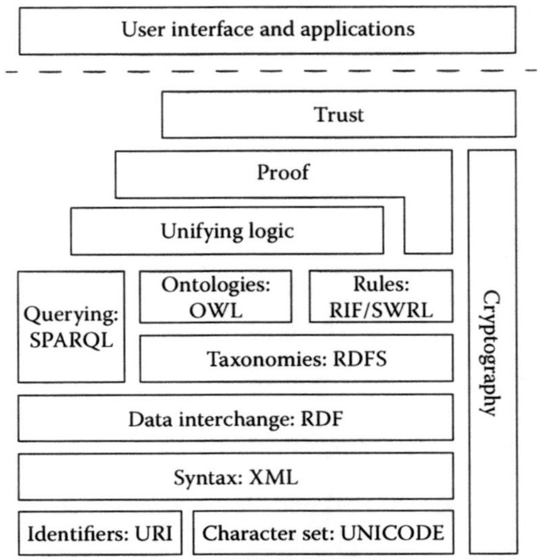

Figure 12.1 The Semantic Web stack. OWL = Ontology Web Language; RDF = Resource Description Framework; RIF = Rule Interchange Format; SPARQL = Sparql Protocol and RDF Query Language; SWRL = Semantic Web Rule Language; URI = Uniform Resource Identifier.

Semantic Web. The principal goal of RDF is to add semantics on top of data and documents.

In RDF, data are expressed and defined by means of RDF statements that, in the RDF terminology, are often called *triples*. Triples are in the form of <subject, property, object>. The subjects denote resources, which are expressed by uniform resource identifiers (URIs) and represent mainly data and documents. Resources can have relationships that are defined using specific properties of a vocabulary. In this way, properties (also called predicates) are typed links with a precise meaning; it is therefore possible to express certain semantics within the data. Properties can also connect resources to literal values, that is, descriptive character strings. Accordingly, a resource is described by all its connections and relationships. The RDF framework allows users to combine triples coming from different sources: This provides more information than the separate triples do. Consequently, with metadata and representation of web resources, RDF can enable semantic interoperability across multiple different applications.

A set of RDF triples is called an RDF graph. The nodes of the graph are the subjects (resources) and the objects (resources or literals); the edges connecting the nodes are the predicates (properties) between them. The collection of all existing RDF statements can be seen as a giant RDF database represented in the form of a directed graph, that is, the previously mentioned web of data.

12.2.1.1 Sparql Protocol and RDF Query Language

To fulfill the need of querying RDF data, in literature, query languages have been designed and proposed. Among them is W3C standardized Sparql Protocol and RDF Query Language (SPARQL; SPARQL protocol, 2012), an SQL-like language that allows querying graph-modeled data. SPARQL provides most of the SQL constructs such as *select*, *where*, and *count*.

12.2.2 Linked Open Data

To fully benefit from open data, citizens and the market in general must be able to access data that are structured, flexible, and easy to read and use because only in this way they can be effectively (re)used in mash-ups, crowdsourcing, and services in general. To this end, in 2006, the LOD initiative inspired practitioners and organizations to start publishing, and building from scratch, data sets derived from data that had previously been stored and exposed using traditional models. The LOD initiative relies on the following principles (Berners-Lee, 2012): (a) use URIs to identify things; (b) use http URIs so people can look things up; (c) provide useful data in standard RDF (preferably reusing ontologies); and (d) use RDF to link to other things.

Upon these principles, W3C defined an open data rating system (Berners-Lee, 2012) used to assess the quality of published open data. If open data are fully compliant with the above principles, they are evaluated at the top level, that is, five. The LOD initiative appears to be particularly successful: In September 2011, 295 different LOD data sets of Level 5 have been counted, containing an overall of 31 billions of triples (State of the LOD Cloud, 2012).

12.3 The Italian E-Government Interoperability Framework

Last year, the Italian government applied significant changes to a Law Decree (n. 82/05) named Digital Administration Code (CAD), whose aim is to define the entire Italian legal interoperability framework. The applied changes were necessary to align the rules regarding the digitalization of PAs to the new trends in information and communication technology and to simplify further the complex administrative processes of PAs by promoting a wider cooperation among their information systems. The legal framework is enforced by the development of the Italian information and communication technology interoperability framework SPC (Sistema Pubblico di Connettività e Cooperazione; Public Connectivity and Cooperation System), which CAD itself regulates. The SPC is the actual means through which interoperability and cooperation between PAs can be guaranteed. It consists of two principal building blocks: the application services and national interoperability infrastructures.

The former are owned by both central and local PAs and are mostly deployed within PA's local administrative boundaries. Application services are offered to final end-users such as citizens and enterprises. Examples of these include all web-based services, e-documents management, digital signatures, and, for example, what in Italy is called certified e-mail, that is, a kind of e-mail that provides users with guarantees on the actual sending and receiving (Petrucci, Gennai, Shahin, & Vinciarelli, 2011).

The latter building block consists of the set of infrastructures and services that are used to support secure connections and advanced interoperability and cooperation between PAs. It is through these infrastructures that technical, semantic, and organizational interoperability, as defined by the European Interoperability Framework (European Commission, 2010), can be ensured in SPC. The national interoperability infrastructures can be further classified into three main categories: infrastructures for governance, infrastructures for connectivity, and infrastructures for applications.

In the context of this chapter, we focus our attention on a specific service that falls into the infrastructures for applications category. The next subsection describes this service in detail.

12.3.1 The IPA

The IPA is particularly relevant for all PAs. Introduced in 2000 by a President of Council of Ministers Decree, the role of the IPA has been recently strengthened by CAD. The IPA is clearly defined by CAD as the national registry within which PAs are obliged, at least every 6 months, to include all those up-to-date data that are necessary to establish a, possibly electronic, communication with legal value either among PAs or between PAs and citizens/enterprises. The IPA can therefore be considered one of the pillars for enforcing interoperability and cooperation in the complex e-government context formed by several heterogeneous central and local PAs.

Currently, the IPA includes such data as tax codes, e-mail and certified e-mail addresses, postal addresses, fax numbers, and official web site URLs, as well as people in charge and organizational structures (e.g., offices and departments) of 20,720 Italian PAs. These data are assuming a relevant role in the PA interoperability and cooperation scenario: They are in fact used in other national base registries to uniquely identify PAs (an example is the tax code, which is also used as identifier in the PAs' territorial data base registry). In addition, with the introduction of e-payments, the IPA is expected to become the main "hub" in which most interactions between PAs will pass through.

Originally, the IPA was developed as a Lightweight Directory Access Protocol (LDAP) (Wahl, Howes, & Kille, 1997) registry that could be queried using both an LDAP and a web interface. However, due to the unfriendly nature of LDAP protocol and syntax, the registry has been assisted over time by the well-known relational database management system MySQL, to which access is via a dedicated web-based public portal (Indice della Pubblica Amministrazione, 2012). The portal allows anyone to search previously mentioned contact information of PAs.

The next section describes the work carried out at our agency to convert the IPA data set in RDF and thus create what we called Linked Open IPA.

12.4 Methodology Used in SPCData and the Application to Linked Open IPA

Our work for creating Linked Open IPA has been driven by LOD principles and Semantic Web best practices. As a result, the Linked Open IPA is assessed at the highest level of the W3C open data rating system (Berners-Lee, 2012).

The methodology we have devised consisted of five main phases that are sketched in Figure 12.2. We think this methodology is sufficiently general to be applied for producing linked open data sets starting from any relational DBMS that constitute both SPC and PAs' assets. In fact, we have applied such methodology to other data sets that we have published, for example, a data set about public contracts.

12.4.1 Data Cleansing

The input of our whole process is a relational DBMS dump of IPA. As mentioned before, IPA data were originally managed by using LDAP and subsequently imported into the relational DBMS MySQL. However, this import process, along with a lack of proper modeling constraints, caused a degradation of the quality of the data represented into the database. Thus, a data cleansing phase becomes crucial.

We managed the data cleansing through MySQL itself; in particular, we implemented an SQL script to automate the execution of this phase. An example of IPA data on telephone numbers and URLs of services offered by PAs is the following.

des_srv	tel_srv	uri_srv	nome_srv
1#Il servizio fornisce informazioni sugli atti e la carto(	1#05526743220	1#http://www.adbarno.it/cont/testo.php?id	1#Sportello consulenza Piano Assetto Idrogeologico
1#Associazione all'ACI on line con pagamento trami	1#800313535	1#www.aci.it	1#Associazione ACI
1#tasse automobilistiche	1#032130321	1#null	1#tasse automobilistiche

As can be noticed, in these example data, the "1#" characters are pre-pended; they derive from LDAP and can cause problems when converting IPA in RDF. With a MySQL script, we can successfully clean the data. The cleansing phase can be performed before or after the modeling phase (see Section 12.4.2).

12.4.2 Modeling

In parallel to the cleansing phase, our modeling phase consists of an assessment of the schema of the IPA database where the most important concepts to be published are selected and then renamed if required. This phase also comprises an evaluation

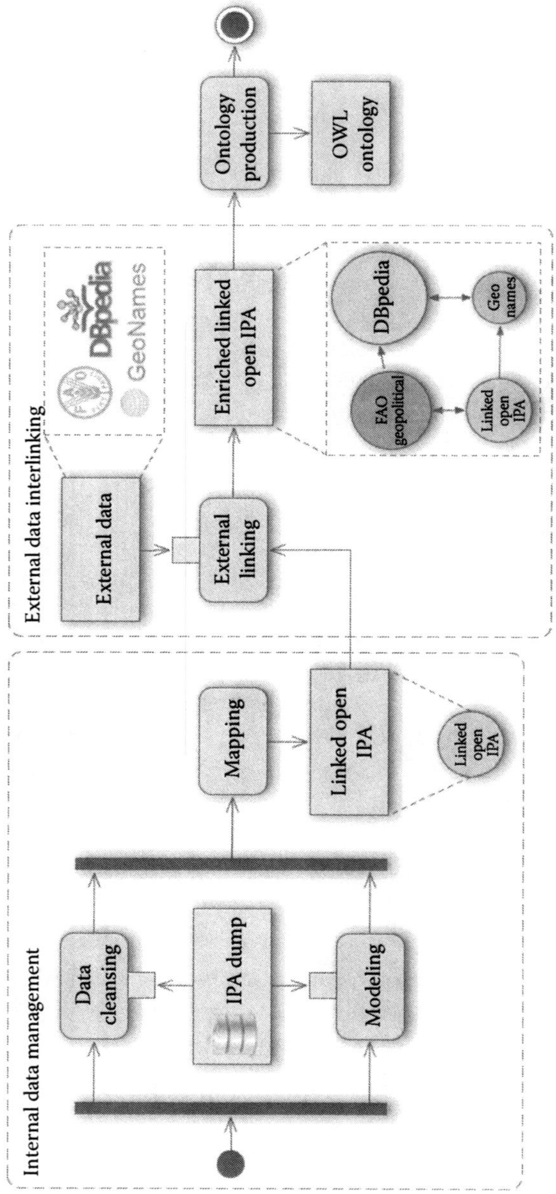

Figure 12.2 UML activity diagram of Linked Open IPA creation. IPA = Indice della Pubblica Amministrazione; OWL = Ontology Web Language.

of the possible connections among the selected concepts. Every connection was named with a property that describes the type of connection between two different concepts. Afterward, we evaluated all the well-known and wide-accepted ontologies and vocabularies to find properties and concepts corresponding to ours. The reuse of common ontologies and vocabularies has the merit of making our data set easier to understand for practitioners and machines and of guaranteeing semantic interoperability with the web of data. For instance, we used the ontology Friend of a Friend (FOAF; 2012) (i.e., the property "`homepage`" to denote all the web site URLs of Italian Pas), the ORG ontology to represent organizations, and Data Catalog to represent catalogues of data sets. In addition, it makes large use of the core vocabularies, recently promoted by W3C and the European Commission. Some of the property we specified in the data set has been directly taken from the standards Ontology Web Language (OWL) (W3C, 2004), RDF, and RDF Schema (W3C RDF Vocabulary Description Language 1.0: RDF Schema, 2012). For instance, we represent the description of the services (`des _ srv` in the previous IPA data example) with the existing property `rdfs:comment`.

12.4.3 Mapping

Once both previous phases have been completed, the RDF triples reflecting the given IPA conceptual model have to be produced. In doing so, we carried out an assessment of current reference documentation (Sahoo et al., 2009) of the W3C RDB2RDF Working Group to survey all the existing approaches that can be applied at this stage. Among them, we opted for D2RQ platform (Bizer & Cyganiak, 2007; Freie Universitat Berlin, Web-based Systems Group, 2012), which is a comprehensive and flexible framework for our purposes. D2RQ includes, among others, a declarative mapping language that allows one to transform data included in a relational database into RDF triples. We implemented a specific D2RQ script that helped us to automate the whole mapping process.

Figure 12.3 shows an example of a D2RQ script clause we used to map the previously illustrated IPA records into RDF triples.

The type of clause (`d2rq:PropertyBridge`) illustrated in this figure serves for connecting literals (see `d2rq:column`) to entities that belong to some concept (see `d2rq:belongsToClassMap`) through a specified property (`d2rq:property`). In our example, we connect web site URLs (contained in `db _ servizio _ amministrazione.uri _ srv`) of PAs' services (ServizioAmministrazione

```
map:ServizioAmministrazioneWebsiteProperty a d2rq:PropertyBridge ;
    d2rq:property foaf:homepage ;
    d2rq:column "db_servizio_amministrazione.uri_srv" ;
    d2rq:condition "db_servizio_amministrazione.uri_srv <> 'null' and db_servizio_amministrazione.uri_srv <>''" ;
    d2rq:belongsToClassMap map:ServizioAmministrazioneClassMap .
```

Figure 12.3 D2RQ script clauses used in the mapping phase.

ClassMap) through foaf:homepage property. The mapping is performed if and only if the specified condition (see `d2rq:condition`) is verified.

It is worth noticing that the generated URIs are formed by a prefix, in our case `http://spcdata.digitpa.gov.it`, the name of the concept, and an identifier.

At the end of this phase, the Linked Open IPA data set was produced, as depicted in Figure 12.2. The following RDF triples are generated by the clause in Figure 12.3 over the above IPA data example.

```
<http://spcdata.digitpa.gov.it/Servizio/abfa-1> <http://xmlns.com/foaf/0.1/homepage> "http://www.adbarno.it/cont/testo.php?id=49" .
<http://spcdata.digitpa.gov.it/Servizio/aci-1> <http://xmlns.com/foaf/0.1/homepage> "www.aci.it" .
```

12.4.4 External Linking and Self-Description Metadata

However, to further enrich the data set (fulfilling the fourth LOD principle), and to better integrate it into the LOD cloud context, we searched for existing public data that represent similar concepts to ours. We found that the IPA contains geographical data (e.g., municipalities and regions) that are also present in well-known data sets of the LOD cloud such as DBpedia (Auer et al., 2007) and Geonames (Geonames ontology, 2012). Therefore, we created external interlinks between them and Linked Open IPA, thus connecting IPA to the LOD cloud (see Figure 12.2). The interlinks are represented by the logic equivalence of two RDF resources and can be expressed and materialized using the OWL property sameAs.

In particular, this phase was driven by string similarity between the municipalities' URIs in Linked Open IPA and those in DBpedia. Thus, we firstly imported DBpedia's URIs of municipalities into our relational DBMS (specifically in the table column "`geodata.dbpedia`"), and then we executed another D2RQ script. The script allows us to produce the RDF triples of the interlinks with DBpedia, and it is illustrated in Figure 12.4.

In practice, the clause in Figure 12.4 connects IPA municipalities' entities to corresponding DBpedia municipalities through the property (`owl:sameAs`). The connection is valid if the specified SQL condition (see `d2rq:condition`) is satisfied to the municipalities.

The direct linking toward Geonames was made exploiting the already existing links between DBpedia and Geonames, that is, using the transitive property of the data.

```
map:ComuneDBPediaProperty a d2rq:PropertyBridge ;
      d2rq:property owl:sameAs ;
      d2rq:uriSqlExpression "geodata.dbpedia" ;
      d2rq:belongsToClassMap map:ComuneClassMap ;
      d2rq:condition "locate(concat('/',db_comuni_codici.denominazione,'>') , geodata.dbpedia) != 0" .
```

Figure 12.4 D2RQ script for producing interlinks with DBpedia.

In this way, we were able to map correctly around 6872 municipalities out of approximately 8000 that currently exist in Italy. Note that we could not map the remaining municipalities due to ambiguities of the names; however, as we aim to provide a comprehensive information content, we are planning to improve this phase using advanced record linkage tools such as SILK (Volz, Bizer, Gaedke, & Kobilarov, 2009).

In addition, we found a Food and Agriculture Organization (FAO) data set (FAO geopolitical ontology, 2012) that is very rich from a government point of view as it describes geopolitical information (e.g., countries' borders and affiliations to international organizations).

Thus, we added the following two triples related to it in our data set that link the identical concept of country "`Italy`."

```
<http://spcdata.digitpa.gov.it/Nazione/Italia> <http://www.w3.org/2002/07/owl#sameAs> <http://aims.fao.org/aos/geopolitical.owl#Italy> .
<http://aims.fao.org/aos/geopolitical.owl#Italy> <http://www.w3.org/2002/07/owl#sameAs> <http://spcdata.digitpa.gov.it/Nazione/Italia> .
```

Finally, we manually added from the data set itself some triples to the Linked Open IPA concerning the description (e.g., licensing, creators, title of the document, and updating date).

At the end of this external linking phase, we produced the whole data set of approximately 717,600 RDF triples.

12.4.5 Ontology Production

Our last phase consisted of extracting the OWL ontology of the Linked Open IPA using an automatic reasoner on the RDF data. A few observations on this phase are in order. First, the ontology production is not required to produce and publish a linked open data set; however, it can be useful for a better understanding of the data and a better semantic interoperability. Consequently, it is easier to produce new data (triples). Finally, this activity can be also performed at the beginning of the conversion process (i.e., during what we called modeling phase) by applying a top-down approach. Nevertheless, because our experience was more data oriented than logic oriented, we preferred to use, in contrast, a bottom-up approach.

12.5 SPC Open Data Portal and Public Services

This section shows the practical results of our work from an end-user perspective. It helps the reader to navigate through our portal and understand the effects of the backend technical work. To make publicly available the linked open IPA, we have designed the SPC open data web portal. This project started at the beginning of 2012 and is still in production (http://spcdata.digitpa.gov.it). The portal is meant to be used for the publication of all the data sets of SPC's web of data. SPCData has often been used as a laboratory for testing the outcomes of European Community

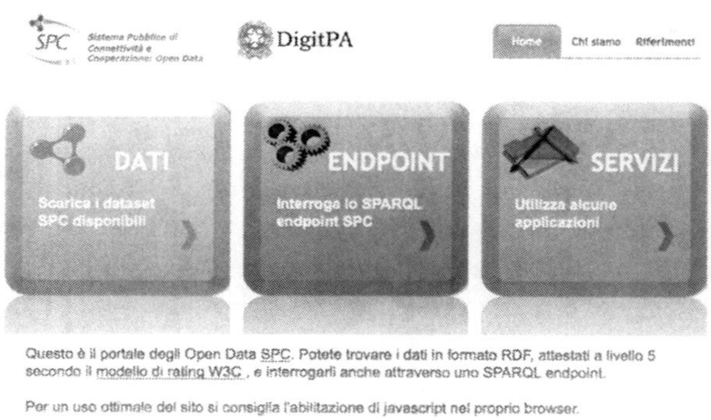

Figure 12.5 SPCData web portal.

working groups on semantic interoperability (European Commission, 2013). At the national level, it was often referred as a demo showcase to show best practices to Italian PAs. The portal exposes the Linked Open IPA data set, its related ontology, and a set of services we implemented for demonstration purposes. The portal is simple in its structure; it consists of three principal sections that enable direct access to the contents: data, SPARQL endpoint, and services* (Figure 12.5). These sections are described in the next subsections.

12.5.1 Data Section

Along with the publication of the entire IPA, we also share the RDF of single pieces of it (Figure 12.6). This allows us to provide end-users with a limited view of the whole data set (e.g., an end-user might be interested in downloading the data regarding PAs' offered services, only). Note that we also publish comma-separate value files, obtained from the IPA database dump, to collect in one single point all types of open data concerning IPA and then simplify the discovery of data sets. In addition to the data, this section of the portal includes the ontology that describes the RDF Linked Open IPA. All the data are subject to CC-BY-SA Creative Commons license, whose peculiar rights are specified within the RDF data set. Note that there exist a variety of licenses for data (e.g., Open Data Commons Attribution License, 2012; Open Data Commons Public Domain Dedication and License, 2012). Our reasons for choosing to use CC-BY-SA are twofold: First, it is compliant to the Italian Open Data license commonly used in our country, and second, it is sufficiently unrestricted to allow us to promote a larger data utilization as possible.

* The portal is currently delivered in Italian as it is intended to offer public services to Italian stakeholders.

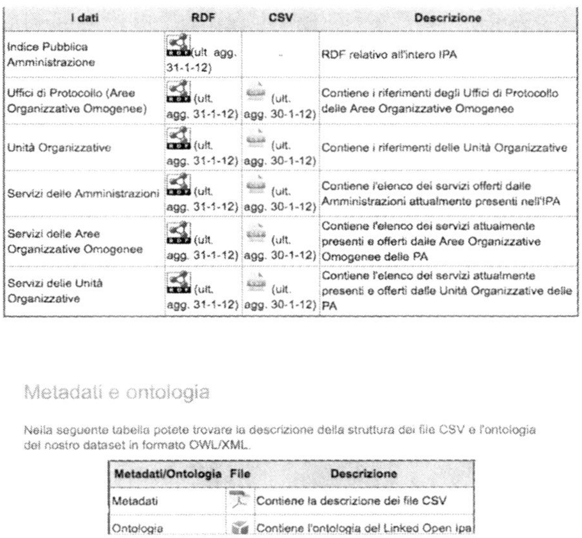

I dati	RDF	CSV	Descrizione
Indice Pubblica Amministrazione	(ult agg. 31-1-12)	-	RDF relativo all'intero IPA
Uffici di Protocollo (Aree Organizzative Omogenee)	(ult. agg. 31-1-12)	(ult. agg. 30-1-12)	Contiene i riferimenti degli Uffici di Protocollo delle Aree Organizzative Omogenee
Unità Organizzative	(ult. agg. 31-1-12)	(ult. agg. 30-1-12)	Contiene i riferimenti delle Unità Organizzative
Servizi delle Amministrazioni	(ult. agg. 31-1-12)	(ult. agg. 30-1-12)	Contiene l'elenco dei servizi offerti dalle Amministrazioni attualmente presenti nell'IPA
Servizi delle Aree Organizzative Omogenee	(ult. agg. 31-1-12)	(ult. agg. 30-1-12)	Contiene l'elenco dei servizi attualmente presenti e offerti dalle Aree Organizzative Omogenee delle PA
Servizi delle Unità Organizzative	(ult. agg. 31-1-12)	(ult. agg. 30-1-12)	Contiene l'elenco dei servizi attualmente presenti e offerti dalle Unità Organizzative delle PA

Metadati e ontologia

Nella seguente tabella potete trovare la descrizione della struttura dei file CSV e l'ontologia del nostro dataset in formato OWL/XML.

Metadati/Ontologia	File	Descrizione
Metadati		Contiene la descrizione dei file CSV
Ontologia		Contiene l'ontologia del Linked Open ipa

Figure 12.6 Data section: the Linked Open Data in Resource Description Framework and comma-separate value.

12.5.2 Endpoint Section

The data we produced have been loaded into an RDF store to which we connected a SPARQL endpoint, that is, a web service that accepts SPARQL queries and returns the related results. To this end, we used the RDF store and SPARQL endpoint of OpenLink Virtuoso (2012) in its open source version. We made this endpoint publicly available within the specific section endpoint of the portal (as shown in Figure 12.5). This allowed us to provide end-users as well as applications with a comprehensive environment to be used to directly query the data sets.

12.5.3 Services Section

To show the effectiveness and power of our approach in the development of applications over the published RDF data set, we provided a number of demonstrative "ready-to-use" services. The services are implemented using Java Server Pages and JENA* APIs and exploit our public SPARQL endpoint. Note that with the availability of the endpoint, these services could have been developed by any other person or organization. Most of the services issue queries on data related strictly to IPA. For instance, we provide a service through which an end-user can obtain the list of

* http://incubator.apache.org/jena/.

```
prefix foaf: <http://xmlns.com/foaf/0.1/>
prefix rdfs: <http://www.w3.org/2000/01/rdf-schema#>
prefix rdf: <http://www.w3.org/1999/02/22-rdf-syntax-ns#>
prefix digitPA: <http://spcdata.digitpa.gov.it/>
prefix geonames: <http://www.geonames.org/ontology#>
select count(?servizio) as ?counter
where
{
        ?cod_prov geonames:locatedIn
                <http://spcdata.digitpa.gov.it/Regione/1> .
        ?cod_com geonames:locatedIn ?cod_prov .
        ?cod_amm geonames:locatedIn ?cod_com .
        ?cod_amm rdf:type digitPA:Amministrazione .
        ?cod_amm digitPA:eroga ?servizio .
}
```

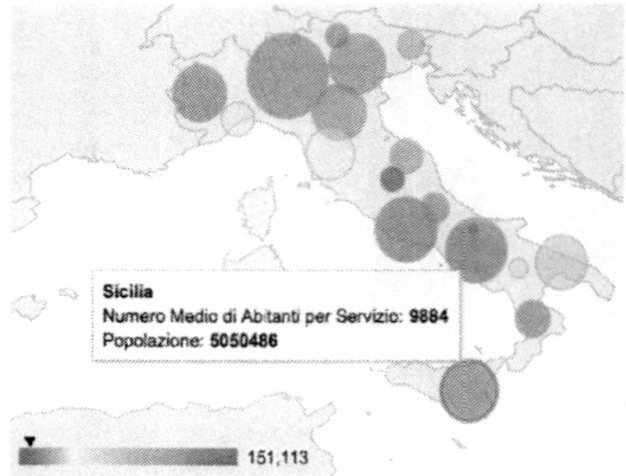

Figure 12.7 Region ranking service: one of the implemented SPARQL query and the final result. SPARQL = Sparql Protocol and RDF Query Language.

all the web site URLs of the PAs of his or her municipality or the list of the certified e-mail references of PA offices. Additionally, we show the effectiveness of the linked data approach by offering a specific service that exploits the external linking we embodied into the data set (see Section 12.4). This service aggregates data from IPA and DBpedia with the aim to display a ranking of the 20 Italian regions on the number of offered services per inhabitant (it is worth noticing that the services are only those the regions have registered into IPA). For the sake of brevity, we report in Figure 12.7 only one of the three queries implemented by the service. The query counts the services of a region. The final result is then shown on a map. We used Google Chart tools (Google Tool Charts, 2012) to display on Italy's map circles whose dimension represents the number of inhabitants of a region and colors that indicate how many services the regions offer per inhabitant (Figure 12.7). As can be noticed, the majority of Italian regions have a good percentage of offered services, included into IPA, per inhabitant. Currently, just one region in the middle of Italy falls into the "red" interval of services per inhabitants.

12.6 Conclusions

In the context of the Italian interoperability framework SPC, a large amount of public government data are generated and used by PAs. These data can be made available online for (re)use purposes. One of the main sets of public data in SPC is represented by the Italian IPA, which contains the reference information of all Italian PAs.

This chapter described the work conducted at our agency to open IPA in a way to produce a standard and human- and machine-readable data set. The work consisted of using standard LOD technologies that allowed us to create interlinks among IPA data and other public data of interest (e.g., DBpedia). We called the resulting data set Linked Open IPA, evaluated at Level 5 in the W3C open data rating system. In addition, we made available an environment to be used for the execution of queries on the data. To show the feasibility and ready-to-use approach we employed, this chapter discussed the design of a currently in production web portal. Within the portal, not only are the data and related ontology published, but also a set of demonstrative applications are provided. These applications can be used by citizens, enterprises, and also PAs to easily obtain contact information of PAs. We believe the experience we reported in this chapter can facilitate PAs in meeting transparency, accountability, and semantic interoperability requirements, and it can serve as a guide for other PAs that wish to effectively open their data. All the administrative processes that now rely on IPA can be, in the future, based on the created Linked Open IPA.

We consider the presented work a starting point toward the creation of a complete SPC "web of data." Therefore, a number of activities are planned in this direction. First, we are currently assessing the possibility to transform in RDF the data of another base registry managed in the context of SPC. This registry includes Italian geodata. The data can be potentially connected to IPA since PAs can feed the registry only once they are registered into IPA. Finally, we are planning to enhance the web portal to (a) add other data sets, (b) develop further applications that demonstrate the added value of the LOD approach and from a user interface viewpoint, (c) address navigational issues of the data sets (e.g., using tools such as ELDA; Epimorphics Ltd., 2012), thus further improving the user experience. At the moment, three different applications (one desktop and two mobile) are developed using SPCData content through its SPARQL endpoint.

Author Biographies

Giorgia Lodi received the PhD degree in computer science from the University of Bologna (Italy) in 2006. After a long period in academia, in 2011 she began working as a consultant for AgID in such topics as digital identities, technical and semantic interoperability, and open data. She is currently a research assistant at the Semantic Technology Laboratory (STLab) of the Institute of Cognitive Science and Technologies of the National Council of Research (CNR).

Antonio Maccioni is a PhD candidate and a computer engineer with several professional experiences in the software industry, government, and academia. In 2011, he started working for the Agency for Digital Italy, focusing specifically on semantic interoperability, big data, and open data. He is currently a research assistant at the database group of Yale University.

Francesco Tortorelli currently works in the Agency for Digital Italy, where he is the head of architecture, standard, and infrastructure department, and the interim head of citizens, enterprises, and technology transfer department. He has worked for the Italian national e-government agency for the last 15 years as a manager of different units. Previously, he worked in the Italian stock exchange commission as a manager of the system and network sector. Before that experience, he worked in the innovation department for an Italian IT company.

References

Archer, P., Dekkers, M., Goedertier, S., & Loutas, N. (2013). *Study on business models for linked open government data.* Retrieved from https://joinup.ec.europa.eu/sites/default/files/85/31/25/Study_on_business_models_for_Linked_Open_Government_Data_BM4LOGD_v1.00.pdf.

Auer, S., Bizer, C., Kobilarov, G., Lehmann, J., Cyganiak, R., & Ives, Z. (2007). DBpedia: A nucleus for a web of open data. *Proceedings of 6th International Semantic Web Conference, 2nd Asian Semantic Web Conference (ISWC+ASWC 2007), 4825*, 722–735.

Auer, S., Lehmann, J., & Hellmann, S. (2009). LinkedGeoData: Adding a spatial dimension to the web of data. *International Semantic Web Conference (ISWC).*

Berners-Lee, T. (2012). *W3C open data rating system.* Retrieved from http://www.w3.org/es/LinkedData.html.

Bizer, C., & Cyganiak, R. (2007). D2RQ—Lessons learned. *Proceedings of W3C workshop on RDF Access to Relational Databases.*

CNR. (2012). Retrieved from http://data.cnr.it/site/.

Data.gov. (2012). Retrieved from http://data.gov.

Dati Camera dei Deputati. (2012). Retrieved from http://dati.camera.it/it/.

Ding, L., Difranzo, D., Graves, A., Michaelis, J., Li, X., McGuinness, D., & Hendler, J. (2010). Data-Gov Wiki: Towards linking government data. *Spring Symposium of the Association for the Advancement of Artificial Intelligence (AAAI10).*

Ding, L., Lebo, T., Erickson, J. S., DiFranzo, D., Williams, G. T., Li, X. et al. (2011). TWC LOGD: A portal for linked open government data ecosystem. *Web Semantics: Science, Services and Agents on the World Wide Web, 9*(3), 325–333.

Epimorphics Ltd. (2012). *ELDA—An implementation of Linked Data API.* Retrieved from http://elda.googlecode.com/hg/deliver-elda/src/main/docs/index.html.

European Commission. (2010). *European interoperability framework for European Public Services (EIF) Version 2.0.*

European Commission. (2013). *D5.2.1—Report on the pilot implementation based on core vocabularies—Pilot on the Organization Ontology and the Registered Organisation Vocabulary.*

FAO geopolitical ontology. (2012). Retrieved from http://www.fao.org/countryprofiles/geoinfo/geopolitical/resource/.

Freie Universitat Berlin, Web-based Systems Group. (2012). *The D2RQ platform—Treating non-RDF databases as virtual RDF graphs.* Retrieved from http://www4.wiwiss.fu-berlin.de/bizer/d2rq/.

Geonames ontology. (2012). Retrieved from http://www.geonames.org/ontology/documentation.html.

Goodwin, J., Dolbear, C., & Hart, G. (2008). Geographical linked data: The administrative geography of Great Britain on the semantic web. *Transactions in GIS, 12*, 19–30.

Google Tool Charts. (2012). Retrieved from http://code.google.com/intl/it-IT/apis/chart/.

Governo Italiano. (2012). *I dati aperti della PA.* Retrieved from http://dati.gov.it.

Halb, W., Raimond, Y., & Hausenblas, M. (2008). Building linked data for both humans and machines. *1st Workshop about Linked Data on the Web (LDOW2008)*, Beijing.

Indice della Pubblica Amministrazione. (2012). Retrieved from http://www.indicepa.gov.it/.

Linked Data Research Centre. (2012). Retrieved from http://linkeddata.deri.ie.

Lodi, G., Maccioni, A., Scannapieco, M., Scanu, M., & Tosco L. (2014). Publishing official classifications in linked open data. *2nd International Workshop on Semantic Statistics. Proceedings of the 2nd International Workshop on Semantic Statistics (SemStats 2014)*, Riva Del Garda, Italy.

Lodi, G., Maccioni, A., & Tortorelli, F. (2012). Linked open data in the Italian e-government interoperability framework. *6th International Conference on Methodologies, Technologies and Tools enabling e-Government (METTEG).*

Open Data Commons Attribution License. (2012). Retrieved from http://opendatacommons .org/licenses/by/.

Open Data Commons Public Domain Dedication and License. (2012). Retrieved from http://opendatacommons.org/licenses/pddl/.

Opening up Government. (2012). Retrieved from http://data.gov.uk.

OpenLink Virtuoso. (2012). Retrieved from http://virtuoso.openlinksw.com/.

Petrucci, C., Gennai, F., Shahin, A., & Vinciarelli, A. (2011). *IETF RFC 6109 Informational, "La Posta Elettronica Certificata—Italian Certified Electronic Mail."*

Sahoo, S. S., Halb, W., Hellmann, S., Idehen, K., Thibodeau, T., Jr., Auer, S. et al. (2009). *A survey of current approaches for mapping of relational databases to RDF.*

SPARQL protocol. (2012). Retrieved from http://www.w3.org/TR/rdf-sparql-protocol/.

State of the LOD Cloud. (2012). Retrieved from http://www4.wiwiss.fu-berlin.de/lodcloud /state/.

Volz, J., Bizer, C., Gaedke, M., & Kobilarov, G. (2009). Silk—A link discovery framework for the web of data. *2nd Workshop about Linked Data on the Web (LDOW2009)*, Madrid.

W3C RDF Vocabulary Description Language 1.0: RDF Schema. (2012). Retrieved from http://www.w3.org/TR/rdf-schema/.

W3C Semantic Web. (2012). *Resource Description Framework (RDF).* Retrieved from http:// www.w3.org/RDF/.

W3C. (2004). W3C OWL Web ontology language overview. *W3C Recommendation.*

Wahl, M., Howes, T., & Kille, S. (1997). *Network Working Group RFC 2251, "Lightweight Directory Access Protocol (v3)."*

Chapter 13

Coupling Public Sector Information and Public-Funded Research Data in Europe: A Vision of an Open Data Ecosystem

Sven Schade, Carlos Granell, and Andrea Perego

Contents

Abstract

Scientific domains and the public sector increasingly open their data repositories. In June 2013, the G8 boosted the topic in a charter, and in April 2014, the Open Government Partnership welcomed its 64th partner to make governments more open and accountable. The U.S. and Australian governments, together with the European Commission, complement this initiative of the public sector with the Research Data Alliance, to ease research data sharing and exchange. Although many objectives are shared, we see different interpretations and developments all over the globe. This chapter highlights the central activities in the European context and anticipates possible developments. It explores the concept of a cross-domain open data ecosystem by (a) explaining the current situation, (b) exploring innovative approaches and recent trends on information and communication technologies, and (c) outlining potential pathways to couple public sector information and public-funded research data. The chapter concludes by summarizing practical implications and outlining directions for future research.

Keywords: open data, science, research, public funding, curation, preservation

13.1 Introduction

Over the past few years, we have witnessed a transition from closed to open data, across scientific domains as well as administrative levels and sectors. In June 2013, the G8 boosted the topic in a charter (G8, 2013), and the Open Government Partnership (OGP) unites over 60 partners to make governments more open and accountable. The U.S. and Australian governments, together with the European Commission (EC), complement this initiative with the Research Data Alliance (RDA)* to ease research data sharing and exchange. Although common goals emerge, we see different interpretations and developments all over the globe.

As a major example, the Europe 2020 strategy—the EC's response to the economic crisis—asks to remove barriers for data reuse to generate economic benefits and lead to greater transparency in public administrations, as well as promotes social and political engagement. At its core, open data are introduced as an engine for innovation, economic growth, and transparent governance (EC, 2011). Recent case studies demonstrate the feasibility of implementing the requested open data strategies and illustrate how they enhance transparency in the public sector (U.K. Cabinet Office, 2012). It can be concluded that concrete actions on critical data-related issues—such as data governance, societal engagement, data licenses and regulations, and data usability—can help governments to unleash the potential of public data (their main asset) to empower a clear and transparent governance model. Likewise, open data can help in preventing errors, abuses, and inefficiencies through continuous collaboration between citizens and the government, and it may be a main driver to increase trust between public administrations and their stakeholders (Domingo, Bellalta, Palacin, Oliver, & Almirall, 2013; Fioretti, 2011). Open data initiatives are thus crucial to stimulate the reuse of government data by public and private stakeholders alike, to create new, innovative, added-value services and applications and to enhance efficiency, effectiveness, and cost savings (Domingo et al., 2013).

This chapter addresses open data in the context of public sector information (PSI) and other government-funded data in conjunction with public-funded research data. It focuses particularly on the case of Europe, including also related programs at a global level. We identify the current status and challenges to derive future pathways for coupling PSI and public-funded research data. As a book chapter, this work is not foreseen to give a detailed academic review of the state-of-play. Instead, it provides an initial overview of major European activities to novices in the field, as well as entry points for further studies to those with a deeper interest in particular facets. It equally addresses practitioners from academia and public administration by highlighting some of the implications that open data strategies might have on their daily work, as well as researchers who are eager to learn about the underlying scientific challenges.

The remainder of this chapter is structured as follows. Section 13.2 reviews major ongoing open data initiatives, in the public sector and in the scientific

* https://rd-alliance.org.

communities, to provide an impression of the rapidly growing and dynamic open data landscape from a European perspective. Recent developments in matching activities in the public and academic sectors are highlighted in Section 13.3, just before introducing the vision of a cross-domain open data ecosystem in Section 13.4. Finally, in Sections 13.5 and 13.6, respectively, we derive major lessons learned—highlighting central consequences for the theory and practice of data management—and indicate directions for future work.

13.2 A Gentle Review

In the following, we present a noncomprehensive review on initiatives, activities, and projects relevant for the scope of the chapter. We organize the list into two categories: (a) projects related to PSI and (b) projects related to research data. In Section 13.3 we then explore initiatives that bridge the gap between both categories.

13.2.1 PSI Initiatives

In the legal context of the European Union (EU), PSI access and reuse are regulated by two directives, having complementary purposes:

1. Directive 2003/98/EC of the European Parliament and of the Council of 17 November 2003 on the reuse of PSI, usually referred to as PSI Directive (EU, 2003)
2. Directive 2007/2/EC of the European Parliament and of the Council of 14 March 2007 establishing an Infrastructure for Spatial Information in the European Community (INSPIRE) (EU, 2007)

The PSI Directive entered into force in December 2003 and focuses on the economic aspects of reuse of information rather than citizens' access to information. A revision to the PSI Directive entered into force recently, in June 2013 (EU, 2013), and it was followed by a public consultation (EC, 2013a), to devise recommendations and guidelines to assist Member States in implementing this directive consistently. Moreover, in October 2013, the EC published a document (EC, 2013b) concerning the EU implementation of the principles set out in the G8 Open Data Charter (G8, 2013) and following the initial communication of the EU Open Data Policy (EC, 2011).

The PSI Directive provides the framework for a number of activities aiming at promoting open data principles and cross-domain interoperability in the public sector. In particular, in December 2012, the EC published the EU Open Data Portal,* which gives access to open data from EU institutions and bodies. Moreover, work is under way for the creation of a pan-European Open Data Portal (EC, 2010;

* http://open-data.europa.eu.

Shadbolt, 2010), which is meant to provide a single access point for the discovery of open data from EU Member States. Finally, the EU program "Interoperability Solutions for European Public Administrations" (ISA)* funds a number of actions addressing interoperability issues across levels and sectors of European public administrations. This includes standardization activities that are meant to support the creation of the federated data infrastructure of the pan-European Open Data Portal, as a common metadata interchange format (EU ISA Programme, 2013).

Whereas the PSI Directive covers public sector and government data as a whole, the INSPIRE Directive defines specific requirements for information relevant to environmental policies or policies that may have an impact on the environment. This is achieved by setting out a legal and technical framework for the harmonization of the representation, access, and sharing of geospatial data and services operated by Member States. Entered into force in May 2007, the adoption and implementation of the INSPIRE Directive should be completed stepwise by 2020. Notably, the thematic scope of INSPIRE includes themes that are also relevant beyond the environmental domain, for example, addresses and buildings, and that can be potentially used as a basis for supporting cross-sector interoperability.

13.2.1.1 Open Government Partnership

Unlike the above, the OGP† is a voluntary initiative at the political level, currently involving over 60 countries (including, at the time of writing, 20 of the 28 Member States of the EU), to stimulate national governments to adopt transparent, accountable, and sustainable government models in response to their citizens. As the OGP's vision claims, this requires "a shift in norms and culture to ensure genuine dialogue and collaboration between governments and civil society."

Each OGP member (country level) must endorse the Open Data Declaration‡ principles, which basically aim to make government data more open and accessible to citizens, and foster the use of standards throughout public administrations. Although these principles are not legally binding, they provide minimal but well-defined guidelines and action plans to reach the objectives. The OGP also conducts regular monitoring activities on a number of member countries.§ These progress reports account for the current implementation status of open data portals, measures to facilitate publications and reuse of public data, promotion of standards, information and communication technology (ICT) support, regulations, strategic plans, and others. They address public administration, institutions, and agencies at national and subnational levels and provide up-to-date information of the status quo of openness, transparency, and accountability of the public sector and governmental data.

---

* http://ec.europa.eu/isa.

† http://www.opengovpartnership.org.

‡ http://www.opengovpartnership.org/about/open-government-declaration.

§ http://www.opengovpartnership.org/independent-reporting-mechanism.

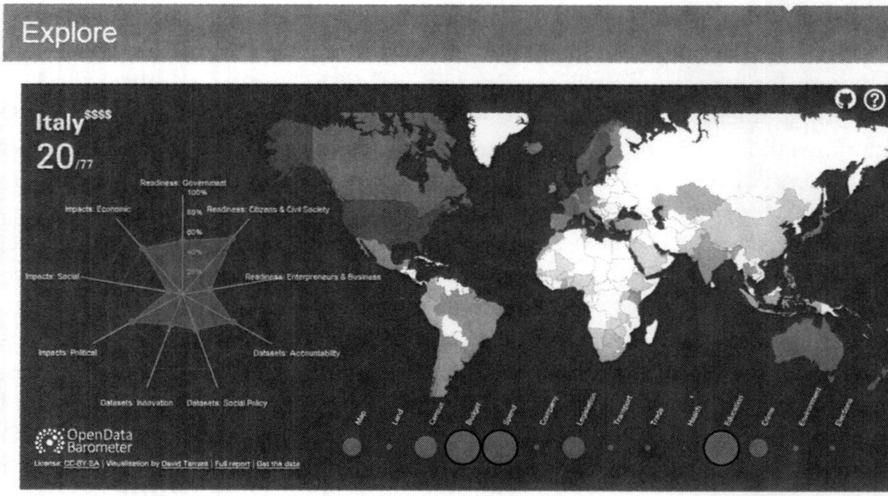

Figure 13.1 Example of the Open Data Barometer visualization tool.

13.2.1.2 Open Data Research Network

The Open Data Research Network* is a collaborative network coordinated by the Web Foundation† and the International Development Research Centre‡ to promote and explore the implementation and impact of open data initiatives across the world. Like OGP, the Open Data Research Network evaluates national open data initiatives. Researchers of the ongoing project study "Exploring the emerging impacts of open data in developing countries,"§ for example, conduct regular studies on the implementation of open data initiatives and strategies and their socioeconomic and environmental impacts in several countries. Along the same lines, the Open Data Barometer¶ explores the degree of adoption of open data initiative in 77 countries, including 17 Member States of the EU (Davies, 2013). The data visualization tool (Figure 13.1) lets users inspect these data per individual country, analyze how a country is investing in the development of national open data infrastructures, and compare different states.

The OGP and the Open Data Research Network play a similar role. Both are seen as trusted, third-party, supranational entities that can evaluate open data developments at country levels. Obviously, countries must commit to deliver

* http://www.opendataresearch.org.

† http://www.webfoundation.org.

‡ http://www.idrc.ca.

§ http://www.opendataresearch.org/emergingimpacts.

¶ http://www.opendataresearch.org/barometer.

up-to-date and verified information on the current status of deployment of open data initiatives to both entities. It remains to be seen how these regular monitoring studies and reports influence national strategies toward the full realization of open government data across countries, also given that geopolitic contexts profoundly vary from country to country.

13.2.1.3 Share-PSI 2.0

Share-PSI 2.0* is a European network for the exchange of experience and ideas on implementing open data policies in the public sector. Unlike the first two initiatives, in which the predominant focus is at national level, Share-PSI 2.0 scales down to also include subnational levels to identify issues and practical solutions for the upcoming implementation of the revised PSI Directive (see earlier this section discussing PSI initiatives).

The majority of European countries belong to this network. Nevertheless the list of partners also includes standards bodies (W3C), national and local government departments, academic institutions involved directly in PSI dissemination and usage, commercial bodies—from microcompanies to small and medium enterprises—and citizen engagement bodies and open data advocates (the Open Data Institute [ODI] and the Open Knowledge Foundation†). The network already organized a series of stakeholder workshops‡ to analyze cases in which open data have improved the efficiency of government departments and institutions. Results and lessons learned provided input to the W3C Data on the Web Best Practices Working Group.§

13.2.1.4 Publicdata.eu

Publicdata.eu¶ provides access to open, freely reusable data sets from local, regional, and national public bodies across Europe (Figure 13.2). It aggregates data sets and centralizes the access to national open data portals that are reachable by following the link "Data Portals" in Figure 13.2. As a one-stop portal, Publicdata.eu lets users discover and access a great variety of open data sets at European and Member State levels. Because Publicdata.eu is built upon the Comprehensive Knowledge Archive Network (CKAN)**—the same technology platform as most relevant national open data portals of European countries (United Kingdom, Austria, Holland, Italy, etc.) —technical interoperability among these open data portals is easily achieved.

* http://www.w3.org/2013/share-psi.
† http://okfn.org.
‡ http://www.w3.org/2013/share-psi/workshop/samos.
§ http://www.w3.org/2013/dwbp/wiki/Main_Page.
¶ http://publicdata.eu/about.
** http://ckan.org.

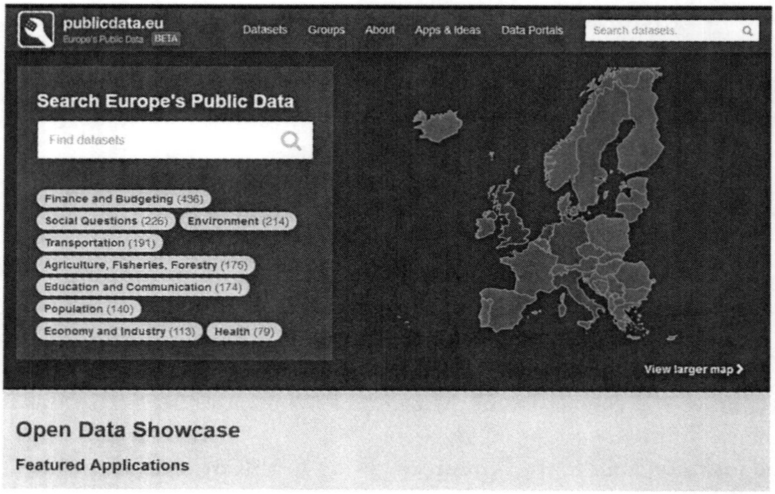

Figure 13.2 Portal of Publicdata.eu.

Notably, Publicdata.eu does not only harvest data sets from national open data portals. Experiments on the integration of advanced Semantic Web technologies to provide Linked Open Data in the form of Resource Description Framework triples are ongoing.* As we will discuss further in Section 13.4, raw, open, and linked data are key to substantially improve the internal and external value for governments at any level.

13.2.1.5 Research Projects

Whereas the previous initiatives were examples of networks at global and European scale to exploring exiting cases studies, experiences, and best practices, this section takes a complementary perspective and briefly introduces as subset of ongoing EC-funded research projects with a much narrower scope and focus on ICT than the aforementioned initiatives.

The Alliance for Permanent Access to the Records of Science in Europe† is an EC-funded Network of Excellence that brings together an extremely diverse set of organizations and experts. It envisions a long-lived Virtual Centre of Digital Preservation Excellence to bring coherence, cohesion, and continuity to research and remove the barriers to long-term accessibility and usability of digital information.‡

* http://eur-lex.publicdata.eu.
† http://www.alliancepermanentaccess.org/index.php/aparsen.
‡ http://www.alliancepermanentaccess.org/index.php/community/virtual-centre-of-excellence.

The second example is the SmartOpenData* project, which investigates the creation of a Linked Open Data infrastructure (including software tools and data) fed by public and freely available data sources. SmartOpenData focuses particularly on data for biodiversity and environment protection and research in European protected areas and national parks. As the focus is on environmental data, the project applies Semantic Web technologies and Linked Data in the context of the INSPIRE Directive.

Finally, the Data Supply Chains for Pools, Services and Analytics in Economics and Finance (DOPA) project† is an interesting example of opening financial and economic data. The DOPA project aims to semantically link large amounts of economic and financial data and make these publicly available as standardized data sets for further data analysis and reuse. In this sense, the integration of distinct platforms by linking contained data sets, either public environmental, economic, or financial data, within each platform seems to be a promising way to make advances toward the coupling of government and research data.

13.2.2 Research Data Initiatives

We now summarize a selection of the many already ongoing initiatives and research projects for making public-funded research data publicly available, discoverable, and accessible to the entire scientific community. Past studies in the literature already highlighted the fact that most researchers would be willing to share their research data and reuse others' research data for their own research projects, but unfortunately, the majority do not know how to do so or simply do not follow the required practices to share their output data and make these available to others in a convenient way (Piwowar, 2011; Tenopir et al., 2011). Along the same lines, the EC (2012a, 2012b) promotes multidisciplinary collaboration and the sharing of scientific findings. In July 2013, the EC held a public consultation (EC, 2013c) on open access to research data to get some insights for the definition of the open data policies in the current EU Framework Programme for Research and Innovation (Horizon 2020). Some obvious considerations came out of the consultation, as different stakeholders, such as private companies, researchers, industry, ICT and data professionals, publishers, and libraries, have distinct needs with regard to use of and access to research data. Of course, open access to scientific publications is a shared requirement, but equally important is to access and reuse related research results from scientific research projects. In this context, we overview some of the major initiatives that put the emphasis not only on the openly access and share of scientific publications but also on related research resources, such as data, models, algorithms, and software.

* http://www.smartopendata.eu.

† http://www.dopa-project.eu.

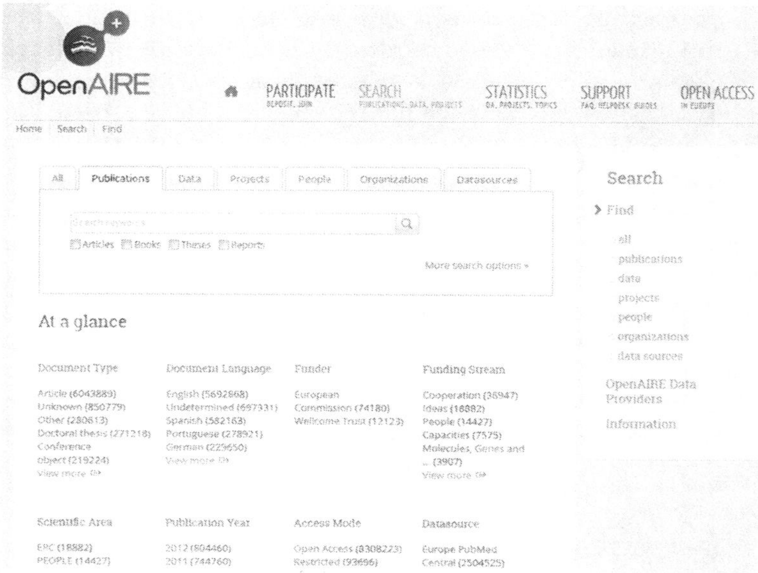

Figure 13.3 OpenAIRE portal (https://www.openaire.eu/search/find/publications?noscroll=true#publtab).

13.2.2.1 Open Access Infrastructure for Research in Europe

The EC supports initiatives to ensure that research results funded by European research programs are made freely available to citizens. In this way, open access is a driver to improve the EU's return on research and development investment, which has been reinforced in Horizon 2020. In 2008, the Open Access Infrastructure for Research in Europe (OpenAIRE)* project was launched as an open access pilot under the previous funding program of the EC (the Seventh Framework Programme, FP7). It aimed to foster open access practices for researchers by building support structures, agreements, and alliances to depositing scientific publications derived from FP7-funded research projects.

As a result, such "support structures" are now realized as a network of digital repositories that enable free access to scientific content across academic disciplines. To this regard, the Digital Repository Infrastructure Vision for European Research (DRIVER) initiative† sets the background scene to create a cohesive pan-European infrastructure for open access digital repositories. The OpenAIRE portal (Figure 13.3) is built upon the DRIVER infrastructure to offer open access to scientific publications across an interrelated network of digital repositories.

* http://www.openaire.eu/en.

† http://www.driver-repository.eu.

Figure 13.4 RECOLECTA portal, an open science harvester for Spanish digital repositories (http://recolecta.fecyt.es).

Digital repositories at universities are increasingly using the OpenAIRE and DRIVER guidelines and claim compliance. Taking the example of Spain, the national portal Recolector de Ciencia Abierta (RECOLECTA [Spanish for Open Science Harvester]) (Figure 13.4) is an open science harvester at national level that collects research data in form of scientific publications from digital repositories that are provided by Spanish universities. Scientific publications are harvested using Open Archives Initiative–Protocol for Metadata Harvesting (OAI-PMH; Open Archives Initiative, 2002), which is both a widespread standard and an interoperability tool to enable communication and exchange information among OAI-PMH-compliant repositories. Those resources from Spanish digital repositories that are compliant with the OpenAIRE and DRIVER guidelines are also reachable from the OpenAIRE portal. In this way, interoperability and integration of different digital repositories are achieved across Europe. The guidelines thus enable sophisticated and added-value services to both researchers and the general public.

13.2.2.2 European Data Infrastructure

Whereas the focus of OpenAIRE is to enable open access and sharing of scientific publications, the European Data Infrastructure (EUDAT)* goes beyond scientific

* http://www.eudat.eu.

publications as the only type of research outputs and aims to establish cost-effective, collaborative data infrastructures to support the access, discovery, and sharing of research data for multidisciplinary research communities. The project has the objective to build a solid foundation—in terms of scale, scope, domain, practices, and others—that allows heterogeneous research communities to share research data and to preserve these in the long-term.

The main users of the EUDAT infrastructure are experts who need to collaborate in demanding research projects or validate research results that were produced by other scientists. To define a set of cross-community, common services for data sharing and management, EUDAT has worked closely with thematic research communities (e.g., Common Language Resources and Technology Infrastructure [CLARIN], LifeWatch, European Plate Observing System [EPOS], European Network for Earth System Modeling [ENES], Virtual Physiological Human [VPH]).*

The project is organized into dedicated working groups to deal with the management of massive generation of real-time data from research facilities, instruments and sensors, the definition of data workflows for sophisticated services, the required semantics to let researchers annotate research data in consistent and interoperable ways, and, most important, the establishment of policies regarding data access and reuse. The latter was extremely challenging, as stated by partners of the EUDAT consortium, due to the need of harmonizing diverse open data and data access policies as well as devising a common licensing scheme to set pragmatic ways for opening, accessing, and sharing cross-domain research data.

13.2.2.3 Research Data Alliance

The RDA is an international initiative launched in March 2013 and jointly funded by the EC, the U.S. National Science Foundation and National Institute of Standards and Technology, and the Australian Government's Department of Innovation. The RDA aims to accelerate and facilitate research data sharing and exchange, use, and reuse; standards harmonization; and discoverability. This will be achieved by quickly creating and adopting focused pieces of code, policy, infrastructure, standards, or best practices that enable data to be shared and exchanged now.

To achieve its objectives, the RDA provides the framework for the creation and coordination of working groups, proposed by its members, to address interoperability issues across disciplines. Such groups are meant to help build consensus and establish partnerships across communities on objectives of common interest.

Due to its worldwide participation and support, the RDA can greatly contribute to the adoption of consistent data management and sharing principles across the scientific communities. However, the effective and consistent governance of an initiative of the size of the RDA poses a number of issues, for example, preventing duplication and/or fragmentation of efforts and monitoring whether and how RDA

* http://www.eudat.eu/eudat-communities.

groups coordinate their work and address cross-domain interoperability issues. At the early stage, it remains to be seen how the RDA will cope with these challenges.

13.2.2.4 Research Projects

Complementing the earlier text this section on research data initiatives, several ongoing research projects share the common vision of enhancing the sharing of, integration of, and access to research data, but from different perspectives. Here, we present a small subset of the current EC-funded research projects that seem promising and worthwhile to follow over the next months and, in some cases, years.

The Policy Recommendations for Open Access to Research Data in Europe* project will leverage existing networks, communities, and projects to address challenges within the open access and data dissemination and preservation sector and produce policy recommendations for open access to research data based on existing good practice. For doing so, besides taking into account infrastructural and technological issues, the Policy Recommendations for Open Access to Research Data in Europe project puts special emphasis on legal, ethical, institutional, and policy issues by analyzing cross-domain case studies (e.g., physics, health, earth sciences) to enable open access for research data. Even though the project will address different interoperability levels related to open access, it aims primarily to incentivize professional researchers to deposit their data on open access repositories. The project wants to formulate the required open access polices in an easy, unambiguous, and understandable way.

The following two recently started projects exploit large amounts of open research data deployed in research data infrastructures through the design and development of sophisticated analytical tools and data visualization services. The Maximizing the Exploitation of Linked Open Data in Enterprise and Science† project will develop advanced services to maximizing the potential of Linked Open Data for private and public sector alike, whereas the OpenCube‡ project will focus on publishing and enriching Linked Open Statistical Data for the development of data analytics and enhanced visualization services. Like research data gains value when it is open and accessible, data analysis and visualization tools will follow the same tenet and will thus be regarded as a type of research output that needs to be open and accessible to the broad scientific community.

The Co-ordination and Harmonization of Advanced e-Infrastructures for Research and Education Data Sharing§ project focuses particularly on promoting collaboration between scientists facilitating interoperability between thematic data infrastructures.

* http://recodeproject.eu.
† http://www.melodiesproject.eu.
‡ http://www.opencube-project.eu.
§ http://www.chain-project.eu.

13.2.3 Summary

Regardless of the data as such, PSI and research data initiatives share the grand vision of making data more open, public, and accessible. Many technicalities are common in PSI and open research data initiatives, for example, the use of semantic technologies, CKAN platform, Linked Data principles, controlled vocabularies, and persistent identifiers (Willmes, Kürner, & Bareth, 2014). However, the nature of the data, the incentives of making data openly available, and the target user communities vary between PSI and research data initiatives. It remains unclear if further similarities can be derived, for example, in reuse mechanisms or minimum requirements for data description.

Having these differences in mind, one might wonder whether the coupling of government data and research data makes any sense at all. What are the benefits of coupling existing infrastructures and how could they be built to overcome the aforementioned issues? We reflect on these questions and related challenges in the next section.

13.3 Emerging Approaches for Coupling PSI and Public-Funded Research Data

This section outlines existing approaches and arising needs to combine PSI and research data. We present early steps for establishing a rich space of open data as a foundation for the innovation ecosystem, which is requested by the Europe 2020 strategy and reemphasized in the G8 Open Data Charter, the OGP, RDA, and most lately by the EC (2014) communication on a data-driven economy. Activities on exemplary synergies between the two fields are introduced, and prominent examples for new types of joint initiatives are detailed.

13.3.1 Addressing Synergies in Collaboration

Although the projects and initiatives described in the previous section differ in scope and objectives, it is yet possible to identify a number of common issues and requirements. This section highlights a few possible synergies and the potential benefits of collaboration across different domains and communities. We focus our discussion on three specific use cases concerning key issues as cross-domain interoperability, the scientific principle of reproducibility, and the notion of context.

13.3.1.1 Interoperability through Reuse

A first use case of enabling interoperability through reuse practices is provided by the EU ISA Programme, which hosts on its online platform (Joinup), a catalogue* of "semantic interoperability assets" (i.e., resources like meta/data schemas, code lists,

* https://joinup.ec.europa.eu/interoperability/search.

thesauri), which are used by the different sectors of European public administrations. Notably, a similar activity is underway in the framework of RDA, where the Metadata Standards Directory working group aims at creating a catalogue of metadata schemas used in the different scientific communities.

Despite the specific requirements to be addressed by these initiatives, they share the same rationale: promoting the reuse of existing semantic interoperability assets and the development of new ones, to fill the existing gaps. Interestingly, both initiatives recognize that to achieve their objectives, semantic interoperability assets should be associated with information about their context, for example, what was their intended use, where and how much they are actually used, and existing reference implementations. Work in this area is underway in the EU ISA Programme and in RDA (this is the specific scope of the Data in Context Interest Group).

13.3.1.2 Curation for Reproducibility

The notion of context introduces another use case, concerning the ability to assess the quality of an information resource. This is a fundamental requirement in the scientific community, where research outputs should be associated with all the information needed to verify their scientific soundness and, in case of an experiment, to reproduce their results (e.g., a data set obtained by processing other data through a given model). In some cases, this requires methodologies for data curation and preservation. For example, versioning support may be required for both the data and the models and the analytics used to process them.

Notably, such requirements are going to become important also in the public sector, especially when both data and studies are being used as a basis for policy-making processes. In such cases, the ability to repeat an experiment and, in most cases, to reproduce its result is a fundamental part of policy assessments; for example, when a policy does not obtain the expected results, it is necessary to verify each step followed for its formulation, to verify whether and where errors have been made. In such a scenario, the context of public sector data consists of a constellation of heterogeneous resources, including, for example, legal acts, that should be available persistently.

Expertise on data curation and preservation is available both in public administrations and in the scientific community but is limited to given sectors and disciplines. A typical example is the library and archival science community, which provides the common background for experts working in libraries and archives of public administrations and research institutions. It is worth noting that this community addressed very early issues related to archiving, preservation, and access to digital artifacts, and digital libraries are currently a consolidate technology, able to deal with a huge amount of heterogeneous multimedia objects (Candela, Castelli, & Pagano, 2011). Other examples are provided by specific institutions, as those dealing with the preservation of data concerning DNA (Benson et al., 2013) and particle physics experiments (Akopov et al., 2012). All these experts can then contribute useful expertise to other scientific communities and sectors of public administrations, taking into account the specific requirements of the data to be maintained. This may apply to the selection criteria determining what to preserve and how, against different criteria for data acquisition and publications, as well as limitations concerning the available resources for data storage and maintenance.

13.3.1.3 Information Integration from Multiple Sources

The last use case we would like to discuss concerns the integration of data from different sources and, in particular, the integration of crowd-sourced data, such as data mined from social media platforms, with research and government data. The ability to make available different views of the same phenomena not only is beneficial to their better understanding but also may help the effective use of data for practical purposes. For instance, many interesting examples are available of Smart Cities applications able to integrate official data with real-time feedback on traffic, accidents, emergencies, and others (Craglia & Granell, 2014). However, a number of issues prevent the adoption of data integration as a common practice for data publishers. One of the main obstacles is related to the risk perceived by data publishers of being considered accountable for the information coming from other data sources, using different criteria for data acquisition, processing, and validation. For instance, public administrations are, in general, reluctant to integrate other data sources because they can be considered as "official" data. A similar concern is shared by the scientific community.

The ability to express the context of data—in particular, their provenance and attribution—may contribute to lower the barriers to data integration. This would provoke a more effective use of crowd-sourced data, which are nowadays recognized by both the public sector and the scientific community as an increasingly important source of information, not only for statistical analyses but also for knowledge acquisition (see, e.g., the relationship between folksonomies and ontologies; Mika, 2007), data validation, generation, and quality control, that is, as an input for the data curation process.

The RDA recently highlighted the particular relevance of location information to shape contextual information.* The capacity of geospatial data to offer a common reference frame and, thus, to enable the integration of previously decoupled PSI or scientific data will be further investigated in this context.

13.3.2 Exemplary Solutions

Synergies such as those described previously in this chapter stimulated the development of new solutions across both domains. We are now going to highlight three specific examples that illustrate central innovations in this emerging landscape.

---

* https://www.rd-alliance.org/bof-geospatial-information-meeting-rda3.html.

13.3.2.1 Europeana

Linked Data is being adopted by libraries and cultural information projects like Europeana.* The "Library Linked Data" W3C Incubator Group's final report (LLD XG, 2011) points the way forward to highlight the potential of Linked Data to interrelate and make sense of large amounts of diverse social and cultural data as well as scientific data. Europeana's goal is to become a European digital library that digitally collects and provides open data access to a vast amount of European cultural heritage resources hosted by several institutions such as libraries, archives, museums, galleries, and audiovisual collections. It is indeed an ambitious goal that needs a widely consensus-based approach for the establishment of political and technological agreements among all the involved institutions.

Similar to the OpenAIRE portal, Europeana acts as a unique entry point to give users access to all metadata of digital resources that are collected and harvested by Europeana. It has converted into a big metadata repository, harvester, and provider of cultural and scientific resources. To accomplish this, Europeana specifies a simple set of technical and legal requirements that any contributing institution must adhere to. From the technical point of view, contributing institutions must provide their metadata records in compliance with the Europeana Data Model† and the stipulated syntax and semantics of the established persistent identifiers, as well as to ensure compatibility with the OAI-PMH standard. From a legal perspective, each contributing institution must sign an agreement on the exchange and maintenance of the source metadata records in the realm of the Europeana Licensing Framework.‡

13.3.2.2 Open Data Institute

The ODI§ carries out mostly training, education, and promotion activities about the consumption and publication of public open data, which are equally applicable practices to public-funded research data. For example, ODI offers a variety of training courses and educative programs in the area of Open and Linked Data Technologies. Such programs are targeted to developers and technical-skilled users who can transfer open data technology know-how to public and private organizations like small and medium enterprises.

The ODI shares commonalities and objectives with the OGP and Open Data Research Network (Section 13.2.1) in the promotion of and awareness raising on open data. Unlike the latter two, which carry out their activities mostly at country level, ODI addresses especially individuals on the acquisition of technical skills and

---

* http://www.europeana.eu.

† http://pro.europeana.eu/edm-documentation.

‡ http://europeana-photography.iprguide.org/lang_en/page/documents/Europeana-licensing-framework.

§ http://theodi.org.

tools to put in practice open data solutions at local level. That is, it fosters open data literacy as a means to promote economic growth and innovation by facilitating the exploitation of open data capabilities, along the same line of the EC's vision on data-driven economy (EC, 2014). Indeed, far from being overlapping and mixed initiatives, they complement very well, as the success and impact of the ODI's training activities could be measured in the mid-term through the monitoring studies of open data use cases conducted regularly by the OGP and Open Data Research Network at national level.

13.3.2.3 The ENGAGE Project

The ENGAGE project* aims at the deployment and use of an advanced service infrastructure, incorporating distributed and diverse PSI resources as well as data curation, semantic annotation, and visualization tools. The provided services support scientific collaboration and governance-related research across disciplines while also providing open governmental data to citizens.

A key aspect of ENGAGE is the bottom-up approach to using and exploiting public sector data. Empowering and engaging the research community and nonscientific community alike to devise new ways to use, exploit, and make sense of the wealth of data are central. Community building is a promising connection to unite PSI and research data and find innovative services based on both worlds.

13.4 Toward an Open Data Ecosystem

The sheer number of initiatives and projects presented previously in this chapter provides a first insight into the topic and illustrates the richness of already ongoing activities. A few cutting-edge examples already indicate the current thinking and initial convergence between the public sector and research communities. But where are we going next?

Searching for an appropriate way ahead, it is not surprising that methodologies for data archiving and practices for data access, curation, and preservation in the research and library domains are more mature and widespread than in the public sector. For example, some EU Member States already have a network of recognized data archives in the research domain (typically funded by the national research councils). OpenAIRE and Europeana are scaling this up to the European level and successfully exploit best practices, standards, and technology in the library domain. The same holds for ongoing initiatives for sharing and preserving research data like EUDAT.

Nevertheless, the current status of open government data portals is still far away from the examples in library science. As commented in Section 13.1, the trend may change in the immediate future as several international and national networks are

* http://www.engage-project.eu.

making quick progress to disseminate best practices in terms of standards and technologies for increasingly opening PSI. From a technical perspective, it is a matter of time before open government data initiatives will reach the level of maturity of library and research data initiatives.

Ultimately, the main issues are not technical (persistent identifiers, quality, controlled vocabularies, registries, etc.) but institutional/organizational. How can we develop a network of interrelated, recognized public sector data repositories (or data archives, or data infrastructures, regardless the label used) that take on the responsibility of long-term management of PSI across Europe?

Public institutions are characterized by not being able to get full return on investment from their valuable internal data sets for boosting innovation, due to several factors such as the lack of human and economic resources as well as the lack of proper stimuli and future vision. Fortunately, acts like the PSI Directive (EU, 2003, 2013) and some ongoing initiatives are trying to raise awareness of the economic value of opening data because data are the new raw material for the current knowledge-based economy. Government data can be "opened" in the sense of publishing freely available data to the entire society but can also be "opened" with no restrictions to create social utility, new services, and added value to stimulating stakeholder ecosystems and economic growth based primarily on open data. These two meanings of "open" encompass the idea of Open Innovation formulated originally by Chesbrough (2006) in the economic and business sectors, in that business boundaries become permeable to external actors' influence, promoting then a new wealth of innovate ideas and synergies, thus creating a data-driven economy.

An open data ecosystem could bridge different open access solutions (platforms, approaches, etc.) in an interoperable way because a one-fits-all solution seems unrealistic due to the extraordinarily great diversity of fields, disciplines, and areas of knowledge. It should build on central lessons that can be learned from past and ongoing activities in the PSI sector and research communities and the derived recommendations.

13.5 Practical Implications

From the examples in the previous sections, and crossing many of the guidelines and recommendations of individual projects and initiatives, we can extract several practical implications when establishing an open data ecosystem—each beginning with a lesson learned, followed by supporting examples, and ending with a derived need for action (highlighted in italic):

1. The distinction between PSI and research data is not definite; for example, statistical data from public authorities might be used for scientific analysis, and data that are acquired as part of a public-funded research project might be considered PSI. Accordingly, the coupling of PSI and research data implies

cross-institutional management. This is exemplified by the MELODIES, OpenCube, and ENGAGE projects mentioned earlier. *New organizational structures and business models have to be cocreated by all involved parties—overcoming the "artificial" barriers between public sector and research communities.*

2. Whereas culture, motivations, and approaches between the public sector and academics and other research environments are extremely diverse, ICT requirements are similar. Accordingly, available technical solutions from one sector can meet the needs of the other. The remaining barrier is in sharing and awareness raising. The example of Europeana, which combines technology from diverse contexts (Linked Data, taxonomies, distributed catalogues, and OAI-PMH), demonstrates that ICT requirements across communities and domains are not as critical as the will to share data. *Diverse culture and attitudes should not block the sharing of available technical solutions, that is, data models, software, and example applications. Practitioners in the public sector and research have to leave their comfort zones and collaborate.*
3. As much as open data are about technology, they are about governance. Transparent licensing schemes and sustainable governance models for the access, management, and maintenance of data are fundamental. This lesson is illustrated by the lack of common open data policies and licensing scheme, which have been identified as main obstacles in OpenAIRE and EUDAT. *Dialogues between the public sector and academics and other research environments have to be increased, starting from an identification of the communalities and differences from a librarian's viewpoint.*
4. Collection management is central to librarians. They are the connectors between both worlds; although they are essential enablers for both the public sector and research communities, their activities largely overlap. In addition, each administrative unit hosts at least one librarian. For example, an early attempt to build federated networks of connected libraries has been proven in Europeana. *It is the time to practically exploit the ubiquity of libraries and the established network of interrelated libraries (e.g., public, private, and university libraries), along with their inherent expertise on data management, to go beyond the traditional library services to reach the overall goal of transparent handling and management of all kinds of data without boundaries.*
5. Several possible solutions have been and are currently exploited, but the overall scenery is still fragmented. Commonalities have to be identified, and the interoperability between available ICT architectures and developed components and services has to be improved, with and between both domains. For example, projects like DOPA (economic and finance open data) and EUDAT (network of thematic research infrastructures) expose similar services that could, by large, serve other projects and initiatives. *Operational success will depend on the projection of case studies and experiments. Initial deployments should be launched to identify common technology that could be used by both for content discovery, management, and handling. The potential of general-purpose*

data platforms, using mainstream technologies (e.g., CKAN), should be particularly investigated.

6. While establishing the required tools, the definition of the responsible parties and the partnerships to host the required infrastructure/platform/framework remains a major challenge. Most of the projects in the central part of this chapter acknowledged the need to seek sustainable models to ensure the feasibility of the underlying infrastructures beyond the project duration. *Public–private partnerships and consortium have to be explored. Furthermore, practitioners in public and research sectors should, by principle, consider also open tools.*
7. Context is essential to be able to use (open) data. It will be fundamental to find a common way to describe, manage, and handle context commonly. This is widely recognized both in PSI and research data projects and initiatives, such as RDA and Europeana. *Given that location is particularly relevant to shape contextual information, we expect an increase in geospatial information linked to PSI and scientific data, which will impact future data management systems.*

13.6 Areas for Future Research

As the previous text illustrates, the creation of an open data ecosystem including both public sector and research data is an evolving process. To continuously progress on the diverse items identified above, we see a particular need to address the following research questions (in italics):

1. First, we need to move away from open access as a "yes/no" question to a question of "how." *How do we benefit from experiences and lesson learned across disciplines?*
2. All of the mentioned initiatives imply long-term curation and preservation needs. *Which organizational structures are needed to set up and maintain such data infrastructures?*
3. The open access to databases is necessary, but not sufficient, to reach transparency and to empower diverse audiences to use the unlocked resources. *How might public application programming interfaces allow not only access to but also understanding and proper usage of open data sets?*
4. The citizen component is expected to get a more prominent role—in relation to science, as well as policy making. *How could we foster civic engagement based on open data and, as part of that, which privacy and ethical issues have to be considered?*

Apart from these burning issues, we should continuously remind ourselves of two central questions: Once all of the items discussed so far have been resolved, who would want to use the new ecosystem? And who will be able to use it? In the future, developed thematic expertise will have to be increasingly complemented with skills in data science so that the full depth of the newly available information can be successfully exploited. Building this expertise will be one of the central challenges for the next decade.

Author Biographies

Dr. Sven Schade is a geospatial information scientist with a special interest in knowledge extraction from (big) data, multidisciplinary interoperability, and open innovation. Since September 2013, he has been working as a scientific officer for the Digital Earth and Reference Data Unit of the European Commission's Joint Research Centre (JRC), where he already worked for three years as a postdoc (2009–2012). Before rejoining the JRC, he was employed by the European Environment Agency (EEA) as a project manager for the Shared Environmental Information System (SEIS). In his early career, Sven spent ten years with the Institute for Geoinformatics (IfGI) of the University of Muenster, where he completed his diploma and PhD studies and contributed to numerous national and European-level research projects in the area of semantic interoperability. Sven is the editor-in-chief of the International Journal for Spatial Data Infrastructure Research (IJSDIR), is a council member of the International Society for Digital Earth (ISDE), and has coauthored more than 80 publications in the fields of geospatial semantics, observation web, and next-generation knowledge infrastructures.

Dr. Carlos Granell is a postdoctoral fellow at GEOTEC (UJI). His research lines are centered on spatial data interoperability, geoprocessing services, composition and reuse of geographic services, and more recently architectural styles for integrated environmental modeling and data visualization. He has carried out short research stages at SINTEF (Norway), at the Faculty of Geo-Information Science and Earth Observation of the University of Twente (The Netherlands), at the University of Nottingham (UK), and at the University of Milano Bicocca (Italy). Prior to joining GEOTEC again in 2014, he spent three years as a postdoctoral researcher at the European Commission–Joint Research Centre (EC-JRC), Ispra (Italy). Granell has a PhD degree in computer science from the University Jaume I of Castellón, Spain.

Dr. Andrea Perego is a researcher at the Digital Earth and Reference Data Unit of the European Commission's Joint Research Centre (JRC). His research interests cover semantics, knowledge engineering, cross-domain interoperability, and Open Data. Andrea is a member of the JRC team in charge of the technical coordination of the INSPIRE Directive of the EU, and he is involved in European and international initiatives on Open Data and cross-domain interoperability—as the Research Data Alliance (RDA) and the EU program on "Interoperability Solutions for European Public Administrations" (ISA). Before joining JRC in 2011, Andrea worked for ten years in academic institutions, where, as a PhD student and then as a postdoc researcher, he contributed to national and EU research projects on knowledge engineering, distributed systems, and the social web, addressing scenarios concerning content personalization, trust, and privacy protection.

References

Akopov, Z., Amerio, S., Asner, D., Avetisyan, E., Barring, O., Beacham, J., von der Schmitt, H. (2012). *Status report of the DPHEP Study Group: Towards a global effort for sustainable data preservation in high energy physics* (Technical Report DPHEP-2012-001). CERN, arXiv:1205.4667.

Benson, D. A., Cavanaugh, M., Clark, K., Karsch-Mizrachi, I., Lipman, D. J., Ostell, J., & Sayers, E. W. (2013). GenBank. *Nucleic Acids Research, 41*(Database issue), D36–D42.

Candela, L., Castelli, D., & Pagano, P. (2011). History, evolution, and impact of digital libraries. In I. Iglezakis, T. Synodinou & S. Kapidakis (Eds.), *E-publishing and digital libraries: Legal and organizational issues* (pp. 1–30). Hershey, PA: Information Science Reference.

Chesbrough, H. (2006). *Open business models: How to thrive in the new innovation landscape.* Boston, MA: Harvard Business School.

Craglia, M., & Granell, C. (Eds.). (2014). *Citizen science and smart cities.* (JRC Scientific and Technical Reports, EUR 26652 EN). Luxembourg: Publications Office of the European Union.

Davies, T. (2013). *The Open Data Barometer 2013 Global Report: World Wide Web Foundation and Open Data Institute.* Retrieved from http://www.opendataresearch.org/dl/odb2013/Open-Data-Barometer-2013-Global-Report.pdf.

Domingo, A., Bellalta, B., Palacin, M., Oliver, M., & Almirall, E. (2013). Public open sensor data: Revolutionizing smart cities. *IEEE Technology and Society Magazine, 32*(4), 50–56.

European Commission. (2010). *Technical workshop on the goals and requirements for a pan-European data portal.* Report. Retrieved November 3, 2010, from http://ec.europa.eu/information_society/newsroom/cf/dae/document.cfm?action=display&doc_id=1192.

European Commission. (2011). *Communication from the Commission to the European Parliament, the Council, the European Economic and Social Committee and the Committee of the Regions. Open data: An engine for innovation, growth and transparent governance.* COM(2011) 882 final, 12/12/2011.

European Commission. (2012a). *Communication from the Commission to the European Parliament, the Council, the European Economic and Social Committee and the Committee of the Regions – Towards better access to scientific information: Boosting the benefits of public investments in research.* COM(2012) 401 final.

European Commission. (2012b). *Commission recommendation on access to and preservation of scientific information.* COM(2012) 4890 final.

European Commission. (2013a). *Results of an online consultation on the guidelines on recommended standard licences, datasets and charging for the re-use of public sector information.* Final report. Retrieved December 11, 2013, from http://ec. europa.eu/digital-agenda/en/news/results-online-survey-recommended-standard-licensing-datasets-and-charging-re-use-public-sector.

European Commission. (2013b). *EU implementation of the G8 Open Data Charter*. Retrieved October 31, 2013, from http://ec.europa.eu/digital-agenda/en/news/eu-implementation-g8-open-data-charter.

European Commission. (2013c). *Report of the European Commission public consultation on open research data*. Retrieved from http://ec.europa.eu/digital-agenda/en/results-consultation-open-research-data.

European Commission. (2014). *Communication from the Commission of the European Parliament, the Council and European Economic and Social Committee and the Committee of the Regions—Towards a thriving data-driven economy*. COM(2014) 442 final.

European Union. (2003). Directive 2003/98/EC of the European Parliament and of the Council of 17 November 2003 on the re-use of public sector information. *OJ L* 345, 31/12/2003, pp. 90–96.

European Union. (2007). Directive 2007/2/EC of the European Parliament and of the Council of 14 March 2007 establishing an Infrastructure for Spatial Information in the European Community (INSPIRE). *OJ L* 108, 25/04/2007, pp. 1–14.

European Union. (2013). Directive 2013/37/EU of the European Parliament and of the Council of 26 June 2013 amending Directive 2003/98/EC on the re-use of public sector information Text with EEA relevance. *OJ L* 175, 27/06/2013, pp. 1–8.

EU ISA Programme. (2013). *DCAT application profile for data portals in Europe*. Retrieved from https://joinup.ec.europa.eu/asset/dcat_application_profile.

Fioretti, M. (2011). *Open Data, Open Society: A research project about openness of public data in EU local administration*. DIME network. Retrieved from http://www.dime-eu.org/files/active/0/ODOS_report_1.pdf.

G8. (2013). *Open Data Charter*. Retrieved June 18, 2013, from https://www.gov.uk/government/publications/open-data-charter.

LLD XG. (2011). *Library Linked Data Incubator Group final report*. W3C Incubator Group Report. Retrieved October 25, 2011, from http://www.w3.org/2005/Incubator/lld/XGR-lld-20111025.

Mika, P. (2007). Ontologies are us: A unified model of social networks and semantics. *Journal of Web Semantics, 5*(1), 5–15.

Open Archives Initiative. (2002). *Protocol for metadata harvesting. Version 2.0*. Retrieved from http://www.openarchives.org/pmh.

Piwowar, H. A. (2011). Who shares? Who doesn't? Factors associated with openly archiving raw research data. *PLoS One, 6*(7), e18657.

Shadbolt, N. (2010). *Towards a pan EU data portal—data.gov.eu. Version 4.0*. Retrieved December 15, 2010, from http://ec.europa.eu/information_society/newsroom/cf/dae/document.cfm?action=display&doc_id=1191.

Tenopir, C., Allard, S., Douglass, K., Aydinoglu, A. U., Wu, L., Read, E., Frame, M. (2011). Data sharing by scientists: Practices and perceptions. *PLoS One, 6*(6), e21101.

U.K. Cabinet Office. (2012). *Open Data White Paper: Unleashing the potential*. HM Government. Retrieved from http://data.gov.uk/sites/default/files/Open_data_White_Paper.pdf.

Willmes, C., Kürner, D., & Bareth, G. (2014). Building research data management infrastructure using open source software. *Transactions in GIS, 18*(4), 496–509.

Chapter 14

How Public Administration Restructuring Can Contribute to Greek Educational Reform

Konstantinos Zacharis, Pandelis Ipsilandis, and James O'Kane

Contents

Abstract

For the last two decades, information technology professionals and public servants have witnessed a rapid coevolution of information and communication technology (ICT) artifacts, which have invaded all major sectors in public administration. These tools offer valuable help in managing everyday operations of the organization but at the same time incubate a whole mindset of innovation and novelty. Especially in the educational area, the combination of large-scale automation projects with prolific didactic theory interventions (that reinforce the sociocultural theories of learning as the most dominant today) shapes contemporary high-level priorities in education as well.

It is always useful to have comparative examples with other countries that have also tried to reform their systems. Two cases are presented, Finland and England, where the changes had measurable results. These examples represent two different approaches that can contribute fruitful outcomes to the discussion.

The degree of pragmatologic approach of ICT in the educational system is being ascertained. This procedure stirs between two poles: the technocratic and the holistic. Having defined the current status, we investigate the role of innovation as a catalyst for change. Then, some critical research questions come up: Under which circumstances can technology-driven innovations emerge and flourish inside educational organizations? What performance indices could measure the impact of change? Can we model systemic transformation based on ICT?

A solid transformation model that consists of four standard modules (namely, Management by Objectives, Common Assessment Framework, International Organization for Standardization, and e-government) is suggested. Each one of them is being explicated. The combination of all comprises an operational framework, suitable for the restructuring of an educational organization in terms of improved efficiency and performance. A simple modeling tool that uses metrics vector is applied, after careful selection of a series of attributes that make up a school unit's "eigenvector." This can lead to the calculation of a final numerical expression representing the whole process.

Keywords: restructuring, reform, innovation, education, public administration, ICT, management by objectives, assessment

14.1 Introduction

The contemporary function of Greek public administration (PA) is significantly compromised by legal labyrinths and ramifications that strongly determine both its bureaucratic character and its resistance to modernization and resilience to change. The existence of a large number of detailed laws and regulations not only makes it difficult to monitor their implementation but also creates a rule-based instead of a result-oriented operational background. Thus, the existing legal and regulatory framework is, to a significant degree, responsible for the cachectic administrative culture and its subsequent problems that dominate PA operation in the country today (Matei & Flogaitis, 2011).

Meanwhile, the economic crisis that broke out in the last 5 years has revealed all the malfunctions of PA in its full spectrum. The same applies to the educational area. It is now obvious that the situation cannot be reversed until the importance of other deviant factors is realized (e.g., innovation). Not only should anyone understand the conditions for the development of innovation, but one should also coordinate multiple initiatives to foster it in both private and public sectors. Formulation of such a strategy contains, sine qua non, the accurate description of a national vision and its coherent master plan.

The emergence of innovation as an economy driver is a relatively recent phenomenon. It is due to the rapid growth of local environment based on synergies among three key factors: (a) the establishment of research and excellence inside educational nodes, which become capable of producing state-of-the-art results, (b) the active participation of large companies that are willing to immediately finance innovative ideas, and (c) the emergence of start-up companies that provide flexibility and efficiency in the exploitation of scientific results in everyday practice. These three cooperating factors constitute a self-sustained symbiotic system that has the ability to drive the country away from the slippery recession track (Babones, 2012).

In this work, the authors try to isolate factors that lead to success stories when changing or transforming educational systems in large scale. First, some elements that come out of the international experience should be considered. The cases of Finland and England are examined. The next section describes the idea of "innovation," when taking policy actions, followed by a section that introduces an applicable model of the transformation process in education using four generally accepted PA standards (Management by Objectives [MBO], Common Assessment Framework [CAF], International Organization for Standardization [ISO], and e-government [eGov]).

14.2 Relevant International Cases

The first case that comes to mind when the discussion refers to successful educational systems is Finland. Indeed, this country has been a model case for sustainable school improvement since it has rapidly climbed to the top of relevant international tables, after it started its long-term transformation process a few decades ago. Once lowly ranked educationally, struggling around a centrally guided system that produced poor quality services, today it stands on top among the Organisation for Economic Co-operation and Development (2012) nations, qualified against the "PISA" test assessment.

What were the key elements to this effective change? What advantages helped Finland with this remarkable achievement? Can other countries share this paradigm? The basic pillars of Finland's reform policy were teacher quality, widespread equity–autonomy, and public financing. This led to elegant school buildings embracing happy children and highly qualified and devoted teachers. As one analyst notes: "...Most visitors to Finland...recognize the large autonomy that schools enjoy, little interference by the central education administration in schools' everyday lives, systematic methods to address problems in the lives of students, and targeted professional help for those in need" (Sahlberg, 2009, p. 324).

The analysis summarizes the core principles as follows: (a) highly qualified staff, (b) resourcefulness, especially for those in need, (c) special needs high standards, (d) system evaluation and assessment, and (e) balanced centralization versus decentralization policies (Laukkanen, 2008).

Starting four decades ago, Finland's first priority on educational investment was in personnel. They considered that improving teacher qualifications would help in the implementation of wider policies concerning school autonomy, free content curriculum formation, and instruction design. This was actually proven correct. All teachers are required to have a master's degree on research-based (field) education before their assignment to school units. This knowledge helped them develop diverse high-order skills, like problem-solving and critical thinking. Therefore, they can easily convey this cultivation of independence and active learning to their students (Buchberger & Buchberger, 2004).

The notion of personalized caring for students is a central principle at Finnish schools. Schools are generally small (containing about 250 students) and the same applies to classroom size (about 20). All students receive free daily meals, healthcare, transportation, and learning resources, so that an environment of equal opportunities is actually molded.

The focus on teachers' professional development has led to an increased prevalence of effective teaching methods, resulting in innovations and lateral capacity building. Much attention has been given to voluntary in-service training programs (obligatory programs have been canceled), thus increasing the quality of professionalism, efficiency, and healthy competition among teachers. According to Sahlberg (2007) "...This shift in teachers' learning conditions and styles often reflects ways that classroom learning is arranged for pupils. As a consequence of strengthened professionalism in schools, it has become understood that teachers and schools are responsible for their own work and also solve most problems rather than shift them elsewhere."

The second case worth studying is United Kingdom, where school policies were developed in quite a different way. The Education Reform Act (1988) made considerable changes to the entire system. These changes were aimed at creating a "market" in education with schools competing with each other for "customers" (pupils). The theory was that "bad" schools would lose pupils to the "good" schools and would either have to improve, reduce in capacity, or close.

The reforms introduced at that time were (a) a national curriculum, which made it compulsory for schools to teach certain subjects and syllabuses, (b) national curriculum assessments, (c) league tables that began showing performance statistics for each school (which are regularly published in newspapers and on the Internet), (d) formula funding (which meant that the more children a school could attract to it, the more money the school would receive), and (e) open enrollment and choice for parents was brought back, so that parents could choose or influence which school their children went to. These arrangements have slightly changed and are still applied today.

UK education secretary Michael Gove has defended free schools, tougher examinations, and performance-related pay as needed to improve "stagnant" international rankings after figures published by the Organisation for Economic Co-operation and Development showed a persistent gap between British schoolchildren and their Asian peers (Guardian, 2013). One can easily note that UK policies are in the opposite direction from Finnish ones. This argument contains also nuances of the dominant political philosophy in those countries.

Mainstream politics in United Kingdom today is neoliberal. This reflects the majority belief that minimizing the public sector can resolve all pathogenic situations. The counter objections are that completely "free" schools would hire untrained teachers, which is a good recipe for further decline. One solution could be an optimal combination of freedom with accountability and autonomy with minimum standards and teacher qualifications, but that still remains a sparking political debate.

14.3 Is Innovation the First Step?

Innovation is defined as the implementation of a new or improved product, service, or method; a new marketing practice; even a new business pattern, whose adoption actually leads to, a large extent, restructuring of the organization. Innovation creates economic as well as social added value (Windrum & Koch, 2008). Innovation is a strategic element in the change process because it creates a fine thin line that connects present to future status. It comes as a complement to the process of "standardization" or "prototyping": Whereas the first one is a little chaotic by nature, the second puts things in order in a realistic manner.

It is clear, at first, that innovative ideas are generated by people. Innovation stems from the peculiar way of one's thinking, one's mindset, or world view.

To flourish, the idea must rely on two major factors, preparation and discipline (Laperche, Uzunidis, & Von Tunzelmann, 2008). What people call "inspiration," "serendipity," "brilliant idea," or "intangible essence" is usually nothing more than the expected outcome of a well-prepared mind that took place at the right moment. That is why another factor must be introduced, called "momentum." Most ideas are actually "under our nose," but it is too difficult for a naive eye to recognize.

Innovation is not a privilege of large countries. Small countries such as Norway, Finland, Switzerland, and Taiwan claim international supremacy. Achievements of small countries show that a Greek vision for innovation is not impossible, at the beginning. But it cannot be founded solely on dry fiscal data, without taking into account mainly the human factor. In this area, the situation in Greece today is rather disappointing—Turbulent economic phenomena have progressively caused a moral and deeply ideological malady: crisis of values, depreciation of elites, mitigation of the sense of responsibility, disruption on social issues, and lack of vision and spiritual values (Matsaganis & Leventi, 2013). Those principles and values should be restored as soon as possible. New strategies based on the country's comparative strengths and opportunities should be formulated.

The development of innovation ecosystems in education requires government support and intervention to create infrastructures and to establish adequate legal frameworks. Sufficient funding is also important for the development of start-up businesses. This process requires continuous and steady effort to bloom and to achieve harmonious results (Von Stamm, 2008).

What are the characteristics of people who are able to create high rates of growth and innovation inside their organizations? These people have the ability to think "outside the frame," which means that although they understand the importance of standard managerial factors (like efficiency and effectiveness), they embrace uncertainty, ambiguous research, and unorthodox methods. They realize that innovation follows its own rules, which do not coincide with the conventional strategy of rewards and reprimands that most organizations follow (Osborne & Brown, 2005). Generally summarized, we conclude that innovation

- Does not necessarily rely on a great idea, but it can be a refined or small improvement of a trivial everyday routine (with possible future impact)
- Is a leap to the future, so no previous or past organizational records can forecast advancement of the next day
- Can canonically arise from the adaptability to change, something that most organization cannot achieve because they need to protect their status quo, which is a normal reaction for every organized system

The third factor is critical. Resilience to change is one of the most difficult attributes to evade. Most managing directors and principals would reject that dilemma in favor of stagnation because they feel threatened and out of their comfort zones (Berkun, 2010). It is, after all, quite difficult to adopt an ever-changing way of thought, to keep trying to follow unpredictable and unstable conditions, consuming lots of mental energy, waiting for the next innovation "ghost" to show up. It requires large amounts of expertise, stress control methods, empathy in relationships with others, and a perpetual appetite for knowledge and creativity (Boutellier, Gassmann, & Zedtwitz, 2008).

14.4 Basic Restructuring Procedures

In this section, we propose mainstream procedures that aim at reshaping the educational system from its contemporary state toward a strategic element of a modern state (field of interest is the Greek case). The process should emphasize the redesign of high-order policies that actually reestablish PA by facing drastically the causes rather than the symptoms of its pathology (Von Stamm, 2008). It should be characterized by the following principles:

- Transparency in all activities, especially in those undertaken by high standing authorities
- Quality of produced services
- Inclusiveness and participation
- Meritocracy among students and staff
- Trust in cultural and schooling institutions
- Effectiveness, concerning the learning outcome
- Student-oriented approach behind every initiative
- Skill reinforcement by targeted use of information and communication technology (ICT)

In the past, there have been enough reformatory programs, zealously proclaimed by the Ministry of Education and other institutions, that have shown little success in the course of their implementation. The difference with the current proposal is the operational—targeted and result—oriented approach it uses to improve the expected outcome.

We chose to apply four different methods, each one representing a fundamental concept in the international public management literature. They are MBO (for planning), CAF (for assessment), ISO (for standardization), and eGov (for innovation). They correspond to different facets of the transformation process and are implemented sequentially, to achieve the best result.

14.4.1 The System MBO

The term refers to the process of determining specific targets and goals, emanating from the upper administration levels and diffusing to the lower levels of hierarchy, in the form of specialized actions (MBO). Directors and principals at all levels, as well as teachers, commit themselves to the implementation of these targets and attainment of quality and quantity outcomes on an annual basis (Ministry of Interior, 2004).

Therefore, objective selection is extremely important. This is, in fact, a tedious multistep process that involves the communication and cooperation of all the departments up and down the hierarchy, to set objectives and action directions at the strategic level and specialize them in operational goals and detailed action plans, appropriate for each level. Chosen actions must reflect the mission and the vision of leadership, which is clearly reminiscent of a top-down approach. On the contrary, the aim of the implementation follows the bottom-up pattern because all the partial action plans finally will sum up to the fulfillment of the aspirations at the top managerial level.

What are the prerequisites for effective target setting? One should summarize these in the following list (McNabb, 2006):

- Prioritization of objectives, identification of critical key activities
- Diversification into individual targets
- Identification of the main and alternative modes of action and allocation of responsibilities and powers by organizational unit and employee
- Description of connections and interdependencies among specialized activities
- Determination and securing of the required resources (manpower, financial assets, ICT)
- Scheduling of a precise timetable for fulfilling the objectives

In recent years, the gradual adoption of the MBO system by Greek authorities is becoming evident by two observations: (a) For the first time in Greek PA, official documents of the Ministry of Education refer to strategic targets (Ministry of Education, 2014a). This official document clearly describes 20 objectives (at the higher policy level) that need to be achieved in the current year. And (b) there is dissemination of the action of self-evaluation of the educational work in all school units inside the Greek territory (Observatory of Evaluation, 2013). This project is a rather ingenious way to mobilize the teacher community at primary and secondary levels, to develop, implement, and follow their own action plans, inside the school unit, scheduled on an annual basis. Plans must be based upon sectors (or quality factors) that measure different aspects of the educationally produced results inside school classrooms. The whole work is connected to innovation activities through testing performance metrics inside these sectors. Until now, results of self-evaluation are indeed promising, although there are reactions of mistrust.

The implementation of the MBO approach yields multiple benefits that arise from better internal functioning of PA and from the relationship between PA and external environment. It improves interactions between tutors and students. It aims at continuous improvement through citizen's satisfaction. It leads to better short-term results and also to overall competitiveness and service reliability.

14.4.2 The CAF Model

A lot of discussion has been made on this issue. The so-called CAF is a model of quality management, which has been effected by the Excellence Model of the European Foundation for Quality Management (2014). According to the model, the best results in organizational performance, customers (students here), and society depend on leadership, strategy and planning, human and nonhuman resources, partnerships, and administrative procedures. The CAF is a holistic analysis of organizational performance approaching it from different angles simultaneously (Engel, 2002).

The CAF is offered as an easy-to-use tool to assist public sector organizations across Europe to apply quality management techniques and improve their performance. The CAF provides a framework set pertaining to the following abilities:

- To respond to citizens'/learners' needs and accountability
- To communicate on an institutional level
- For stakeholders (the educational community, i.e., teachers, students, and parents) to participate and balance their needs
- To provide excellent services (teaching and learning)
- To achieve objectives (conquer knowledge, acquire skills, get moral values and ethos)
- To modernize, innovate, and change

We graphically depict the functionality of the CAF model in Figure 14.1. It consists of two parts: inputs and outputs. These are connected with the causal relationship of transformation processes, whereas the feedback loop is called "learning and self-adjusting." In that way, this model takes care of these two catalysts to improve its overall performance.

In September 2012, the Greek Ministry of Education participated at the 5th International Conference for CAF users in Oslo, Norway, titled "The CAF as a Driving Force for Innovation" (Ministry of Education, 2012). The conference was organized under the auspices of the European Institute of Public Administration (EIPA) CAF Resource Center. The Greek presentation demonstrated how CAF results can be combined with an emerged innovation (in the sector responsible for organizing and conducting national-level examinations). It showed how a well-designed partial restructuring can lead to more efficient methods in PA. The Greek presentation was chosen to be first among the 18 "best administrative practices" in public services and institutions within the European Union presented in that conference.

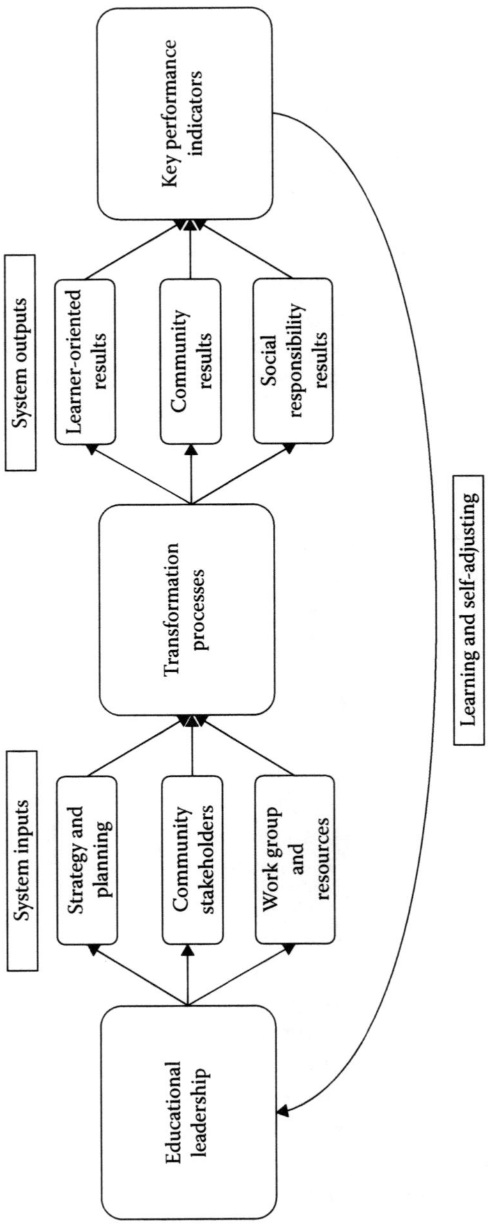

Figure 14.1 The Common Assessment Framework model.

14.4.3 ISO Certification

The acronym ISO stands for the International Organization for Standardization. It is the largest international organization for standardization. Founded in 1947, the ISO developed standards to cover many different industry sectors (ISO, 2014). Today, it comprises the national standardization bodies of 148 countries. The ISO certification procedures are followed by 2981 technical corporate bodies (technical committees, subcommittees, working groups, and study groups). In Greece, the ISO is represented by the Greek Organization for Standardization.

There are more than 14,000 international standards of the ISO, and among them is the ISO 9000 series, which includes standards related to the creation (development) and implementation of quality management for organizations of all types.

In the area of education, international standardization helps bring technological, economical, and societal benefits altogether. It helps the harmonization of educational products and services, thus fostering sustainability and social welfare. A lot of systematic work has been done in different areas, such as the repository of resources and teaching materials, good practices, and cooperation between different educational institutions and bodies. As we noted in Section 14.2, it is a common belief that standardization can be the starting point of resurgent innovation (Blind & Gauch, 2008).

A common standard in education is ISO 9001:2008, which is strongly focused on the continuous improvement of overall performance and efficiency of an organization, aiming to serve the needs and expectations of its stakeholders in the best manner. The ISO 9001:2008 is based on applying a continuous improvement cycle (the Deming cycle) consisting of four distinct parts: "plan–do–check–act" (Deming & Edwards, 1982). It is applicable to educational organizations as well.

To anticipate and accept results in the area of vocational education and training, Greece has created a national, statutory coordination point for qualifications named EOPPEP (2012). This organization develops and implements the National Accreditation and Certification System for nonformal education (including initial and continuing adult education) and provides scientific support, vocational guidance, and counseling services in Greece. Among its principal fields of activity and responsibility are the accreditation of providers of nonformal education (institutes, teacher profiles, etc.), the development of national qualifications framework (e.g., Europass), and the certification of personnel and the career development for youth and adults.

14.4.4 Employment of the eGov Tool

According to the United Nations, the term *governance* includes laws and procedures through which the state, individuals, and social groups promote their views and interests, resolve their differences, and fulfill their responsibilities (Fukuda-Parr & Lopes, 2013). This strong interdependence among state power, private sector, and civil society is a vital element of the type of cooperation that promotes nationwide welfare and justice.

One of the biggest challenges that PA faces today is the deep economic and social impact due to new technologies. The ICTs can indeed help PA to face numerous problems. However, ICTs cannot themselves be a goal. The epicenter should be the exploitation of ICT in combination with structural changes and emergence of new skills to improve public services, democratic processes, and policies exercised by PA. That exactly is the subject of electronic governance (Grönlund & Horan, 2005).

Good practices in many different countries have shown that eGov is a powerful instrument for the provision of improved quality public services, reduction of waiting times, improvement of economic efficiency, increase in productivity, and improvement of transparency and accountability. The eGov strategy, as a central element of the Action Plan e-Europe, should focus on achieving Lisbon goals, reducing barriers in the internal European Union common space and increasing mobility across Europe while effective implementing national policies and regional or local development. This will lead to a global integration in the area of PA (European Commission, 2010).

The basic target of eGov is the electronic support of the whole life cycle of public policies, from diagnosis and understanding of society's needs to planning and implementation of appropriate policies to cope with. Therefore, eGov should be used on the following (good governance) cases: (a) increased communication with clients/students, (b) cooperative implementation of planned programs and services based on workflow management systems, (c) immediate delivery of services through multiple physical and electronic channels, and (d) evaluations of produced services by using e-democracy software. In general, eGov tools can be a catalyst for a massive wave of cultural change in the whole PA area.

Inside the realm of the Greek Ministry of Education now, a massive program toward the digital school has been implemented for the last 3 years (Ministry of Education, 2011). This program is fruitful enough: It comprises modules like interactive school books, the national repository of teaching material, a digital educational platform, and a library of digital content and context (called light-tree). In parallel, a large-scale project of digitization of standard bureaucratic processes, like files, protocols, and student registers, is being carried out, under the name "myschool." It is very interesting to observe the acceptance of these new artifacts throughout the school community. The reactions vary from moderate acceptance to intense rejection. At the same time, several ethical problems arise, like eligibility from the upper hierarchies to gather and maintain sensitive data, security, and authentication. The whole operation has been physically based upon the optical backbone of Panhellenic school network.

14.5 Putting Things Together

The main developmental choice in the area of improving the managerial capacity of PA can be summarized in the creation of a citizen-centered effective, open, and flexible government. This will affect all the subsidiary areas, including education, molding upper- and lower-level policies. The initiatives aiming at PA restructuring actually involve the strengthening of "back-office" (which is the old structure) to "front-office" (the modern open, professional perspective) relationship, as we can see in Figure 14.2.

The preconditions of effective restructuring can rely upon the following pillars (Hargreaves, 2003):

- Accountability: Authorities are willing to show the extent to which decisions are consistent with clearly formulated objectives.
- Transparency: Actions, decisions, and decision-making processes of the government are subject to appropriate control of other parts (auditors), the civil society, and in some cases, by external bodies and governments.
- Efficiency and effectiveness: The master plan aims at producing quality public services for citizens at the lowest cost, ensuring that the results meet the original intention of the policy makers.
- Responsiveness: Upper hierarchies have the ability and flexibility to respond quickly to social changes, taking into account the expectations of civil society when it determines the general public interest and is willing to review critically the role.
- Insight: Authorities anticipate future problems based on current data and trends (forward vision) and formulate policies that take into account the future costs and expected changes (e.g., demographic, economic, environmental).

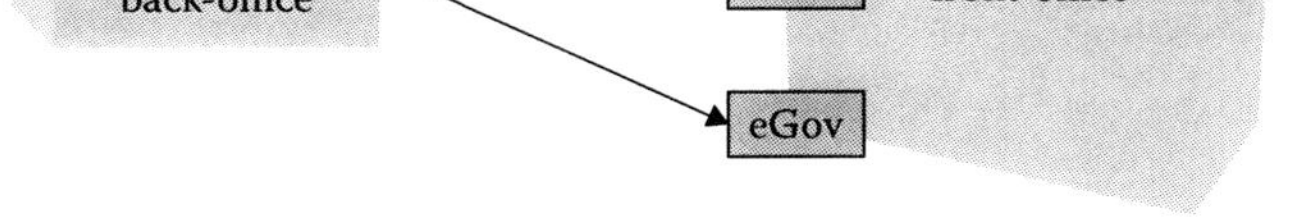

Figure 14.2 Putting things together. CAF = Common Assessment Framework; eGov = e-government; ISO = International Organization for Standardization; MBO = Management by Objectives.

Upon these pillars, we apply the aforementioned transforming processes, consequently the four modules (MBO, CAF, ISO, and eGov). The whole pallet of these techniques can initiate an effective organization reform. To computationally measure the impact of change and the qualities of the transformation, first, some concrete metrics will be needed. They must be picked up carefully so they represent the whole spectrum of attributes. Once they are defined, they must be simple enough to evaluate to ascertain the impact of transformational appliances. An assembly of all these variables creates a unique vector (Kerzner, 2013). Analyzing these metrics is up to any different school unit to measure, modify, or augment the specific criteria according to their particular conditions. In Figure 14.3, we list a skeleton of the most important ones (Solomon, 1998).

The model here is selected among the many in the international literature (Kerzner, 2013). It consists of a set of 16 variables (metrics) that depict the main characteristics of a school unit (namely, 1a, 1b, 2a, 2b, 2c, 3a, 3b, 4a, 4b, 4c, 5a, 5b, 6a, 6b, 6c, and 7a), as shown in Figure 14.3. They were chosen based on their expressed purpose, provision of specific information, reflection of the true status of the organization, and likelihood to gauge as performance indicators. They constitute the "eigenvector" of the unit as follows:

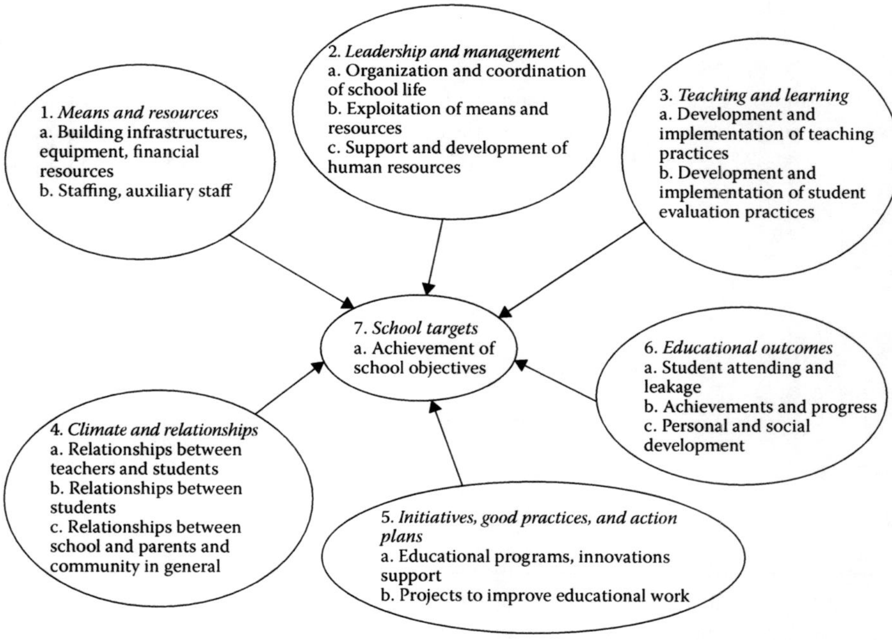

Figure 14.3 Simplified school unit model.

$$G = [m_1, m_2, m_3, \ldots, m_{15}, m_{16}]^T, \tag{14.1}$$

where m_i is a real normalized value ($i = 1, \ldots, 16$), corresponding to each metric, and G is a 16×1 vector matrix (shown here is its transpose). In the next phase, we try to sequentially apply the four transformation factors as functions on vector G:

$$\lambda = tr_{\text{eGov}}\,(tr_{\text{ISO}}\,(tr_{\text{CAF}}\,(tr_{\text{MBO}}(G)))). \tag{14.2}$$

The resulting value of λ should be a single real value representative of the transformation index of the whole process. The aim is to maximize expression λ over all (possible viable) instances of the G vector.

14.6 Toward the Holistic Model

Discussing the ICT model in education usually brings to mind new settings of hardware equipment or software packages that facilitate teaching and learning inside a modern school classroom. But the process of embedding ICT inside a classroom goes (according to the holistic model) deeper than that: It involves changing the whole mindset of stakeholders. The successful integration should work as an immersion experiment, when a virtual space is created and a set of new tools are available in a "lean-back," as opposed to "lay-forward," configuration to demonstrate use of more intrusive techniques (Rakib, Bronstein, Bronstein, & Devictor, 2009). This implies rather extensive changes in the knowledge management process, as shown in Table 14.1.

A vivid example of a contemporary program of an enormous scale is the one recently approved by the European Regional Development Fund–Operational Program "Digital Convergence" Action titled "Digital Services in Greek Ministry of Education," with a total budget of about 140 million euros, which will "arrive" soon in classrooms (Ministry of Education, 2014b). The aim of the project is the massive digital upgrade of educational processes, both in primary and secondary education. It is about a substantial change in means and pedagogy that will make a fertile ground for innovations and future novelties.

The procedures for the procurement and deployment of interactive systems (IWBS) and mobile computer labs in elementary and secondary schools throughout the country have already started. These systems will be placed, as a priority, in high and vocational high schools, gymnasia, and primary schools. In absolute numbers, this program will cover about 200,000 pupils, 315,000 gymnasium students, 230,000 high school students, and 85,000 vocational high school students and colleges across the country, according to ministry executives. Important issues are also the networking of all this hardware and the services it will run. It will be surely based on the Greek school network backbone, already running fiber optics technology (Kalochristianakis, Paraskevas, Varvarigos, & Xypolitos, 2007).

Table 14.1 Knowledge Management Process and Supporting ICT Tools

Knowledge Management	*ICT Tools*	*ICT Enablers*	*Example Platforms and Settings*	*Adopted Model*
Creation	Data mining, e-learning tools	Creation and combination of new sources of knowledge, online learning	Knowledge practices, blueprints, stories, dialogues, and discussions	Technocratic
Storage and retrieval	Blogs and bulletin board systems (BBS), digital repositories, databases, search engines	Support of individual and organizational memory	Groupware and communication technologies	Technocratic
Combination and transfer	Discussion fora, directories, web 2.0 tools, social media, interactive whiteboards (IWBs)	Intranets, faster access to corporate sources	Digital marketplace, communities of practice and expertise, cloud	Holistic
Application and feedback	Expert systems, workflow systems	Faster application of new knowledge, startups	Knowledge management systems, knowledge repositories, new paradigms for learning	Holistic

Source: Alavi, M., & Leidner, D. (2001). *MIS Quarterly 25,* 107–136.

At the same time, interactive digital educational content is being created, enriched, and archived in large databases. The central e-service of the Ministry of Education is the light-tree (the so-called photodendro, in http://ebooks.edu.gr). It is accessible by the entire community as freeware and hosts a collection of approximately 3500 learning objects (experiments, interactive simulations, inquiries, images, educational games, videos, 3D maps, exercises) of rich interactive textbooks. Meanwhile, the knowledge-eager user will find all the teaching material (textbooks, exercises, teacher books, specifications) of all lessons in the syllabus of the 12 years of formal compulsory studies.

A third feature of the system is bureaucracy. A lot of work that is being done by paper and pencil is now fully digitized. Some of the most important documents are individual student sheets (which contain certificates and scores of every student in each quarter), registry of students (which contains a hardcopy of score transcripts), school leaving certificates (most important documents that accompany the student for the rest of his or her productive life), directory of students (includes all students who attended this school in the past), absence registration books, the main proceeding book, mail protocols, and others. All this paperwork is now becoming digital, thus saving a lot of cost, money, and system reliability.

Procurements on this scale have been attempted a few times in the past two decades. The outcomes of these projects are still under intense debate (Orwa, 2009). Many people support the idea that such large-scale, top-down, massive initiatives are, by default, too bureaucratic and tend to degenerate quickly. On the other hand, clever small-scale interferences can lead to more flexible and creative conditions.

14.7 Perspectives

The term *public sector educational organization* refers to that part of the economy that is concerned with providing governmental or state services in the realm of education (provisionally including all its three levels: primary, secondary, and tertiary). According to this definition, it is among the fundamental missions of a state to provide the necessary conditions for economic prosperity and social justice for all its citizens (Freire, 1985). These two pillars define the function of the state as a strategic node that designs and applies policies, to achieve its long-term visions and goals (Christensen, Lægreid, Roness, & Rovik, 2007).

However, Greece needs more than that: a national strategy that defines a general framework, priorities, and roles for achieving long-term goals. The frame is determined by a number of necessary structural reforms aimed at restructuring research and facilitating cooperation among all key factors in creating innovation ecosystems.

In the area of applied research, the country must accelerate and augment the restructuring of the fragmented potential to create nodes of excellence with a critical mass. To achieve excellence, a series of measures are necessary. It is needed to (a) introduce the evaluation criteria of both researchers and research institutions, (b) offer researchers and tutors better working conditions, such as recognition of creativity and elimination of bureaucratic procedures, and (c) create the appropriate institutions for the implementation of the national strategy plans (Haynes, 2003).

A national strategy should identify specific roles for each of the developers of innovation and provide the necessary measures and incentives for seamless collaboration. It will not be dictated by upstream bureaucratic decisions unrelated to economy or technological progress. Should this initiate a composition dialogue between the main stakeholders, it would replace the decadent practices of client relationships that cause bad influence. Those practices of the past need to be annihilated (Boutellier et al., 2008).

It is presumed that the development of innovation is the country's long-term bet. It requires not only multiple structural changes but also continuous effort and radical sanitization of public life. Let us start with education and research restructuring by establishing centers of excellence. There should be an R&D Ministry inauguration operating closely beside the Ministry of Education. In addition, a broad dialogue between state bodies, institutions, industrialists, and entrepreneurs should be initiated. The aim is the creation of business clusters, appropriate for the realization of strategic technology objectives (Corallia, 2014).

The whole effort should be supported by investments of venture capital. Despite the significant decline in labor costs and the availability of well-qualified scientific and technical potential, the country is still unattractive for investment capital. That happens because investors remain unconvinced about the stability and orderly function of state administration and everyday services.

In the end, the crisis offers us a valuable opportunity, but as the famous Greek historian Thucydides noted in his famous work "Peloponnesian War, Thuc.1.142": "Time waits for no one" (Hobbes, 1942). Now is the time to redesign the future of the education system.

Author Biographies

Konstantinos Zacharis holds a diploma in Computer Engineering and Information Science from Patras Polytechnic School (1991) and an MSc in Studies in Education from the Greek Open University (2003). He has two decades of experience in working within educational organizations in various positions and levels (mostly in secondary and tertiary, formal and nonformal, and also adult education) in Greece and Germany. Currently, he serves as principal in 5th Lyceum Karditsa. He is a PhD candidate in Staffordshire University.

Pandelis Ipsilandis is a professor of Operations Research in the Department of Project Management in the School of Business and Economics and is also vice president of TEI of Larissa. He holds a first degree in Mathematics from the National University of Athens, Greece, and he did his postgraduate studies at the Polytechnic Institute of New York, United States, where he received both an MSc and a PhD degree in Operations Research. He is K. Zacharis' second PhD supervisor.

James O'Kane is a director of the Business School at Edge Hill University, United Kingdom. He studied physics at Hull University then joined Hatfield Polytechnic as a research assistant doing work on simulation. He then joined Marconi Radar Systems, based in Chelmsford in Essex, where he worked on mathematical modeling and simulation in radar signal processing. His academic career began at Staffordshire University, where he taught and did research in Operations Management in the School of Engineering. He is K. Zacharis' first PhD supervisor.

References

Alavi, M., & Leidner, D. (2001). Knowledge management and knowledge management systems: Conceptual foundations and research issues. *MIS Quarterly, 25*(1), 107–136.

Babones, S. (2012). *To end the jobs recession, invest an extra $20 billion in public education.* Retrieved April 1, 2014, from http://truth-out.org/opinion/item/11031-to-end-the-jobs-recession-invest-an-extra-$20-billion-in-public-education.

Berkun, S. (2010). *The myths of innovation.* Paris, France: O'Reilly Media.

Blind, K., & Gauch, S. (2008). Trends in ICT standards in European standardization bodies and standards consortia. *Telecommunication Policy, 32*(7), 503–513.

Boutellier, R., Gassmann, O., & Zedtwitz, M. V. (2008). *Managing global innovation. Uncovering the secrets of future competitiveness* (3 Rev.). Berlin: Springer Verlag.

Buchberger, F., & Buchberger, I. (2004). Problem-solving capacity of a teacher education system as a condition of success? An analysis of the "Finnish Case." In F. Buchberger & S. Berghammer (Eds.), *Education policy analysis in a comparative perspective* (pp. 222–237). Linz, Austria: Trauner.

Christensen, T., Lægreid, P., Roness, P. G., & Rovik, K. A. (2007). *Organization theory and the public sector: Instrument, culture and myth.* London: Routledge.

Corallia. (2014). *Hellenic technology clusters initiative.* Retrieved February 28, 2014, from http://www.corallia.org.

Deming, W. E., & Edwards, D. W. (1982). *Quality, productivity, and competitive position* (Vol. 183). Cambridge, MA: Massachusetts Institute of Technology, Center for Advanced Engineering Study.

Education Reform Act. (1988). England, UK: Her Majesty's Stationery Office.

EFQM. (2014). *European Foundation for Quality Management.* Retrieved February 28, 2014, from http://www.efqm.org.

Engel, C. (2002). Common Assessment Framework: The state of affairs. *Eipascope, 2002*(1), 1–5.

EOPPEP. (2012). *National Organization for the Certification of Qualifications and Vocational Guidance.* Retrieved February 28, 2014, from http://www.eoppep.gr.

European Commission. (2010). *The European e-Government Action Plan 2011–2015. Harnessing ICT to promote smart, sustainable & innovative Government.* COM (2010) 743 Final, Brussels.

Freire, P. (1985). *The politics of education: Culture, power and liberation.* Westport, CT: Bergin and Garvey.

Fukuda-Parr, S., & Lopes, C. (Eds.). (2013). *Capacity for development: New solutions to old problems.* London: Routledge.

Grönlund, Å., & Horan, T. A. (2005). Introducing e-Gov: History, definitions and issues. *Communications of the Association for Information Systems, 15*(1), 39.

Guardian. (2013). *Michael Gove stands by school reforms.* Retrieved July 1, 2014, from http://www.Theguardian.com/politics/2013/dec/03/gove-defends-education-reforms.

Hargreaves, D. H. (2003). *Education epidemic: Transforming secondary schools through innovation networks.* London: Demos.

Haynes, P. (2003). *Managing complexity in the public services*. Berkshire: Open University Press.

Hobbes, T. (1942). Thucydides. *Historiae in two volumes*. Oxford: Oxford University Press.

ISO. (2014). *International Organization for Standardization*. Retrieved February 28, 2014, from http://www.iso.org.

Kalochristianakis, M. N., Paraskevas, M., Varvarigos, E. A., & Xypolitos, N. (2007). The Greek School Network: A paradigm of successful educational services based on the dynamics of open-source technology. *Education, IEEE Transactions on, 50*(4), 321–330.

Kerzner, H. R. (2013). *Project management metrics, KPIs, and dashboards: A guide to measuring and monitoring project performance*. New Jersey: John Wiley & Sons.

Laperche, B., Uzunidis, D., & Von Tunzelmann, G. N. (Eds.). (2008). *The genesis of innovation: Systemic linkages between knowledge and the market*. Cheltenham: Edward Elgar Publishing.

Laukkanen, R. (2008). Finnish strategy for high-level education for all. In N. C. Soguel & P. Jaccard (Eds.), *Governance and performance of education systems* (pp. 305–324). The Netherlands: Springer.

Matei, L., & Flogaitis, S. (Eds.). (2011). *Public administration in the Balkans—From Weberian bureaucracy to new public management* (Vol. 1). Bucharest, Romania: Economica Publishing House.

Matsaganis, M., & Leventi, C. (2013). The distributional impact of the Greek crisis in 2010. *Fiscal Studies, 34*(1), 83–108.

McNabb, D. E. (2006). *Knowledge management in the public sector: A blueprint for innovation in government*. Armonk, NY: M.E. Sharpe.

Ministry of Education. (2011). *Digital educational platform, interactive books and learning object repository*. Athens. Retrieved February 28, 2014, from http://dschool.edu.gr.

Ministry of Education. (2012). *Web-entrance application procedures for higher education in Greece*. 5th European CAF Users' Event. Oslo, Norway.

Ministry of Education. (2014a). *Defining strategic targets for the Ministry of Education, year 2014*. Official Document 23076/K/18-02-2014. Athens.

Ministry of Education. (2014b). *Recording classroom data in the frame of co-financed action 'Digital services'*. Official Document 2785/07-03-2014. Athens.

Ministry of Interior. (2004). *Management by objectives, measurement of efficiency and other provisions*. Law 3230 (Official Gazette FEK 44A-11/2/2004). Athens: National Printing Office.

Observatory of Evaluation. (2013). *Self-evaluation of educational work*. Operational Program ESPA 2007–2013. Retrieved February 28, 2014, from http://aee.iep.edu.gr.

Organisation for Economic Co-operation and Development. (2012). PISA 2012 Results in Focus. *Program for International Student Assessment*. Retrieved July 1, 2014, from http://www.oecd.org/pisa.

Orwa, W. O. (2009). Reconstruction of education for the 21st century. *Innovative Teacher Education and Classroom Practice in the 21st Century*, Nairobi, Kenya, July 8–10, 2009, 8, 427.

Osborne, S. P., & Brown, K. (2005). *Managing change and innovation in public service organizations* (Vol. 1). London: Psychology Press.

Rakib, S. S., Bronstein, A., Bronstein, M., & Devictor, G. B. M. (2009). *U.S. Patent Application 12/423,752*.

Sahlberg, P. (2007). Education policies for raising student learning: The Finnish approach. *Journal of Education Policy, 22*(2), 147–171.

Sahlberg, P. (2009). Educational change in Finland. In A. Hargreaves, M. Fullan, A. Lieberman & D. Hopkins (Eds.), *International Handbook of Educational Change* (pp. 323–348). The Netherlands: Springer.

Solomon, J. (1998). Trends in the evaluation of education systems. *School (self) evaluation and decentralization, papers, reports, discussion outcomes*. European Workshop, Athens December 1997. Athens: Pedagogical Institute.

Von Stamm, B. (2008). *Managing innovation, design and creativity*. London: John Wiley & Sons.

Windrum, P., & Koch, P. (2008). *Innovation in public sector services*. Cheltenham: Edward Elgar Publishing Limited.

Index

Page numbers followed by f and t indicate figures and tables, respectively.

H

I

P

Q

R

S

T

£76.99